THE
FUTURE
EATERS

OTHER BOOKS BY THE AUTHOR

Mammals of New Guinea

Tree Kangaroos: A Curious Natural History
with R. Martin, P. Schouten, and A. Szalay

Possums of the World: A Monograph of the Phalangeroidea
with P. Schouten

Mammals of the South West Pacific and Moluccan Islands

Watkin Tench, 1788 (ed.)

The Life and Adventures of John Nicol, Mariner (ed.)

Throwim Way Leg

The Birth of Sydney

*Terra Australis: Matthew Flinders' Great Adventures in the
Circumnavigation of Australia* (ed.)

The Eternal Frontier

The Explorers

A Gap in Nature with P. Schouten

THE
FUTURE
EATERS

**An Ecological History of the
Australasian Lands and People**

TIM FLANNERY

GROVE PRESS
New York

First published in 1994 by Reed Books Australia

This edition is published by arrangement with
Reed New Holland, an imprint of
New Holland Publishers (Australia) Pty Ltd.

Published simultaneously in Canada

Library of Congress Cataloging-in-Publication Data

Flannery, Tim F. (Tim Fridtjof), 1956–
    The future eaters : an ecological history of the Australasian lands and people / Timothy Fridtjof Flannery.
        p. cm.
    Originally published : Australia : Reed Books Australia, 1994.
    Includes bibliographical references (p.).
    ISBN 978-0-8021-3943-6
    1. Natural history — Australasia. 2. Nature — Effect on human beings on — Australasia — History. 3. Australasia — History. 4. Extinction (Biology) — Australasia.   I. Title.

QH196.8 .F63 2002
508.9 — dc21   2002016445

Grove Press
an imprint of Grove Atlantic
154 West 14th Street
New York, NY 10011

Distributed by Publishers Group West
groveatlantic.com

DEDICATION

I dedicate this book to the Australasians who are trying to forge nations out of the chaos of colonial history. It is my hope that it will help them understand that they are special, for they inhabit unusual lands. It was in their countries that future eating began. If they are to preserve their unique natural heritage, their newly forged nations must cease to be the realms of the future eaters.

ACKNOWLEDGMENTS

This book owes its existence to a large number of people. During the time I was writing it many people, including my family and friends, saw far too little of me. Others, including friends and colleagues who are busy themselves, doubtless felt that I was seeing them far too often, asking questions, proposing ideas and generally being a drain on their time. To all of the individuals who have suffered because of my obsession with this work, I ask forgiveness and give my most heartfelt thanks.

Inspiration for this book sprung in large part from my experiences in the Australasian bush. Many people have shared their knowledge with me, including specialists with whom I have spent time in the field, expert hunters of the mountains of New Guinea, and ordinary people throughout the region. Although it seems unfair to select a few individuals for special thanks, I must mention Dr Geoff Hope of the Australian National University, who introduced me to both Papua New Guinea and Irian Jaya; Dr Don Gardner of the same university, who gave me my first real insights into

Melanesian lifestyles; Murrabudda of Groote Eylandt, and Anaru, Miyanmin elder, both of whom opened doors to worlds I never imagined existed.

Jonathon Kingdon, Selwa Anthony and Bill Templeman all encouraged me to publish this work, while Mary Halbmeyer and Alexandra Szalay provided excellent and good-natured editorial expertise. Without their confidence in its value I doubt I would have completed it.

A number of my colleagues and friends read drafts of the manuscript, providing invaluable criticism and correcting embarrassing errors. They are Nick Allen, BHP; Bob Beale, Sydney Morning Herald; Dr Harold Cogger, Deputy Director, Australian Museum; Dr Bill Hurditch of the NSW Chamber of Mines; Professor Rhys Jones, Research School of Pacific Studies, Australian National University; Dr Ronald Lampert; Dr Betty Meehan, Australian Heritage Commission; Dr John Paxton, Ichthyology Section, Australian Museum; Dr Harry F. Recher, Department of Ecosystem Management, University of New England; Dr Patricia Rich, Department of Earth Sciences, Monash University; Dr Tom Rich, Curator of Vertebrate Fossils, Museum of Victoria; Ronald Strahan, Book Publishing Services, Australian Museum; Mr Haydn Washington; and Professor Peter White, Department of Prehistoric and Historical Archaeology, University of Sydney. To them all I give my most sincere thanks.

Finally, I must thank those people—many of them anonymous—who came to my aid when I lost my house and library to an ENSO-induced bushfire while writing this work. To those who gave me books I am especially grateful, for my referencing system was destroyed and I doubt that I would have been able to pick up the traces of this study had it not been for the inspiration I received from their gifts.

CONTENTS

Part One

AN INFINITY BEFORE MAN

1 The New Lands / 20

New Caledonia, New Holland, New Guinea and New Zealand—European discovery: irony in their names—Separate 'experiments' in evolution—Diverse peoples—Continental drift—Zoogeography.

2 Australia in Gondwana / 30

The age of dinosaurs in south-eastern Australia—A picture of Australasia 120 million years ago—Chicken-sized dinosaurs of Antarctica—Late-surviving labyrinthodonts—Australia's oldest mammal fossils—Break-up of Gondwana.

3 Land of Geckos, Land of Flowers / 42

Tasmantis: its creation and break-up—New Caledonia: geological history and discovery by Cook—A unique flora from toxic soils—Assemblage of ancient trees—Reptiles win out over birds and mammals— A perverse Garden of Eden.

4 Land of Sound and Fury / 52

New Zealand: cold climate and fertile soils—Frogs and their importance in past ecosystems—The Moa: diversity, ecology and appearance—An extraordinary array of birds—Monotremes and other mammals: failure to thrive.

5 Meganesian Enterprises / 67

Meganesia: its creation and composition—Antarctic and South American land-bridge—Origins of Australasian marsupials—*Dromiciops.* A Chilean link with Australia—A South American platypus—Australia loses its early placentals and marsupials win the race.

Part Two

ARRIVAL OF THE FUTURE EATERS

Part Three

THE LAST WAVE: ARRIVAL OF THE EUROPEANS

*When this publication was nearly ready
for the press; and when many of the opinions it
records had been declared, fresh accounts from Port
Jackson were received. To the state of a country, where so
many anxious trying hours of his life have passed, the
author cannot feel indifferent. If by any sudden revolution
of the laws of nature; or by any fortunate discovery of
those on the spot, it has really become that fertile and
prosperous land, which some represent it to be, he begs
permission to add his voice to the general congratulation.
He rejoices at its success.*

Watkin Tench, preface
to *Settlement at Port Jackson* (1793).

INTRODUCTION

My childhood world was a very different one from that enjoyed by young people today. I learned that Australian history consisted of gallant, but often ignorant and unsuccessful attempts by Englishmen to explore our continent. The Aborigines, where they rated a mention, were just one of the natural hazards. They were universally held to be among the world's most primitive people who were now, in a sad although inevitable process, making way for a superior people. Some heroic Europeans, such as Daisy Bates, were kind enough to devote their lives to 'smoothing the pillow of the dying race'. During English lessons I was given the works of one of modern Australia's first poets, Judge Barron Field, to read. In 1825 he wrote:

> *I can therefore hold no fellowship with Australian foliage but will cleave to the British oak through all the bareness of winter.*[24]

Only later did I discover that his balding head had been described by a contemporary as 'a barren field indeed'!

I was also taught that our continent was inhabited by inferior animals. The marsupial kangaroos, wombats, koalas and the like were quaint but, in a remarkable parallel with our Aborigines, unable to cope with competition from introduced sheep, cattle and foxes. They would pass away in the natural course of things, to make way for a new, vigorous and somehow more fitting European Australia. I read books that told me that Australian rainforests, what little there were of them, were recent invaders from the north, an impoverished and unimportant appendage of the great forests of South-East Asia. I learned also that Aborigines had not had any impact on Australian flora and fauna, that we had no dinosaur fossils of note, that Australia had always inhabited a fixed position on the face of the Earth and that our wildflowers, although strange and wondrous, could never attain the grace and beauty of an English rose. I am not talking of a distant age, for these views were common in the 1960s when I attended school.

I took a great interest in all of these things as a child and eagerly absorbed these 'facts'. It was only when I reached university that, one after another, the pillars supporting my world view began to crumble. I was finally convinced only during the writing of this book that Australia has a history of human occupation extending back at least 60 000 years. It is a history of one of the world's most unusual and highly specialised people. Their impact on Australia was enormous and I now see virtually all the continent's ecosystems as being in some sense man-made. Finally, far from passing away, Aboriginal peoples and the things that we can learn from their culture are daily becoming more significant to all Australians. Even more importantly, I now believe that changes in technology and thought undergone by the Aborigines changed the course of evolution for humans everywhere, for they were the world's first future eaters.

As for our 'inferior animals', now that the sheep has faltered, Australians ride more and more upon the marsupial's back. This is partly because Australia's tourism industry, now the third strut of our economy, is based in no small part upon the attractions of our wildlife. To a large extent, but more difficult to quantify, our fauna and flora are being used as a unique resource. In scientific disciplines from reproductive physiology and evolutionary biology to medicine, our native species are hailed as a unique and priceless heritage. They are providing insights into the way the world, and we ourselves, work.

The whole concept of inferiority is now being questioned as people realise that context is all-important in determining what is inferior or superior. Given certain conditions marsupials cannot compete with placental mammals, but alter those conditions and the reverse is true.

Our rainforests—those 'unimportant appendages'—are now widely acknowledged as being the most ancient of our land-based ecosystems, which gave rise to most others. It is also becoming increasingly accepted that rainforests arose on the southern continents and that Australia has some of the most ancient rainforests on Earth. Our rainforests are thus filled with primitive plants. Botanical discoveries of world importance are being made in them every year. Australian botanists have recently completed a catalogue of Australian plants, in which they list 18 000 species. Their taxonomic work over recent years has resulted in a 50 per cent increase in the number of species in the groups examined. Yet they estimate that about 7000 undescribed plant species still exist in Australia. Many surely inhabit Australian rainforests and are members of ancient and bizarre families, like the southern pine (*Podocarpus* species) recently found growing in a steep valley in Arnhem Land, thousands of kilometres distant from its nearest relatives.

Research on newly discovered Australian dinosaur faunas is challenging previous conceptions of what dinosaurs were like. So important are these discoveries that an Australian dinosaur recently made it onto the cover of *Time* magazine. It was discovered in one of only two deposits in the world which was laid down near the South Pole during the age of dinosaurs. The chicken-sized species survived three months of darkness each year in a refrigerated world. Study of these fossils is teaching us much about the greenhouse effect as well as the lives of the dinosaurs themselves.

Far from being fixed on the Earth, we now know that Australia has wandered over the face of the planet for billions of years, sometimes lying in the northern hemisphere, sometimes in the south. For 40 million years, after finally cutting the umbilicus with Antarctica, it has slowly drifted northwards, in isolation, at about half the rate at which a human hair grows.

Our splendid native blooms have also now come to the fore and every day fashionable Europeans, Americans and Japanese pay premium prices for banksias, waratahs, dryandras and many other spectacular Australian flowers. Even Australians have overcome their cultural cringe. The incomparable waratah bloom can command five times the price of a fine rose in Sydney's florists.

Another most vital change is occurring in Australia. It is a growing realisation of the way in which nature works there. For biologists are finally understanding that evolution in Australia is not driven solely by nature 'red in tooth and claw'. Here, a more gentle force— that of coadaptation—is important. This is because harsh conditions force individuals to cooperate to minimise the loss of nutrients, and to keep them cycling through the ecosystem as rapidly as possible. Thus, entire ecosystems have evolved in Australia that, when untampered with, recycle energy and nutrients in the most extraordinarily efficient ways. Aboriginal people have long understood this and have shaped their culture accordingly. Even the Europeans, with their code of mateship, are perhaps being shaped by these same forces.

I will argue that these changes to our world view are an early, yet crucial step in the process of adaptation of an essentially European people to life in Australasia. These changes have come so rapidly and are so profound and irrevocable, that many older Australians have become lost in the whirlwind of changing perspectives that presently constitutes our world view. Other Australians have arrived so recently that they are unaware of the importance of the changes.

By virtue of my profession, I passed through the revolution a little in advance of most, for the powerhouse of the change has been centred upon discoveries in biological science, archaeology and anthropology. Through this personal experience I have learned how important histories are to people, including myself. They define our place in the world and validate our claims to inheritance, both individual and national. The radically changed world view that many Australians possess today means that Australians can now define themselves through things that are uniquely Australian.

The revolution continues. It promises to become more profound—and perhaps more painful—as more fundamental assumptions are challenged. The passing of Federal legislation to recognise native title in Australia on 22 December 1993 is one of the first and perhaps one of the most important legal changes to flow from our new and very different view of our place in the world. In a sense, it brings to a close a period in our history when we possessed a purely European view of land.

There is no doubt that conflict between people holding different world views will become more forceful. The debate as to whether Australia should become a republic or remain a monarchy reflects this. While on some levels it is a legal debate concerning constitution and Parliament, it is also a debate between those who see themselves as people of British descent in Australia, as opposed to those who see themselves as becoming a people unlike any other.

These same sentiments drive many arguments concerning resource use in Australia. Should we, for example, eat kangaroos? How should we manage our water resources? Should we 'manage' wilderness areas? Where should we allow mining and how should we conduct agriculture and grazing? It is also critical to the greatest question of all, which concerns how many Australians there should ultimately be. This question is inextricably interwoven with our views concerning Australia's racist past, multiculturalism, ongoing immigration and government attitudes towards family size. It is also partly dependent upon the technologies that we possess, and the standard of living to which we aspire. As difficult as these factors make it to address the question of population, it more than any other, will define Australia's future.

The Australian nation came into existence less than 100 years ago as a result of the last successful invasion of a continent. The invasion has brought together people from virtually every corner of the Earth in a land unlike any other. Fully one-quarter of all living Australians grew up in other lands, while the very earliest non-Aboriginal Australians have ancestries that extend back only three human lifetimes in Australia. The traditional histories and folk wisdom of the new settlers are as different and diverse as those found anywhere. But they are all, like the essentially British world view of my childhood, the products of other places, other ecologies and other times.

How are Australians, then, to adopt, develop and feel comfortable with a world view that will help them to survive in this strange land? For me, the solution has come through an increasing understanding of the way in which our continent works. As a research scientist studying Australasian wildlife, I am daily amazed at what I learn. In a sense, this book is a summary of all that we—that tiny proportion of humanity privileged to call themselves Australasians—have learned about their Australasian homelands and of how the land has, often in ways not perceived, shaped its people.

As might be surmised from the above, this history is far from an objective telling of past events. It argues a case and interprets events and evolutionary trends from a particular viewpoint. Only time will tell whether that viewpoint is a valid one.

It is not without some reserve that I have emulated the title of Bede's great work *A History of the English Church and People* (731AD) for my own book. But I do so because these books share a fundamental similarity. Bede wrote at a time when the English Church was a new thing; yet it was destined virtually to define the English people to themselves. The Australasian people too are mostly newcomers. They and their land must form a bond as great and lasting as that between the English and their Church. Otherwise we will always remain poor, confused strangers in our own lands.

By necessity, the compilation of this history has involved research in many disciplines, in few of which can I claim expertise. I can do no better that borrow Bede's own words to the readers of his overarching work:

> *Should the reader discover any inaccuracies in what I have written, I humbly beg that he will not impute them to me, because, as the laws of history require, I have laboured honestly to transmit whatever I could ascertain from common report for the instruction of posterity. I earnestly request all who may hear or read this history of our nation to ask God's mercy on my many failings of mind and body.*

Part One

AN INFINITY BEFORE MAN

The mind seemed to grow giddy
by looking so far into the abyss of time ... [and]
we became sensible how much
farther reason may sometimes go than
imagination can venture
to follow.

Playfair 1805, quoted in McAdam 1986: 111

CHAPTER 1

THE NEW LANDS

To write a history of Australia without reference to its geographic neighbours would be as senseless and uninformative as to tell the story of Antony without Cleopatra, or Romeo without Juliet. This is because Australia is a member of a great family of lands. Distant relatives include India and Africa, more recent ones New Zealand, New Caledonia and South America. Our intimate family, however, consists of Antarctica, New Guinea, the eastern part of the Indonesian island of Sulawesi and a few other smaller Indonesian islands. We know this because we know that at one time all of these landmasses were joined into one, which we now call Gondwana. We now know the timing and sequence of separation of the pieces of this once great land and we know that, by and large, those fragments that separated earlier show less similarity with the rest than those that stayed together longer. By examining some of these fragments of Gondwana we can see how different histories and environmental conditions have shaped each land by modifying its common heritage. The insights we can gain from such comparisons are enormous, for in a sense each land represents a separate experiment in evolution. Because of some accidents of history the most important of Australia's family of lands, for the sake of this comparison, are those of Australasia.

New Holland (as Australia was once known), New Zealand, New Caledonia and New Guinea; these are the lands that make up Australasia. At a superficial level they share a common history in that they were all 'new' lands, 'discovered' and named by Europeans during a brief period when it was fashionable to name new discoveries after places close to home, if not home itself. Yet there is a wonderful irony in these names. New Holland is in fact one of the world's most ancient landmasses, including rocks over 4.5 thousand million years old. In contrast, much of 'old' Holland is remarkably new. A Dutchman once said to me: 'God gave the land to most people, only the Dutchman had to make his own'. He was referring to the fact that fully one-third of his country has been reclaimed from the sea over the past few centuries. Almost all of the rest is composed of quite recent river sediments.

The name New Guinea is equally ironic, for it was named after Guinea in Africa, as both places are home to a black-skinned, crinkle-haired people. It has recently been suggested, however, that New Guinea has been inhabited by these people for at least as long, if not longer than 'old' Guinea. There are also ironies in the naming of both New Zealand and New Caledonia, for again these are ancient lands which have continuously supported unique and diverse faunas and floras for more than 80 million years. In contrast, both 'old' Zealand (the Dutch province of Zeeland) and Caledonia, as Scotland is still sometimes called, are newly settled by absolutely everything. A mere 15 000 years ago both were entirely devoid of life and Caledonia was groaning under ice sheets hundreds of metres thick. Thus, everything that lives on 'old' Zealand and 'old' Caledonia today is new.

Despite these marked and wonderful ironies, many Australasians continue to see things from much the same perspective as their exploring ancestors. Ask an Australian, for example, which land has been inhabited by modern people longest, Western Europe or Australia? You would be fortunate to get the correct answer from one in a thousand, for despite extensive scientific research, most people are still unaware that Australia has been inhabited by modern humans for longer—indeed possibly nearly twice as long—as Western Europe.

Most people are also unaware of the different histories of other Australasian people. Many, for example, think of Australian Aborigines and New Zealand Maoris as both being indigenous people with similar origins. Yet who is more similar to whom? Aborigines arrived in Australia from South-East Asia at least 40 000 and more probably 60 000 years ago. They travelled on the most basic of watercraft and arrived without domesticated plants or animals. Maoris arrived in New Zealand from elsewhere in Polynesia between 1000 and 800 years ago aboard superb ocean-going vessels, which made landfall after a long and deliberate voyage of discovery. They brought along their domestic plants, dogs and rats, which had been gathered from such diverse places as China, South-East Asia and South America. Maoris were followed some 650–450 years later by Europeans who, while they possessed inferior ocean-going craft (Cook himself admired the faster, longer and superbly manoeuvrable Polynesian catamarans when compared with his own *Endeavour*), also arrived on deliberate voyages of discovery. Within 200 years they too had settled New Zealand and populated it with their own diverse domesticated plants and animals.

While the differences between the 'new' lands and their people are of great interest in themselves, their true significance cannot be understood without examining the deeper histories of the lands, for they have much more in common than recent naming and discovery by Europeans. This is because 80 million years ago these lands formed a single mass. Thus, it is necessary to go back this far in time to gain a full understanding of the common natural heritage of the Australasians.

The fact that the Australasian lands share a long common geological history first gained wide acceptance in the scientific community as late as 1966. In that year two young geophysicists presented a scientific paper explaining the results of a rather esoteric experiment they had carried out on the International Indian Ocean Expedition of four years earlier, along with the results of a similar experiment carried out several years earlier in the northern Pacific Ocean. The research entailed towing a device across the ocean floor from east to west, which could record the magnetism of rocks.[99]

Researchers were interested in the magnetism of rocks because it had been discovered that the Earth's magnetic field reverses occasionally. You can imagine the Earth as an enormous battery whose magnetic field is formed by the movement of molten metal (caused by the Earth's rotation) around the solid metal inner core. At present the positive terminal is at the North Pole, but in times past the orientation has switched so that the positive terminal is at the South Pole. Over the past 76 million years the Earth's magnetic field has reversed 171 times. No-one knows exactly why this happens, but at present the magnetic field is weakening a little each decade. Many researchers think that this presages another shift. Just how long before the reversal happens, however, is unknown.

The relevance of this magnetism to unravelling Earth's history became apparent when scientists discovered that rocks preserve traces of the planet's magnetism as it was at the time the rocks formed. This is because many rocks contain tiny iron particles. As a lava cools, or tiny sedimentary particles fall to the bottom of a lake, for example, these iron particles settle out to align themselves with the Earth's magnetic field. As the lava or sediment hardens, the iron particles are trapped in their original orientation. Thus trapped, they can retain their record of magnetic change for millions of years. Rocks that formed when the positive terminal of 'battery Earth' was at the South Pole have a reverse magnetism, while those that formed when it was at the North Pole (as it is today), have the magnetic particles with normal magnetism.

It is interesting to speculate, incidentally, what will happen when the Earth's magnetic field finally does flip again. No-one knows how long it takes for the flip to occur, or even for certain what events might accompany it. It is obvious, however, that all of our complex electromagnetic technology would be affected. This could result in more than minor inconvenience for the aviation industry. Because the electromagnetic field also shields the Earth from certain kinds of radiation, a flip, if accompanied by a weakening of the field, may result in a temporary increase in the amount of solar radiation entering the atmosphere at low latitudes.

By examining fossil magnetism preserved in rocks, an estimate of the age of the rocks can also be gained. This estimate is particularly accurate if scientists can examine a stacked series of rocks, such as can be seen in some cliffs or the rock cores taken from drill holes by geologists. Then, the records of magnetic reversals, as they are called, can be read much like a supermarket check-out counter can read the bar code on a box of cereal.

Knowing this, the scientists aboard the research vessel HMS *Owen*, which was taking part in the International Indian Ocean Expedition, doubtless felt pleased when they began to record the presence of broad strips of variously magnetised rock on the ocean floor. But it was only later, when they examined their records in detail in the laboratory and compared them with other data, that they realised what the magnetic strips implied. They found that the magnetic record from one side of a structure called the mid-ocean ridge was an exact mirror image of the pattern recorded on the other side of the ridge! Furthermore, they discovered that the oldest rocks were found farthest from the mid-ocean ridge and the most recent ones at the ridge itself.[99] That, they realised, could mean only one thing—the ocean floor was being created at the mid-ocean ridge. This meant that the ocean basin was growing by getting wider. This, in turn, had important implications for the continents—they must be getting further apart!

These astonishing discoveries opened an old wound in science, for some 55 years earlier, a German meteorologist and explorer of great renown, Alfred Wegener, had proposed just such a theory: that the continents wandered across the face of the Earth, sometimes getting closer together but at other times moving apart. But when he first presented his work to an international meeting of geologists in 1912, he was virtually shouted off the stage.[99]

Undeterred, Wegener elucidated his revolutionary theory in a book, reprinted in various editions between 1920 and 1929. His work was a *tour de force*, the first attempt ever to explain the evolution of world geography. Yet it convinced few researchers. Only those few living in the southern hemisphere (including Australia's incomparable Sir Douglas Mawson), where the evidence is so clear and where the political storms of Europe are somewhat distant, took up his ideas in earnest.

Far from winning him the acclaim he deserved, Wegener's discovery was to bring him enormous suffering. From the moment he descended the podium of the international meeting in 1912, he was dogged by criticism so full of vitriol that it cannot be explained adequately given the quality or methods of his science. For example, the geologist Philip Lake said:

Wegener is not seeking the truth, he is advocating a cause and is blind to every argument and fact that tells against it.[99]

The President of the American Philosophical Society called

Wegener's work 'utter, damned rot'. Admittedly, there was no plausible mechanism then known which could move continents, so Wegener's hypothesis was weak in this regard. But it is now clear that two non-scientific factors were more important in working against him. First, he was an outsider (a meteorologist), who had just shown the world's most eminent geologists a brilliant model of how the Earth evolves; and second, he was a German.

The feelings of defensiveness and xenophobia aroused by researchers who stray from their particular speciality remain strong in science even today. This is unfortunate, for the most brilliant insights are often produced by those outside the immediate field concerned. It is easy to forget the enormous animosity felt towards Germans by English-speaking people at about the time of the First World War. Throughout Australia and New Zealand, towns with German-sounding names were rechristened, sometimes to celebrate the site of an ignominious German defeat. Many fine and patriotic Australians of German origin were imprisoned and harassed, while many more were forced to conceal their German ancestry by changing their names. My Norwegian grandfather changed his name from Fridtjof Thiggurson to the more acceptable Fred Scott at this time.

Wegener did not let the vicious attacks destroy him. Somehow he rose above them, despite the fact that he had to live the rest of his life with a reputation as 'that meteorologist with the absurd theory' and was denied several key positions because of it. He died at the age of 50, while leading an exploring expedition high on the Greenland icecap. Only he and a young Greenlander had been courageous enough to carry supplies to a team facing starvation, stranded in winter in the centre of the vast ice desert. Tragically, both men perished on the return journey, Wegener apparently of a heart attack. When his body was located in the spring, his colleagues found him stitched into a cover of reindeer skin and sleeping bags, looking peaceful, a slight smile on his face.[99] Perhaps he was granted a vision of the rout of his enemies, many of whom were still alive to face the irrefutable evidence supporting his theory, presented in 1966. But then, he was probably too great a man to find solace in the discomfort of his detractors.

As a result of the research of Wegener and others, it is now almost universally accepted that the continents are not fixed in their place, but have drifted across the face of the Earth for millions of years. By using sophisticated satellite and laser technology, scientists have now actually measured how fast the continents move. They have found that Australia is moving north at the rate of approximately 6 centimetres per year. Researchers also think that they have discovered the forces that drive the continents. This is important, for one of the great charges laid against Wegener was that no force could move such enormous masses. Much of the Earth is formed of molten rock, which has convection currents within it. Similar currents are readily seen in a pot of boiling water. They result from the liquid below being hotter than that above. We can imagine the Earth as being somewhat like an enormous pot of boiling pea soup. The continents are the thin scum on the top. They move ceaselessly over the surface, driven by the boiling maelstrom below.

These discoveries, all made over the last 30 years, have revolutionised our concept of the world. Armed with them, scientists have been able to chart the positions of the world's continents in times past. They have discovered that Gondwana, that long hypothetical great southern land, was once real and most importantly for our story, they have found that Australasia, that family of 'new' lands, formed a single landmass some 80 million years ago. Remarkably, they have also discovered that Gondwana straddled the South Pole and that part of what is now Australia lay within the Antarctic circle, a peninsula at the end of the great southern land.

Eighty million years ago the world was a very different place from what it is today, yet it nurtured within it the seeds of our modern world. Dinosaurs still ruled the planet at that time and would continue to do so for a further 15 million years. But birds, mammals and many familiar plants were already in existence.

It is amazing just how much evidence still exists of that ancient time. In the deserts of southern Western Australia lie some ancient and now dry river channels. They correspond precisely with some equally ancient drainages found in Antarctica. Before great Earth movements split Antarctica and the western part of Australia asunder some 100 million years ago, they formed a single drainage. On a world scale, the survival of such trivial surface features as small river channels over such a long period of time is highly unusual. Even more remarkable is the fact that growing along the Australian parts of those river channels are plants belonging to ancient southern families such as Proteaceae and Myrtaceae. Australia's near ubiquitous grevilleas and hakeas, so beloved by the bush gardener, are members of the Proteaceae family, while the eucalypts are our best known Myrtaceae.

While the Antarctic ice sheet has long obliterated their Antarctic cousins, plants belonging to these families still inhabit many of the fragments of what was once Gondwana. The well-known South African proteas, clinging to life on the cool mountains at the extreme southern tip of Africa, represent one such group of survivors. In the mountains of Chile, New Zealand and New Caledonia there survives even closer relatives of the Australian species. Tree-waratahs of the genera *Oreocallis* and *Alloxylon*, whose magnificent red flowers brighten the rainforests of northern New South Wales and Queensland, also grace the cool, misty forests of Chile. Species of the genus *Macadamia*, which includes Australia's delicious Macadamia nut, thrive in New Caledonia and Sulawesi, while geebungs (genus *Persoonia*), so common in eastern Australia, also grow in New Zealand.

Study of the evolution of such plants provides evidence that their ancestors inhabited Gondwana. Furthermore, most still cling to the cool, moist environments inhabited by their Gondwanan ancestors. Salt water barriers are absolutely impermeable to them. Thus there is little doubt that they reached their present distribution by clinging atop various fragments of Gondwana.

Every year brings more research in this area of science called biogeography and each year more startling evidence of the Gondwanan origins of our fauna and flora is found. Take our holly plants, for example. We are all familiar with the members of the genus *Ilex* for the contributions they make to our Christmas decorations. My own familiarity with them began in earliest childhood, when, I remember, our family had a particularly splendid plastic wreath with its spiny, dark green leaves and beautiful red berries. It always sat proudly atop the hot Christmas pudding, with my family all around, overheated, stuffed with rich food and more than a little tipsy, for the only dietary concession made to the 40° Celsius heat of an Austral summer was a copious supply of cold beer.

My mother, flushed with the heat of the oven and her own efforts, would carry an enormous turkey into the small, darkened room; blinds drawn against the intense heat and light outside. Formally dressed, we would all face the huge, intolerably hot bird. After that came a hot and dense pudding. Then, in forced acknowledgment that the ancestral mid-winter feast had shifted hemispheres, ties would be loosened, glasses filled and everyone would seek the shade of an old pear tree that stood in the typically spacious Australian back yard. It was, for many years, the only tree in the yard, unless the Hills Hoist be counted.

It was only when I visited England in late autumn and experienced the chilling cold and skeletal trees, that these occasions really made sense to me. After seeing holly growing in its native land I also realised why we had to make do with a plastic imitation. Even in those few parts of Australia where European holly can grow, it quite sensibly refuses to produce its beautiful red berries at the height of the torrid summer.

Today, only two native species of this predominantly northern hemisphere holly family maintain a toehold in Australia. Both are restricted to the far north of the continent. One can hardly imagine a more quintessentially European plant than holly and perhaps in part because of this, our two Australian species were long dismissed by botanists as Johnny-come-latelies from Eurasia. That was until several years ago, when a palaeontologist found fossilised holly leaves in sediments formed in an ancient lake in a rainforest some 60–40 million years ago at Maslin Bay near Adelaide, South Australia. At the time that those leaves formed part of a dark green rainforest canopy, Australia was still in Gondwana. Many other discoveries of fossil holly pollen and leaves have now been made in ancient Australian sediments. Because of these discoveries we now know that the holly family has an 80-million-year-long history in Australia and that it is a venerable member of our rainforest plant assemblage. Indeed, holly pollen is the oldest fossil pollen ever to be found in Australia. Thus, our few surviving hollies are as Gondwanan as geebungs; and, what is more, the entire group is probably Gondwanan in origin. The

Johnny-come-lately may well be that one species which has adapted to life in the icy north of Europe and with which our ancestors were wont to garnish their Yuletide feast.

As a result of these and similar studies by palaeontologists, zoo-geographers, climatologists and other researchers, an image of the Australia of 120–80 million years ago is slowly beginning to emerge. It is an absolutely fascinating image of a now vanished world that was destined to give birth to Australia as we know it today; and to give birth to many northerners—floral and faunal—as well.

CHAPTER 2

AUSTRALIA IN GONDWANA

I have always had a terrible curiosity to know what things were like in Gondwana some 100 million years ago, before the islands of Australasia broke apart. At one level it is a keen, non-intellectual desire to *know* the landscape in the way one knows the body of a lover. I satisfy this need every time I split a rock to reveal a pebble lying just as a small eddy in a stream had left it 100 million years ago; or the impression of a leaf, chewed by an insect before it fell to the bottom of an ancient muddy pool, where it was covered and pre-served for aeons. But in addition to satisfying this emotional need, I study fossils because I have a strong and urgent intellectual need to know about them and their lives. The key to understanding Australia—and even ourselves—lies in the distant past.

Unfortunately, nature has not endowed Australia with a rich fossil record. The period about 80 million years ago, when New Zealand and New Caledonia split from Australia, is particularly poorly repres-ented. For this reason we must go back a little further in time, to 120–100 million years ago, for evidence of what Gondwana was like before that break up.

Every scientist makes a few fortunate discoveries during his or her career and I, thankfully, am no exception. As a child I was fascinated with dinosaurs to the extent, so my indulgent mother informs me, that when the makers of the breakfast cereal 'Weeties' began to include a small plastic dinosaur in each pack, I embarked upon a Weeties-only diet for several weeks. Unfortunately, I have never grown out of my obsession and although I would not again subsist happily upon Weeties in order to obtain plastic dinosaur models, I have always been willing to devote any spare time I had to a pursuit of all things dinosaurian.

To my great chagrin, I had been born and raised in Victoria and, as every schoolboy knew at the time, almost all of the few dinosaurs discovered in Australia had been found in distant Queensland. Victoria—to my intense shame—had produced but a single pathetic relic; a four-centimetre-long bone dubbed by some wag as 'The Cape Paterson Claw' after the place where it had been found in 1900. Although childishly embarrassed by the tiny size of our State dinosaur treasure, one of my most valued possessions when seven years old had been a slender leaflet produced by the curiously named National Museum of Victoria, entitled *Fossils of Victoria*. Written by the Reverend Edmund Gill, who was perhaps one of Australia's last clergyman naturalists and certainly the only one I ever met, it included an obscure photograph of the doleful dinosaurian digit itself.

One of the highlights of my life came when, at age 10, I was admitted, awestruck, through the great doors of the Palaeontology Department, so that the State Palaeontologist could confirm my identification of a strangely shaped stone as some astoundingly valuable fossil (I had my hopes pinned on a petrified dinosaur brain). In consolation he opened a drawer of the vast collection and placed a small box in my hand. Doubly awestruck, I left swearing never again to wash the hand that had held The Cape Paterson Claw!

Many years later, while studying for a degree in Earth Sciences at Monash University, I had the great good fortune to meet Rob Glennie, one of Victoria's most knowledgeable and affable geologists. Rob has a unique knowledge of the history of geological exploration of Victoria. He and I soon discovered a common interest in The Cape Paterson Claw. I found to my great delight that Rob has a wealth of knowledge both about the man who found The Claw and the geology of the Claw-bearing rocks themselves.

The Claw had been discovered by one of the State's all-time great geologists, William Hamilton Fergusson. He was, by all accounts, an extraordinary man. Meticulous in maintaining standards, he did all of his work in frockcoat and waistcoat and every morning, regardless of where he was, he raised the Union Jack over his humble camp. Glennie found his geological work exemplary, for his maps were utterly reliable and detailed. Fergusson had been sent to Cape Paterson because the rocks in that area had coal-bearing potential. He had made a characteristically detailed map of the coast there, at a very small scale. Indeed, my eyes almost popped from my head when Glennie, while explaining this, casually unrolled an antique geological map of the Cape Paterson coast, upon which a large 'X' marked a spot labelled 'Claw found here'. It was a copy of Fergusson's original map, and it showed every detail of the geology of the area, right down to fossilised logs protruding from the shore platform.

The very next weekend saw Rob, myself and my cousin John Long (another dinosaur fanatic and now Curator of Vertebrate Fossils at the Western Australian Museum) standing on top of the high cliff that is Cape Paterson, examining the glum scene below. The tide was high, all but covering the rock outcrop. The weather, uncharacteristic for November, was bitterly cold, the rain slanting in on us almost horizontally. Rob finally said, 'I think it must have been about there' and we all descended the steep and perilous slope (we only found out afterwards that there was a walkway down the cliff a few yards to the west). Within seconds of arriving on the beach John called, 'Here's a bone' and we rushed over to see the cross-section of a tiny yet unmistakable bone showing in a pebble. It was the first dinosaur bone to be discovered in Victoria for nearly 80 years and while small, we all saw in it the potential to enter the world of dinosaurs.

From that moment I became obsessed. Every weekend would see me at Cape Paterson, combing the rocks for a taxonomically useful dinosaur bone. Within a few weeks I had one—the ankle bone of an *Allosaurus*. But it soon became clear that if the rocks were ever to yield more than a glimpse of Australia as it was some 120 million years ago, then a major effort at exploration and excavation of similar rocks around Victoria would have to be undertaken.

Fortuitously, the State of Victoria had a few years earlier appointed a most determined palaeontologist to its staff. Dr Tom Rich, Curator of Vertebrate Fossils at the now more appropriately renamed Museum of Victoria, had once told me that palaeontologists must have the will to fail. Tom Rich has that will on a massive scale. As a volunteer on one of his early 'digs', I and a crew of five or six others had worked for over a week, shifting tonnes of basalt and ancient soil without finding a single fossil. Sick at heart at this failure as only a teenager can be, I had sworn to quit the desolate diggings more than a dozen times. Finally, late one evening, one of us found a single tiny tooth cap, less that 2.5 millimetres long, turning despair into delicious victory.

Through his sheer 'will to fail' Tom has, over the years, turned this fossil site near Hamilton in western Victoria, into one of Australia's most important fossil localities. Through his effort, it has provided a glimpse of life as it was in a Victorian rainforest some 4.5 million years ago, complete with tree-kangaroos, tiny diprotodons and forest wallabies. Without this work, we would never have known that the rich faunas of possums and tree-kangaroos that presently characterise the rainforests of north Queensland and New Guinea, were once found as far south as Victoria.

Tom Rich and his wife Pat, also a palaeontologist who lectures at Monash University, were to need every ounce of the will to fail in their struggle to uncover the lives of Victoria's dinosaurs, for the effort needed was immense. Before anything else, the entire coast which included rocks of the right age had to be prospected. This involved a team of expert palaeontologists combing every inch of rock on some of Australia's most dangerous and least accessible coastline.

Experts from around the world were invited to participate in various aspects of the research work. Climatologists studied ancient ratios of oxygen isotopes, while palaeobotanists discovered the world's oldest flower and studied the adaptations of plants to polar conditions. Sedimentologists studied the flow direction and sediment deposition in the ancient rivers and lakes, while dinosaur experts Tom and Pat Rich studied the dinosaur bones themselves. Indeed, Victoria's dinosaur bones are perhaps the most travelled of any, for they have circled the globe several times as the Riches have carried them from collection to collection, looking for a match close enough to allow an identification to be made.

It is only now, after more than a decade of hard work, that a picture of southern Australia as it was 120 million years ago is emerging.[105] We now know that the sediments containing the fossils were laid down in a rift valley similar in structure to the great rift valley of eastern Africa today. Rift valleys form where a continent is being torn apart by continental drift. It is surprising to find that such a valley was forming between Victoria and Tasmania 120 million years ago, as these landmasses are still joined. It is quite likely that the ancient rift was part of a system that was eventually to tear Australia from Antarctica. Initially Australia and Antarctica began to separate to the north of Tasmania, but for some unknown reason this rift failed and another rift began to form to the south of Tasmania. This meant that Tasmania was destined to follow the rest of Australia north, rather than remain in the frigid south with Antarctica.

As in most rift valleys, there were plenty of active volcanoes. These volcanoes were rapidly eroded and the rock grains transported into the valley in seasonally active streams. As the sediments built up, the valley floor sagged, allowing more sediment to flood in. Eventually sediment built up to a depth of over three kilometres. The great sediment pile enclosed the remnants of ancient river beds, ponds, lakes and many other topographic features. These sediments are now exposed most spectacularly along the southern coastline of Victoria. There, the greenish rock with its large, brown concretions has weathered into fantastic forms. Moonlight Head, a huge promontory formed entirely of this rock, is Australia's tallest coastal cliff. In places all along the coast one can see pebbles and pieces of clay that mark the course of previous stream channels. It is in these channels that most of the dinosaur fossils are found.

One of the greatest surprises to come from the study of these sediments is just how cold southern Victoria was 120 million years ago. Palaeoclimatologists have discovered that the region then lay well within the Antarctic Circle. Various methods for estimating the mean annual temperature have been tried. One method, which examines the ratio of two oxygen isotopes, suggests that the mean temperature might have been zero degrees, but perhaps reaching as high as eight degrees above zero. Such temperatures are today found at Hudson's Bay and Toronto! It is exceedingly difficult to imagine dinosaurs and turtles living in such settings. Perhaps a more realistic estimate is derived from the study of plants, which suggests a mean annual temperature of a (still very chilly) 10° Celsius. Whatever the precise temperature, the wealth of fossil remains of plants and animals clearly indicates that the polar temperature at the time was considerably warmer than it is today.

While the relatively cool temperature was one problem for cold-blooded creatures, an equally serious one would have been the lack of light during winter. Depending upon the precise position of southern Victoria at the time, they would have had to endure between six weeks and four months total darkness each year.

Some understanding of climatic conditions 120 million years ago has come from studies of modern global pollution. We now know that as the amount of carbon dioxide in the atmosphere increases, the world becomes warmer, particularly at the Poles. This is known as the greenhouse effect. Scientists now tend to describe the world of 120 million years ago as a full greenhouse world. Thus the Rich study is useful not only for understanding the evolution of Australian environments but for telling us a little about the kind of world that our grandchildren may inherit.

The plant and animal fossils retrieved from the 120-million-year-old rocks of southern Victoria are fascinating. The plants are dominated by ancient types such as conifers, ginkgoes (today found naturally only in China), ferns, cycads and mosses and lichens. There were only a few flowering plants. Scientists have discovered what they think is the world's oldest flower in the rocks—a diminutive magnolia-like bloom, perfectly preserved between two layers of clay like a flower in a plant press. Species ancestral to the grevilleas and waratahs evolved after this time. It was to be another 40 million years before the 'new' lands were to go their own ways, so there was plenty of time for such plants to evolve and spread.

The animals, however, were unlike any living today. Dinosaurs dominated, but most of the Victorian species were curiously small. The most common kinds were chicken- to dog-sized, bird-hipped dinosaurs that ran on their hindlimbs and ate plants. One large-dog-sized species has been named *Atlascopcosaurus* (because, without the support of the hire company Atlas Copco, the Riches could never have excavated the bones). Another is named *Leaellynasaura*, after their daughter Leaellyn. *Leaellynasaura* is one of the few species known from a partial skeleton, which shows some curious adaptations. At about the size of a chicken, it was one of the smallest dinosaurs ever to exist. It had very large eyes and large optic lobes in the brain. It is likely that it used its acute vision to remain active during the winter darkness. Thus, it may have been active through the very coldest time of the year.

The very largest of the Victorian dinosaurs was a carnivorous species of the well-known genus *Allosaurus*. At about two metres high at the hip, it was dwarfed in comparison with its seven-metre-long North American relatives. It also lived long after its relatives from other continents had became extinct, for *Allosaurus* is known only from rocks of Jurassic age elsewhere (208–147 million years old), while the Victorian *Allosaurus* is about 120–110 million years old.

Other dinosaurs included an as yet unnamed, sheep-sized ceratopsian (a group of compact, heavily armoured, rather rhinoceros-like dinosaurs) and an ostrich-like species named *Timimus*. (I was, incidentally, rather pleased to hear that this dinosaur had been named after myself and the Riches' son Tim. The only thing previously named for me was far less noble—a New Guinean tapeworm which infests the bowels of possums!)

These two species are of particular interest, because both represent the earliest occurrence of their respective groups. Ten million or more years after these Victorian species existed, the rhinoceros-like and ostrich-like dinosaurs had evolved into many species which populated the landmasses of the northern hemisphere in vast herds. Along with holly, they may be gifts that Australia, or at least Gondwana, has given the rest of the world.

This curious dinosaur assemblage has provided some very unexpected insights into the lives and extinction of the dinosaurs in general. They are very different to the only dinosaur assemblages known from the Arctic region. There, the dinosaurs are larger and may have migrated seasonally in and out of the Arctic Circle. An ancient arm of the sea cut off the Victorian dinosaurs from access to the north and they would have had to stay put all year. The limitations that this placed upon their food resources may explain their small size. Because of this enforced isolation we also know that they were truly cold-adapted—not just seasonal visitors.

These discoveries are a strong challenge to those who believed that a deteriorating world climate led to the ultimate extinction of the dinosaurs some 65 million years ago, for it would have been difficult for the entire Earth to have become as cold as southern Victoria was 120 million years ago. Even if it had, we know that dinosaurs could survive in climates at least that cold.

In addition to dinosaurs, the bones of pterosaurs (flying reptiles), lungfish, turtles and crocodiles have also been found. The very largest animals to inhabit the area were not reptiles at all, but enormous amphibians known as labyrinthodonts. These great creatures, perhaps three metres or more long, superficially resembled huge newts and may have filled an ecological niche similar to that of modern crocodiles. Their discovery in 120–110-million-year-old rocks in Victoria was a great surprise, for they had become extinct elsewhere in the world some 80 million years before. Indeed, I well remember the sceptical reaction when I found the first bone that looked as if it might belong to a labyrinthodont. It was the back of a jaw which lacked teeth. Tom Rich and I were puzzled by it, but both of us felt that it matched nothing else but a labyrinthodont. Because the remains were so fragmentary and the survival of labyrinthodonts into the Cretaceous period seemed to be so outlandish an idea, we were reticent about pushing our claims in the face of scepticism from the experts. Instead we christened the fossil GOK (for 'God Only Knows', the most common refrain from the palaeontologists we asked to identify it) and hid it away in a museum drawer. It was nearly a decade later that another jaw was found. This one preserved some teeth and they revealed unequivocally that labyrinthodonts had inhabited Victoria's great rift valley.

Tom Rich suspects that the unlikely survival of these strange creatures for over 80 million years after their extinction elsewhere may be explained by the difference between reptilian and amphibian physiology. Amphibians can generally better survive in colder water than reptiles and it may be that the cold polar climate of southern Victoria offered a refuge to the amphibian labyrinthodonts that the otherwise superior crocodiles could not penetrate. A few million years later the Earth warmed and we find the first crocodile remains in the rocks of southern Victoria. By then, the labyrinthodonts were sadly absent.

The remains of more ordinary animals are also preserved, including several species of herring-like fish. Their bodies have been exquisitely preserved in lake sediments found near Koonwarra in southern Victoria. They are in such good condition that every scale and fin ray stands out in brown against the fine greenish sediments that enclose them. Along with these fish, the perfectly preserved larvae and adults of many insects have been found. These include cockroaches, cicadas, fleas, horseshoe-crabs (now extinct in Australia but surviving in coastal waters off North America) and water beetles.

The method of preservation of these fossils is rather curious, for their bodies were trapped in lake sediments which are divided into layers called varves, each of which is a few millimetres thick. The varves alternate, one being composed of very fine green clay, the next with slightly coarser and browner sediment. Scientists think that the lake was shallow and froze over every winter, killing many of the lake inhabitants. During the winter when the lake was frozen, fine green sediment settled to the bottom. When the lake thawed in spring, coarse sediments were deposited with the spring floods, covering the bodies and preserving them. Thus, every year one fine green and one coarser brown layer was formed, leaving us an exceptionally detailed, year by year picture of how life was in Victoria 120 million years ago.

Some rather enigmatic fossils have also been preserved in these sediments. These include five exquisitely preserved, rather tiny (one- to two-centimetre-long) feathers. These have long been considered as evidence that birds inhabited the area. The fossilised remains of birds are certainly known from equally old rocks elsewhere, but these feathers worry me. I find it difficult to imagine chicken-sized dino-saurs surviving in such a cold climate, particularly if they were active over winter. If they were able to do this they must have been warm-blooded. And if warm-blooded, they must have had ways of retaining body heat. Given their size and rather primitive structure I think that the fossilised feathers of Koonwarra may have belonged to the dinosaur *Leaellynasaura*. This is not really so surprising, as birds are the only surviving members of the dinosaur family. All birds have feathers, so it is not unreasonable to suspect that some of their rela-tives had feathers also.

There is also some evidence that mammals were present, for the sediments from Victoria have yielded the fossilised impression of a flea, of a kind specialised to live among hairs. Mammals are the only living group of vertebrates to possess hairs, so ever since this flea was described in the 1960s the presence of mammals in Australia 120 million years ago had been suspected. It was not proven, however, until 1985, when Mr David Galman, an opal miner from Lightning

Ridge, New South Wales, came to the Australian Museum with a small box of fossils he wished to sell.

Lightning Ridge is a dusty inland town, as different from wet and hilly Cape Paterson as anywhere in Australia. Yet they both have something very important in common, for the rocks that underlay them were formed about 120 million years ago. At Lightning Ridge the actual fossil bones have long since gone, dissolved by acidic groundwater. The spaces that the bones left in the rock remain and in many instances they have been filled with an unusual form of silica known as opal. Thus, at Lightning Ridge, when a miner sinks his shaft in search of opal, he occasionally runs across the most exquisite natural casts of the bones of animals that existed over 100 million years ago. Many of the creatures whose remains have been found are similar to those found in Victoria although, perhaps because Lightning Ridge lay further from the South Pole than Cape Paterson, the remains of larger dinosaurs are also found there but labyrinthodonts are absent.

By far the most striking fossil ever discovered at Lightning Ridge is a small jaw with three teeth in it. It is beautifully preserved in transparent opal, so that tiny details of the root and nerve canals can all be seen. When examined by palaeontologists it proved to be the jaw bone of an ancient relative of the platypus. Named *Steropodon galmani* in honour of Mr Galman, it remains Australia's most ancient mammal fossil by some 60 million years.[4]

By assembling all of the information gleaned from fossil deposits such as those from Victoria and Lightning Ridge, we can build a picture of what Australasia was like some 120 million years ago. To the east and south lay New Zealand and New Caledonia, still firmly attached to the eastern side of Australia. To the north, New Guinea, if it existed at all, was a couple of islands lying near the north coast of Australia. To the south lay Antarctica, still attached to Australia, but with signs of the developing rift that was to set Australia careering northwards some 80 million years later.

All of these lands except Antarctica lay well to the south of their present position, with part of Australia, all of New Zealand and possibly New Caledonia, lying within the Antarctic Circle. This was a greenhouse world and although chilly within the Antarctic Circle, it was not refrigerated as lands lying at such latitudes are today. Because of the greenhouse conditions, there were no polar ice-caps. Thus the sea-level was considerably higher than it is at present and Central Australia was occupied by a great inland sea.

On this great and alien landmass many of the kinds of organisms that Australians are familiar with today were evolving. These formed the common stock of plants and animals for all of the 'new' lands, but subsequent history was to treat each land differently, resulting in strikingly different faunas. By 80 million years ago, when the 'new' lands began to split apart and drift separately across the face of the Earth, a rich assemblage of plant and animal species had developed.

The patterns of evolution that life followed on each of the 'new' lands tells us much about the forces which were strongest upon them. As these forces continue to shape everything that inhabits the 'new' lands today, the following chapters examine in some detail the patterns of life in each of these four places.

Position at 160 million years ago

The great southern landmass Gondwana. Part of what is now Australia lay within the Antarctic circle.

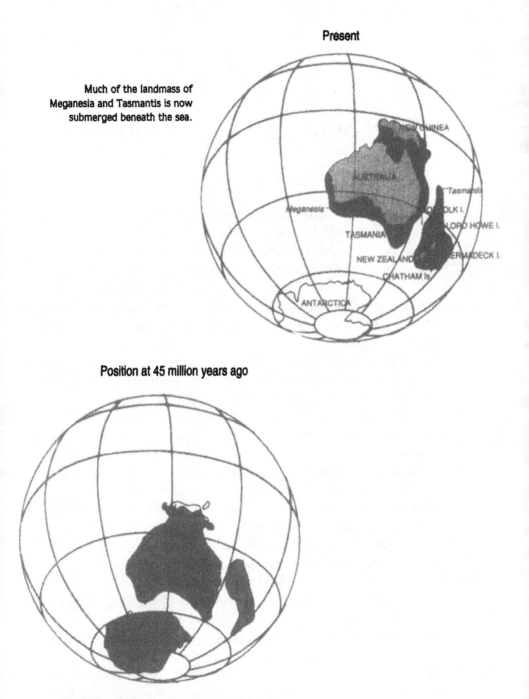

Present

Much of the landmass of Meganesia and Tasmantis is now submerged beneath the sea.

NEW GUINEA

AUSTRALIA

Tasmantis

Meganesia

NORFOLK I.

LORD HOWE I.

TASMANIA

KERMADECK I.

NEW ZEALAND

CHATHAM Is.

ANTARCTICA

Position at 45 million years ago

Separation of the Australian continental plate was complete by approximately 40 million years ago.

CHAPTER 3

LAND OF GECKOS, LAND OF FLOWERS

At between 86 and 82 million years ago a large sliver of land detached itself from eastern Australia. A rift valley resulted from the split probably initiating the formation of the Australian Alps. Dubbed Tasmantis by geologists, this sliver of land was the first part of Australasia to split away from Gondwana. Now largely submerged beneath the sea, it includes New Zealand and New Caledonia, and Lord Howe, Norfolk, Kermadec and the Chatham Islands.

As Tasmantis itself began to fragment and sink, New Caledonia came to lie astride the tropic of Capricorn, between 19° and 23° south, while New Zealand now lies at between 32° and 48° south of the equator. Tasmantis continues to move across the Pacific and geologists estimate that 100 million years from now, the paths of Tasmantis and Meganesia (the name of the landmass including Australia, New Guinea and Tasmania) will again cross. Then, these ancient southern lands will once again amalgamate into one.

With an area of about 17 000 square kilometres, New Caledonia is the smallest of the 'new' lands and is in some ways also the strangest. An elongate island (350 kilometres by an average of only 50–70 kilometres wide), the spine of New Caledonia is mountainous and well-watered. True forest is, however, restricted in distribution and many of the mountains and plains, particularly in the south, support a low, heath-like vegetation that the Caledoche (New Caledonians of French descent) call *maquis*. At a distance, the colour it imparts to the mountain ranges resembles that of heather in the Scottish Highlands. It may have been this that prompted James Cook, who in September 1774 was the first European to see the strange island, to name it New Caledonia. Cook himself noted the great similarity of the flora of New Caledonia and Australia. To the modern biologist, the similarity of the *maquis* to heather vanishes when it is examined at a distance of less than a few miles and similarities with Australia become startlingly apparent.

I remember vividly my first day on the New Caledonian *maquis*. The barren moonscape was an unsettling contrast with the verdant paradise of the other Pacific islands I had visited. It was only when I got out of my car and examined the ground more closely that my sense of alienation was dispelled, for the ground was littered with brown, shiny pebbles, much like those seen on a South Australian gibber plain. Growing isolated among the stones were tiny, almost bonsai-sized bushes. To my astonishment I recognised one as belonging to the same genus (*Araucaria*) as the great klinki pines of New Guinea and the bunya pine of Queensland, both of which are giants towering over lush rainforests. Others were strange-looking relatives of grevilleas and still others had their relatives among the peculiar, primitive flowering plants of north Queensland's rainforest. Eventually I managed to procure a handbook on the New Caledonian flora and learned to my amazement that over 3000 species of indigenous plants are found there. By comparison, New Zealand, which is about 15 times larger, has only 1460 species. Even more extraordinary is the fact that the New Caledonian flora includes some of the most wildly beautiful and primitive plant species to be found on Earth.

The New Caledonian flora is remarkable for including the most comprehensive assemblage of ancient trees in the world. The ancient conifers are particularly well represented. No less than 13 of the world's 19 species of monkey-puzzle trees (*Araucaria*) survive there, as do the majority of the related kauri pine species. This family of trees, now by and large restricted to the southern hemisphere, were a dominant life form everywhere during the age of dinosaurs; the famous fossilised forest of Arizona in North America, for example, is composed of their trunks. There are, in addition, 15 more genera with over 40 species of other ancient pines in New Caledonia. The most unusual species is placed alone in the genus *Parasitaxus*. It is unique among the pines for it is a mistletoe-like parasite which lives on other pines.[120]

There are five families of primitive flowering plants which are unique to New Caledonia, as well as 17 genera of palms (including some of the world's most graceful species), 191 orchids, and eight genera and 42 species of proteaceous plants (the family that includes the banksias, proteas, grevilleas and the like). New Caledonia's Proteaceae rival those found anywhere else on Earth for their beauty and unusual flowers. The species of *Kermadecia*, for example, look rather like huge grevilleas, but have flower spikes 40 centimetres or more long, with each individual flower being relatively enormous compared with Australian members of the family.

New Caledonia also has an exceptionally rich assemblage of myrtaceous plants, although the eucalypts, which are the best known members of the family in Australia, are not present. Among the 200-odd myrtaceous species found there, are some with the most attractive flowers I have ever seen. The species of *Xanthostemon*, for example, particularly *X. aurantiacum* with its brilliant red, green and yellow ti-tree-like flowers, would set any subtropical garden ablaze.

Curiously, many of New Caledonia's plant species do not grow in true rainforest but as natural 'bonsais' on the *maquis*. The reason for the development of this extraordinary plant community lies largely in New Caledonia's unique geological history. The fact that the original plant assemblage was drawn from the rich plant communities of old Gondwana is one vital factor, but another, perhaps more decisive one, is the nature of New Caledonian soils. The rocks from which many New Caledonian soils developed, particularly those of the *maquis*, were formed deep under the sea and originated from the oceanic crust. Only rarely do rocks from the ocean crust appear on land, because, being heavier than the continental rocks, they usually sink below them. Where they do appear they bring a rich mix of metals to the surface of the Earth. In New Caledonia the rocks of the oceanic crust are unusually rich in nickel, magnesium, chromium and manganese. In other parts of the world where similar rocks occur, copper and gold are often more abundant.

In high concentrations, both nickel and copper are toxic to most plant life, yet thin, nutrient-poor, nickel-rich soils everywhere support unique plant communities. These, in turn, support great radiations of insects and other invertebrates, particularly land snails. These rich communities are based on plants which have evolved a tolerance to nickel and/or copper. Despite their tolerance, few grow into trees, but remain as wiry, small-leaved shrubs which grow quite slowly and never support many leaves. As will be explained below, such an environment is very conducive to rapid speciation among plants. This is because variation in nutrients and in nickel and other metal concentrations allows plants to specialise. One species, for example, can grow where nickel is most abundant. Another related species establishes itself where copper concentrations are high and yet another can grow best where concentrations of both are moderate. In good soils, where nutrients are abundant and toxins absent, just a few plant species can dominate because none have to specialise to grow in unusual conditions and those that can use nutrients best outcompete all others. Thus, paradoxically, some harsh environments for plants, such as the *maquis*, support great plant diversity.

While the story of the evolution of the plants of New Caledonia is remarkable, that of its animals is even more strange. We know that at the time New Caledonia broke away from Gondwana quite a variety of animal species were available to populate the newly formed islands. Dinosaurs, turtles, frogs, birds, the ancestors of the platypus and the New Zealand tuatara, geckos and skinks were but a few. For various reasons very few of these groups survived in New Caledonia and it came to be populated by an oddly unbalanced fauna.

Until the arrival of humans about 3500 years ago, New Caledonia was home to crocodiles, lizards, turtles, a few bats (none particularly unusual and all surviving) and birds that had probably arrived by flying over the sea from Australia, Melanesia or New Zealand. All of the larger reptiles—its crocodiles, horned turtles and goannas—became extinct shortly after the arrival of humans.

Given its size and ancient heritage, New Caledonia supported a very modest array of birds. Only 44 species of land birds, including those now extinct, have been recorded from the island.[7] This contrasts markedly with the 169 species once found in New Zealand, or even the 60 living species (excluding extinct ones) found in similar-sized yet more isolated Fiji.

The largest of all of New Caledonia's birds was the giant megapode *Sylviornis neocaledoniae*, a relative of the mallee fowl and scrub turkey of Australia. At about 70–80 centimetres tall and 10–20 kilograms in weight, it was the size of a large goose or turkey and was flightless. Its beak was large and powerful and it had a prominent bony knob on the top of its head. It was probably a herbivore and may have made mounds within which to incubate its eggs, much as the Australian mallee fowl does today. Some researchers think that they have discovered the remains of such mounds, marked by accumulations of pebbles, in the hills of New Caledonia. It seems to have become extinct when humans arrived in New Caledonia some 3500 years ago.

The next largest New Caledonian bird was a flightless species of swamp hen, *Porphyrio kukwiedi*, which would have stood about 50 centimetres tall and weighed less than 10 kilograms. It was perhaps similar in appearance to the takahe (*Notornis mantelli*) of New Zealand. Third largest was a fossil relative of the still living kagou (*Rhynochetos jubatus*), while the kagou itself was next in size. The kagou is the only living member of its family and it is a most peculiar-looking bird. Predominantly grey and rather heron-like in appearance, it has large eyes and a large floppy crest on the head. It has a group of brightly banded feathers used in a striking display, but which are normally hidden in the folded wings. Flightless and ground-nesting, it is vulnerable to introduced predators, and may only have survived because of its extremely secretive habits. Today, it is gravely endangered, its few known nest sites being protected by vigorous rat-trapping programs.

Fifth in size and weight was a now extinct kind of scrub fowl (genus *Megapodius*) somewhat smaller than the scrub fowl inhabiting Australia's Cape York today. The only other prehistorically extinct birds were two large pigeons and two falcons. Within historic times, a rail of the genus *Tricholimnas*, which was related to the Lord Howe Island woodhen, has also become extinct. This very modest assemblage of larger birds stands in distinct contrast with the extravagant feathered faunas of New Zealand, for its largest species weighed less than a tenth as much as a large moa.

Curiously, it was among the reptiles that the finest flower of New Caledonia's fauna is to be found. Hectare for hectare, New Caledonia has one of the most diverse, if not *the* most diverse reptile fauna on Earth. Measured as a proportion of landmass, it is more than 10 times richer in species than Australia, which itself harbours the most diverse lizard communities to be found anywhere.

As is the case with many of Australia's reptiles, it was long thought that New Caledonia's reptiles must have come from elsewhere, having arrived relatively recently by dispersal over the sea. Increasingly, this view is being challenged and it now seems reasonably clear that many have direct Gondwanan origins.

Before people arrived, the top herbivore on the island was undoubtedly the gigantic horned turtle *Meiolania*. These huge, land-dwelling reptiles belong to an extinct group of very primitive turtles which could not draw their heads into their shells. The remains of similar horned turtles have been found in 65-million-year-old rocks from South America, so there is little doubt that they are Gondwanan survivors.

They were heavily armoured, the tail formed a spiked and solid bony mace, while the head sported a pair of horns remarkably similar in position and shape to those seen in cattle. The New Caledonian turtles were small in comparison with those once found in Australia. They may have weighed 100 kilograms or less. No-one knows how many species of horned turtles existed in New Caledonia, but it would not be surprising if there were several, for most lands have more than one species of large herbivore.

The top predator in the New Caledonian ecosystem was probably a goanna rather similar to Australia's mangrove monitor (*Varanus indicus*), but at about 10 kilograms in weight it was perhaps a little larger.[7] It was probably an opportunistic predator, consuming anything that it was large enough to subdue.

New Caledonia was also home to the Caledonian crocodile, an extraordinary and ancient pygmy member of the crocodile group.[7] Known scientifically as *Mekosuchus inexpectatus*, its remains were first discovered only in 1980 and it was named as late as 1987. At about two metres in total length and about 30 kilograms in weight, as far as crocodiles go it was a true pygmy. The Caledonian crocodile is so unusual that some scientists place it in its own family, the Mekosuchidae. Its remains have been found in deep caves well away from water. This, as well as the shape of its bones, indicates that it was a land crocodile. It was distantly related to other hoofed and carnivorous land crocodiles which, 60 million years ago, were found throughout the world. All are now extinct. Some fossilised remains of the Caledonian crocodile date to as little as 1800 years ago. Thus, it was the last of the world's land crocodiles to become extinct. Recently, the jawbone of a Caledonian crocodile was recovered at an old village site, indicating that the first humans to reach New Caledonia hunted it.

Without doubt the most unusual thing about the Caledonian crocodile was its teeth, for while those at the front of the jaw were sharp and pointed as in all other crocodilians, its rear teeth were blunt and rounded. A gap appears to have separated the sharp from the blunt teeth. This suggests that it had a diet very different from that of any living crocodile, for it seems likely that it used its blunt rear teeth to crush mollusc shells.

This extraordinary adaptation tells us much about the New Caledonian environment, for the diverse teeth of the Caledonian crocodile suggest a diverse diet. It is probable that prey species suitable for capture with normal crocodile teeth were so few on New Caledonia that they were insufficient to support even a pygmy crocodile. To survive, the Caledonian crocodile had to add blunt teeth capable of crushing more common food sources such as snails and clams. Only on resource-poor New Caledonia were the adaptable crocodiles pushed to such extremes.

Unfortunately, the only New Caledonian reptiles to survive to the present are two families of lizards: the skinks (with 27 species) and the geckos (with 21 species).[12] While most of New Caledonia's skinks are much like skinks everywhere else, it has one extraordinary and now possibly extinct species that deserves comment. *Phoboscincus bocourti*, or the terror skink, as its name aptly translates from the Greek, is known from a single individual collected by a biologist over a century and a quarter ago somewhere (the exact locality being lost) on New Caledonia. This remarkable lizard was over 40 centimetres long. Externally it looks rather like an elongated version of the familiar Australian blue-tongue lizard. Its teeth, however, reveal a very different lifestyle from any other similar-sized skink, for they are long, recurved and distinctly vicious in appearance. There is no doubt that the terror skink ruled over the smaller reptiles of New Caledonia as effectively as *Tyrannosaurus rex* did over the lesser dinosaurs of its era, or the lion does on Africa's Serengeti today. What happened to the terror skink, or even whether it still survives at some remote location today, remains unknown. It is just one of the many mysteries of the poorly known Pacific.

But it is among the geckos that the strangest surviving reptiles are found, for there is a group of species that form an evolutionary radiation unlike that seen anywhere else on Earth. Fourteen of New Caledonia's gecko species belong to an ancient subfamily whose distribution covers that of much of old Gondwana, including northern New Zealand and Australia.[11] There is no doubt that they have been inhabitants of New Caledonia for the past 80 million years and that they have been fine tuning their ecology to conditions there all that time.

As a result, they have produced some extraordinary species, including *Rhacodactylus leachianus*, the world's largest surviving gecko species. At up to 40 centimetres long and the thickness of a man's wrist, it is, by gecko standards, a titanic beast. As with many geckos, it has the ability to virtually disappear against the mottled trunk of a rainforest tree. I have seen them, lying flat upon a grey-barked tree with their skin and even their eyes, each over a centimetre in diameter, taking on the tone of the bark perfectly. They hug the branch so tightly that even their contour is obscured. It is only when they move their enormous, webbed feet to reveal the bright yellow soles that the magic of their disappearing act is broken.

These huge geckos have a lifestyle as peculiar as their appearance suggests. They are entirely arboreal and nocturnal and emit a strange croaking sound. Their diet consists mainly of fruits and flowers, but they have been reported to eat birds as large as honeyeaters and even the young of their own species. They are thus New Caledonia's ecological equivalent to some of the more omnivorous monkeys and possums found elsewhere in the world, but perhaps have a slightly broader ecological niche.

A related but smaller species, which has not been seen for about a century and may well be extinct, has twin lines of dinosaur-like spines down the back. Yet another species (*Rhacodactylus trachyrhynus*) has adapted to a more exclusively carnivorous lifestyle. At only 17 centimetres long from nose tip to the base of the tail, it is the diminutive leopard of the New Caledonian rainforests, pursuing and subduing other lizards as well as insects. Strangely, its young some-times hide in the centre of water-filled bromeliads.

The species *Rhacodactylus auriculatus* has evolved into a chameleon-like insectivore of the *maquis*. It sports a pair of short horns which emerge, cow-like, from the back of its skull. It lays only two eggs per clutch. Yet other members of this extraordinary group have filled more standard gecko-like niches.

The fauna of New Caledonia is so strange that it begs interpreta-tion. We know that mammals, birds and geckos were all present on the starting blocks when the island broke away from Gondwana. Why was it that the geckos, at first an unlikely looking species to dominate a fauna, became New Caledonia's miniature ecological equivalents of monkeys and possums? Why did a turtle fill the ecological role of an elephant and a crocodile turn to eating mussels and snails?

The answers almost certainly lie in New Caledonia's toxic soils, small size and position just above the Tropic of Capricorn. The tropi-cal climate enjoyed by much of New Caledonia means that cold-blooded species can use the sun almost daily to raise their body tem-perature. The poor soils dictate that overall biological productivity is low. The small size of the island means that species which need large home ranges can sustain only very small populations. These factors work against animals that use lots of energy, such as all warm-blooded animals (including mammals and birds) and against large animals in general.

The general situation can be envisaged this way. Imagine a bird and a lizard of similar size that eat the same prey, in this hypothetical case, a species of grasshopper. Working in favour of the bird is that it is faster, more agile and thus more likely to find and catch prey than the lizard. Working against it is that it must put a very large amount of energy into keeping its body warm and into powered flight (which is generally the most energy-expensive way to travel). Thus, it must eat every day in order to meet these expenses. A further disadvantage is that a given amount of habitat can support relatively few such mobile and energy-demanding creatures.

The lizard on the other hand is slow, restricted to the ground and cannot travel far. But in its favour is the fact that it is an energy miser and that many individuals can be supported by a relatively small patch of habitat. It uses the warmth of the sun to 'hitch a free ride' in raising its body temperature and it can go without food for a long time. If necessary it can 'close down' or aestivate. In this condition, its metabolism just ticking over, it can rest in a burrow for months until favourable conditions return.

In 'normal' parts of the world, where grasshoppers are moderately abundant, the bird enjoys a tremendous advantage and usually out-competes the lizard, for the only restriction upon it—the constant availability of large amounts of suitable fuel—is removed. New Caledonia, however, falls at the extreme low end of the scale as far as energy availability goes. There, poor soils and small land area mean that energy is critically limited. Under such conditions, it may be that no matter how good a species is at finding food, it may some-times have to go for weeks without eating. Furthermore, because the land area is so small, the total populations of any energy-greedy crea-ture may easily fall below the critical level (geneticists reckon it at about 500 individuals) necessary for long-term survival.

It is for these reasons that New Caledonia's land reptiles outnum-ber its land birds—even including its extinct birds—and the reptiles fill such a wide variety of ecological niches. It is also why the Caledonian crocodile was forced to diversify its food choices and why a turtle, rather than a large monotreme, became top herbivore. In a sense, New Caledonia was made for reptiles as much as it was made for flowering plants. It is a sort of perverse Garden of Eden, with a scaly creature hiding behind every bloom, waiting millions of years to tempt the wandering biologist, or to amaze any visitor to this most exotic of lands.

CHAPTER 4

LAND OF SOUND AND FURY

While the flora and fauna of New Caledonia were being transformed by toxic soils under a tropical sun, its southern cousin New Zealand was following quite a different path. The island archipelago that is New Zealand has been a bit of a stay-at-home, for it has drifted the least distance from its Antarctic cradle. Coming to rest in chilly climes at 32–48° south, it has not escaped the influence of its parent the Antarctic continent, for during each of the Earth's last 17 ice ages, New Zealand has been transformed by snow and glaciers. With the exception of a tiny eight-square-kilometre ice-cap in Irian Jaya, New Zealand is the only one of the four 'new' lands to support a glacier at present.

In a way, New Zealand's evolution has been driven by forces almost opposite to those that shaped life in New Caledonia. New Zealand, for example, has not recently suffered a deficit of good soil. Indeed, of all the 'new' lands, it has been most blessed in its extensive and fertile covering of soil. This covering has been formed by all three processes that commonly contribute to soil formation: glacial action, volcanoes and rapid mountain-building. At about 265 000 square kilometres, New Zealand is some 15 times larger than New Caledonia (but only about one-thirtieth the size of Australia) and the climate is much colder. Like New Caledonia, it is among the most isolated of the world's landmasses.

Although it started out with a similar fauna and flora to New Caledonia and Australia, New Zealand's fertile soils, relatively large size, cold climate and isolated position have dictated that it evolve in a very different direction. Its plant communities, for example, are much less diverse (with less than half the species found in much smaller New Caledonia), although they include many species from Gondwana. A conspicuous absence is the ancient tree genus

Araucaria, which is so abundant in New Caledonia and so ancient that it provided shade for the dinosaurs. New Zealand has probably long been too cold to support them, for today these lofty pines are restricted to islands north of Norfolk Island in the Pacific. The beautiful protea family is also extremely poorly represented with only two species, perhaps because New Zealand lacks the really poor sandy soils derived from ancient rocks that members of the family seem to thrive in elsewhere.

New Zealand's cold climate has doubtless played a major role in shaping its reptile fauna. Clues to this lie in the reproductive strategies of New Zealand's 11 gecko species, some of which live at the highest latitudes reached by geckos anywhere in the world. New Zealand's geckos are unique in their family in that all give birth to live young. Even today much of New Zealand is just too cold to leave gecko eggs to the vagaries of nature. During the ice ages, however, conditions were much worse. In such conditions it is critical that the eggs are carried in the body of the mother, to be warmed by the sun whenever possible, if they are to have a chance of survival. Although New Zealand's geckos survived by developing this unusual adaptation, many reptiles could not tolerate the increasingly cold conditions. Today, New Zealand lacks crocodiles, goannas, freshwater turtles and land turtles, even though all were probably part of its Gondwanan heritage.

A major exception to the general rule of reptilian paucity in New Zealand is the tuatara (*Sphenodon punctatus*). This superficially lizard-like animal belongs to the order Rhynchocephalia. One of the more unusual features of the order is the possession of a partially functional third eye, which is hidden under a transparent scale in the middle of the forehead. The last time that rhynchocephalians were a significant force was some 180 million years ago, before even the dinosaurs came to dominance. The survival of the tuatara in New Zealand is thus of great interest. Tuataras are unusual among reptiles in that they can remain active even when the temperature falls to five degrees Celsius, as long as they can sun themselves. They have a slow metabolism.[29] Their eggs take an extraordinary 15 months to hatch. After birth they are very slow-growing and may live for a century or more. These factors may help explain their survival in chilly New Zealand.

The only other reptiles to survive in New Zealand are skinks and geckos. Until the 1980s it was thought that there were about 30 species, but studies of the biochemistry of these reptiles show that in fact 60 species exist. Many are similar in appearance and ecology and species with the same ecological niche replace each other in different locations. Despite the poor record of the reptiles, warm-blooded creatures did not have New Zealand entirely to themselves, for the cold, damp forests were an ideal habitat for frogs—and frogs appear to have once been vital species in New Zealand ecosystems.

In the past, New Zealand was home to six species of frogs.[34] These seem to have played a much larger role in the ecosystem than frogs usually do elsewhere. Unfortunately, the three largest species, which were up to nine centimetres long, have become extinct in the last 800 years, while the three surviving species have been pathetically reduced. One is restricted to a tiny patch of boulders and a rainforest relic on two islands, while the others are restricted to remnant areas on the North Island.

These frogs all belong to a very ancient family called the Leiopelmatidae. All share an unusual and decidedly non-primitive feature of their reproduction, for the entire tadpole stage occurs in the egg, from which hatch perfectly formed froglets which possess a tail. The males brood the eggs and when the young hatch they climb onto their father's back to continue growing there, living on the energy contained in their tails. Reproduction can, therefore, take place quite independent of free water. They are undoubtedly survivors from Gondwana as they belong to an ancient group which has a very limited capacity to cross the sea.

Recent fossil finds indicate that as little as 800 years ago frogs were once a dominant life form in many of New Zealand's forests, for their fossilised bones have been found by the tens of thousands in some fossil deposits. The mossy forest floor probably crawled with them. It seems probable that their dramatic decline was brought about by the kiore (the Maori name for *Rattus exulans*), which reached New Zealand with the first Maori. Kiore may have eaten young frogs, or competed for insects, which are a vital food for both. Some New Zealand biologists think that soon after their arrival, kiore may have formed plagues of such vastness that they have never since been rivalled. In a few brief years they may have stripped the forests of their frogs and other fauna.

Curiously, with the exception of a few bats that arrived by flying over the sea from South America, mammals were never a feature of the New Zealand land fauna. Given their 120-million-year history in Australia, it seems certain that the ancestors of the platypus and echidna were present when New Zealand parted from Australia, but for some reason they became extinct.

The group that really came to dominate this particular corner of Tasmantis was the birds, for until 800 years ago New Zealand had the most extraordinary, indeed unbelievable, assemblage of birds. Nothing like it was found anywhere else on Earth. Examination of the fossil remains of this once great assemblage, as well as study of the few surviving remnants, suggests that in New Zealand birds occupied all of the major ecological niches occupied by mammals elsewhere. One-hundred-and-sixty-four species have been recorded, a very large number of which were flightless.

To a biologist, the most extraordinary thing about the evolutionary radiation of birds in New Zealand is its ecological breadth. Nowhere else on Earth did birds evolve to be the ecological equivalents of giraffes, kangaroos, sheep, striped possums, long-beaked echidnas and tigers. In a sense, New Zealand is a completely different experiment in evolution to the rest of the world. It shows us what the world might have looked like if mammals as well as dinosaurs had become extinct 65 million years ago, leaving the birds to inherit the globe.

Perhaps the most extraordinary—and certainly most striking—of New Zealand's birds were the moas. *Moa* is a Polynesian word meaning chicken. Curiously, it seems that when the Maori arrived in New Zealand they carried domestic chickens with them. But who in their right mind would go to the trouble of tending chickens in that extraordinary land of birds, for in New Zealand the largest 'moa' towered over three metres high and weighed up to 250 kilograms. The domestic 'moa' were probably quickly eaten or neglected once the Maori saw what a 'real' moa looked like.

Unfortunately, we will never know the intimate details of the lives of the moa, for all 12 species are now extinct.[1] They probably belonged to a very ancient group of birds known as ratites. Ratites are today restricted to the southern continents, with the ostrich in Africa, emu and cassowary in Australia–New Guinea, the rheas in South America and kiwis in New Zealand. Thus, everything points to them being yet another group of Gondwanan origin.

Because moas became extinct so recently, some remarkable remains have been found. Feathers, for example, have been discovered on a number of occasions. Moas appear to have been entirely covered in feathers, from the base of the beak to the feet. Perhaps they needed this covering because New Zealand is so cold. The feathers are simple in structure, like the feathers of an emu. Most are reddish-brown and most moas were probably rather like a kiwi in colour. But a couple of quite extraordinary feathers have been found. One, almost 16 centimetres long, was pure white, while another, 5.5 centimetres long, was bright purple. Scientists have interpreted these feathers as evidence of crests, brightly coloured breast patches or colourful tails.[1]

Mummified moa legs, heads and necks have also been found. The neck was extraordinary for its thickness, the base in one of the smaller species being 47 centimetres in circumference and covered in thick, tough skin. Moa eggs and chicks have also been found. It appears likely that moa tended only one or two eggs at a time, which contrasts with the large clutches laid by emus and cassowaries. Even trackways have been found and these, along with the bones of their legs, suggest that the moas were slow-moving compared with the larger living ratites.[1]

Without any mammals to compete against, the ratites were set free to occupy a wide range of ecological niches. The 12 moa species were all herbivores. Some were rather superficially giraffe-like, others filled the ecological roles of rhinoceros and kangaroos. The very largest species were members of the family Dinornithidae, which included three very tall and graceful species that occurred on both the North and South Islands. With its neck stretched up, the largest would have reached over 3.5 metres high, towering twice as high as a man. These impressive birds were browsers of tall shrubs and shorter trees.

The remaining moa species were all much shorter and are placed in the family Anomalopterygidae. The largest would have weighed a little less than 200 kilograms, the smallest a little more than 20 kilograms. The diversity of beak shape suggests some differences in diet and it seems likely that some species were restricted to particular habitats. High elevation beech forest, for example, seems to have been favoured by *Megalapteryx didinus*, a short, 50 kilogram species with a short beak; while *Pachyornis australis*, among the smallest of the moas at 30 kilograms, preferred the subalpine grasslands of the mountains.

Because of their recent extinction and the longevity of certain New Zealand plant species, it is quite likely that some New Zealand plants living today were browsed by moas. An interesting legacy of the moas' long reign in New Zealand is the growth habit of many of New Zealand's smaller trees and bushes. These plants have what is called a divaricating growth habit. In these species the outside of the plant is covered in tough, often leafless twigs, which protect the tender growth inside. This feature is seen in no less that 50 plant species placed in 17 of New Zealand's plant families.[28] Almost all divaricating species are less than four metres high, or, if taller, show this in the juvenile stage. This defence was doubtless sufficient against the toothless moa, but does little to protect the plant against large herbivorous mammals, which are used to dealing with more effective spines and toxins.

The last of the moas became extinct some 400 years ago and today their only surviving relatives in New Zealand are the kiwis. Surely the most unusual of living birds, the four surviving kiwi species (one of which was first recognised as being distinct as late as 1992[29]) are nocturnal and probe for worms and other invertebrates with their long bills. In the northern hemisphere, shrews fill a similar ecological niche, while in New Guinea the long-beaked echidna (*Zaglossus*) does.

One of the most intriguing pieces of evidence that kiwis are descended from moa-sized ancestors concerns the size of their eggs. If a large bird evolves to become smaller, the size of its egg does not shrink as much as its adult body size. Having evolved from a gigantic ancestor, the female kiwi has the unenviable task of laying the world's most outsized egg. At about 420 grams in weight, it is five to 10 times as large as the eggs laid by similar-sized birds. Indeed, it can weigh one-quarter as much as the mother kiwi herself! If humans had gone down a similar evolutionary path, women would be giving birth to babies weighing 15 kilograms (33 pounds) or more.

Curiously, this condition, which at first sight appears to be a mal-adaptation, has allowed kiwis to survive the human invasion more successfully than other New Zealand birds. Because of their reduced size and nocturnal behaviour, they are difficult birds for humans to hunt and so the Maori could not exterminate them as easily as they did the moa. Just as importantly, their large eggs have given them protection from that other enemy of New Zealand birds, the rat, for they are too large to be rolled from the burrow and opened by a rat. The outsized egg also gives chicks a better chance at survival, for when hatched they are large, well-developed and able to defend themselves. The destruction of eggs and young birds by rats continues to be one of the greatest threats to New Zealand's native birds, so these fortuitous adaptations are valuable indeed to the kiwis.

Moas and kiwis were not the only extraordinary birds of New Zealand's past. Eight hundred years ago New Zealand was home to the largest ever member of the eagle family. The males and females of Haast's eagle (*Harpagornis moorei*) differed considerably in size; the larger females, at 13 kilograms in weight, were significantly bigger than the world's largest living eagle (the harpy eagle of South America). There is some debate as to whether these titanic predators were competent fliers, but most researchers suspect that their short, rounded wings were capable of carrying them rapidly through the forest. As might be suspected of an animal capable of killing great bipedal birds such as the largest moa, its weapons were awesome. Richard Holdaway, who has devoted his career to the study of this extraordinary species, has said that it:

> had claws as big as a tiger's and it could strike its prey with the force of a concrete block dropped from the top of an eight-storey building.[65]

Holdaway has found the pelvis bones of moa which have been perforated by the enormous talons of Haast's eagle, showing that it could penetrate the tough skin and bone of the moa's lower back.

Holdaway ponders the nature of interactions between the Maori and Haast's eagle. He notes that, to Haast's eagle, a person dressed in a feather cloak may not have appeared very different from a moa. Perhaps the Maori legend of *Pouakai*, a huge bird that would swoop down and carry off men, women and children to consume in its eyrie, reflects a vague folk memory of the most fearsome of New Zealand's birds.

At least one other large, apparently carnivorous bird existed in New Zealand before the arrival of the Maori. This was the adzebill (*Aptornis otidiformis*). About the size of the smallest moa, this flightless and rather rail-like bird had an extraordinary and fearsome looking adze-like beak. Perhaps related to the kagou of New Caledonia, the adzebill remains very poorly known, its precise role in New Zealand ecosystems being unclear.

One of the most pervasive trends among New Zealand birds is towards flightlessness. It is present, of course, in the moa and kiwi, but almost every bird lineage has some flightless or poorly flighted species. New Zealand was home to flightless or poorly flighted geese and ducks, rails, parrots, owlet-nightjars, wrens and other perching birds.

The reasons for this trend may lie in the energetics of flight. As any aircraft engineer, or indeed any paying passenger can tell you, powered flight is the most energy-expensive way to travel. On the continents, in the presence of mammals, there are very few ecological niches for flightless birds, for they are either eaten or outcompeted by mammals. On islands, however, where (for reasons not entirely clear) mammals are disadvantaged, many niches are open for flightless birds. Energy saved on flight can be put into reproduction and so, through natural selection, the ability to fly is rapidly diminished or lost.

As well as moas and eagles, New Zealand harboured other extraordinary birds, including now extinct pelicans, swans, geese, ducks, falcons, quail, rails, relatives of the New Caledonian kagou, snipes, owlet-nightjars, owls, crows and even wrens. Most of these had diminished powers of flight or were flightless. Among the living or recently extinct species are such unusual birds as the nocturnal, grass-eating kakapo *(Strigops habroptilus)* which, although a parrot, resembles nothing more than an enormous green owl. A thousand years ago it was one of the most widespread of New Zealand birds and its bones have been found in abundance in fossil deposits throughout the length and breadth of the archipelago. Today, it is one of the most endangered birds on Earth, for only 50 individuals survive. They have been translocated onto two tiny islands, where their diet is supplemented. Only 16 of the 50 are female. It is upon their breeding success that the entire future of the species depends. Thus far, only a single young has been hatched in captivity. Born in 1993, it was, thankfully, female.

Another extraordinary survivor is the giant swamp-hen or takahe (*Notornis mantelli*). About 60 centimetres high and coloured vivid blue and green, it has a stout, bright red bill. Remarkably, this brightly coloured bird is a close ecological equivalent of sheep, for it feeds upon alpine grasses until they become covered by deep winter snow. Then it forages inside the beech forest for ferns. It was first described in 1848 from fossilised bones found alongside those of moas on the North Island, by that greatest of anatomists Sir Richard Owen. He presumed that it was as extinct as the moas. Four living specimens were collected in the late nineteenth century but none thereafter. By the 1940s scientists again presumed that it was extinct. But then, in 1948 a tiny relict population was located in a remote grassy valley in Fiordland, South Island.

One of the most extraordinary birds of all time was the crow-like huia (*Heterolocha acutirostris*). At about 45 centimetres long it belonged to a family of birds that is unique to New Zealand, all of which have bright orange or blue wattles at the corners of their beaks. The huia was unique among the world's bird species in that males and females had entirely differently shaped beaks. This was an extraordinary adaptation for obtaining their food, which predominantly consisted of large, witchetty-grub-like larvae of the *huhu* beetle, which bores into rotten wood.[60]

The male, with his short, stout beak would chip into the softer rotten wood and the female with her long, thin, tweezer-like beak would extract the beetle larvae from tunnels in harder wood. There is even some evidence that pairs of huia (the male and female were almost invariably found together) cooperated in obtaining food.

Fossil remains reveal that the huia was once widespread and common in New Zealand. By the time Europeans arrived it had become restricted to a small area of beech forest in the south of the North Island. It was an important species to the Maori, for the beautiful black huia tail-feathers, with their broad, white tips, were used in personal decoration to denote mourning or high social status. Indeed, so valued were they that they were traded, in ornately carved wooden boxes, the length and breadth of New Zealand. The wearing of huia feathers may be a venerable tradition, for in 1642 Tasman saw what was probably a huia feather adorning the hair of one of his Maori attackers.

By the time of European settlement, huia hunting was regulated by tradition. Maori *tohunga* or shamans would place a *tapu* upon a forest area when huia numbers became too reduced. The breakdown of traditional Maori lifestyles with the coming of the European settlers disrupted this practice and hunting on a grand scale began again. In the late nineteenth century, just a few years before the last huia died, a prominent ornithologist recorded that 11 Maori hunters killed 646 huia in a single month. By 1907, the slaughter of one of the world's greatest ornithological treasures was complete.

Despite this great array of fascinating species, the most inherently interesting of all New Zealand birds were, to me, a family of tiny, entirely flightless, wren-like birds. They appear to have been the ecological equivalents of mice. As with so many of New Zealand's birds, they rapidly became extinct following human disruption of their habitat. In this case a few hundred individuals of a single species survived on a tiny, remote and unpopulated island until 1894. Known as the Steven's Island flightless wren (*Xenicus lyalli*), it was to create a scientific sensation when the last population was discovered—and made extinct—all in the same year.[60]

The story began when the New Zealand government decided to build a lighthouse on lonely Stephens Island. The lighthouse keeper, lonely himself, decided to keep a cat. Each day, the cat would bring a few tiny brown birds home. The lighthouse keeper had seen them at dusk, running nimbly like mice through the undergrowth near his lighthouse. He never saw them fly. Puzzled by their unusual appearance and behaviour, the lighthouse keeper sent a few bodies to a museum. But by the time a scientist realised that these tiny birds were the only surviving flightless perching birds (the largest bird group of them all) ever discovered, the lone cat of Steven's Island had exterminated the entire species.

Studies of fossils have now revealed that the Steven's Island flightless wren was but the last survivor of a great group of flightless or poorly flighted, wren-like birds. The remains of several hitherto unknown genera have been discovered in fossil deposits, their tiny and fragile bones previously unnoticed amongst the great bones of the moa. These tiny birds once dashed about the forest floor throughout New Zealand, occasionally perhaps falling prey to the largest of New Zealand's frogs. They appear to have filled the ecological role of mice and voles elsewhere. They are a fragile reminder of just how differently evolution has proceeded in New Zealand.

Unfortunately, it was not only such obviously vulnerable species that vanished from New Zealand with the arrival of humans. The long-extinct New Zealand swan, for example, was quite similar to the Australian black swan. Its nesting habits or restricted habitat may have made it especially vulnerable to hunting by Maori. New Zealand also lost species similar to the Australian Cape Barren goose, Tasmanian native hen, musk duck, freckled duck, Australian sea-eagle and crow. The reason why so many species became extinct in New Zealand, while their relatives survived in Australia, is curious and not readily explained. It may be that there were simply fewer refuges in smaller New Zealand, or that the carnivorous Australian marsupials and reptiles had preadpated the vulnerable Australian birds to the presence of sophisticated predators, including humans.

Why should New Zealand, alone among the world's larger continental fragments, have become a land of birds and frogs in what has elsewhere become, since the extinction of the dinosaurs, an age of mammals? Why, indeed, did it not follow New Caledonia and become a land of reptiles? Several important factors seem to be responsible. First, at 250 000 square kilometres, New Zealand is a relatively large land. This means that it can support large populations of large animals. Second, it has an abundance of superb soil and this, through increased biological productivity, enhances its ability to support large species. Indeed, so productive were New Zealand's soils they allowed the evolution of such energy-hungry species as Haast's eagle. Third, New Zealand is a cool and wet land, which has been refrigerated repeatedly during the ice ages. Amphibians can be advantaged over reptiles in such conditions, which may explain why its frogs were so abundant.

These factors readily explain why warm-blooded creatures should predominate over cold-blooded ones and why frogs should be so successful. But why should birds have won out over mammals? Mammals were certainly present in New Zealand at the time that it broke away from Gondwana, for the ancestors of the monotremes have been present in the region for at least 120 million years. Yet they did they not survive in New Zealand. I must admit that, at present, the failure of New Zealand's monotremes to thrive is a mystery—and somewhat of an embarrassment to those who study mammals and always assume them to be adaptively superior to birds. The only factors I can point to that may help in explaining this is the extreme conservatism of monotremes (probably the only mammals present in New Zealand at the time) throughout their 120-million-year-plus history, and the fluctuating land area of New Zealand through time.

The conservatism of monotremes is best told through their teeth. Remarkably, virtually all known monotreme teeth—whether they are 120-million-year-old fossils from Lightning Ridge, 60-million-year-old fossils from Argentina, or the teeth of the living juvenile platypus—are so similar that they are instantly recognisable. They show much less variability than is present in the teeth of much more recently evolved groups such as the kangaroos, cats or apes. Also remarkable is the extreme conservatism of monotreme body size, for all known monotremes weighed between one and less than 20 kilograms. This great conservatism in tooth form and body size suggests a conservatism in ecological adaptation. This may seem unlikely given the diverse lifestyles of the living platypus and echidna, but a closer look at these species reveals more similarities than might at first be expected. Both, for instance, prey upon invertebrates and although one is aquatic and the other terrestrial, both have an electro-sensitive bill which is used to detect food and both are similar in internal anatomy. It may well be that this great conservatism disadvantaged the monotremes in their competition for survival with New Zealand's birds.

A second important factor relates to land area and geological history. Throughout the last 80 million years New Zealand has been a shifting, unstable archipelago. At various times islands have submerged, while others have risen from the sea. By about 30 million years ago, New Zealand was reduced to a few tiny islands, the rest of the continental rocks being submerged below the sea. This history may have selected for animals that can survive on very small islands (such as energy-hoarding reptiles and frogs), or species that can fly from one island to the other, such as birds (even poorly flighted ones). Thus, the monotremes, unable to do either of these things, were selected against. Although this argument has some appeal, it is not entirely satisfying, for it does not account for the survival of moas and kiwis, which may have been flightless since Gondwanan times. Perhaps the best answer is the one that many scientists are unwilling to accept: that birds are adaptively superior to monotremes and where they come into competition on smaller landmasses (those with less ecological niche space) birds can outcompete egg-laying mammals.

Monotremes were not, however, the only mammals to compete with New Zealand's birds before the coming of humans, for New Zealand's bats provide another fascinating, if frustrating, insight into the way birds and mammals compete on islands. New Zealand has only ever had three bat species. One, a relatively recent immigrant, is related to Australia's lobe-lipped bats (genus *Chalinolobus*), which are common in and around most Australian cities. The other two species are placed in their own family, the Mystacinidae. Their origins were mysterious until 1986, when a team of biochemists compared their DNA with those of bats around the world.[109] They found, to their great surprise, that the mystacinids were derived from a group of South American fishing-bats, which includes one species that eats anchovetta which it catches far out to sea off the coast of Peru. Their work suggests that about 30 million years ago, some South American fishing-bats became lost at sea and were blown to New Zealand. There, they gave rise to the greater and lesser short-tailed bats, as the two mystacinid species are known.

I find it amazing, given that these bats have been present in New Zealand for so long, that they did not give rise to a great radiation of mammals, much as the bird groups have done. Both mystacinid species could fly and the one surviving species is an ecological generalist, eating insects, pollen and nectar from flowers and even scraping the fat and meat off mutton birds put out to dry. No other bat has such broad tastes in food. Unfortunately, we know nothing of the ecology of the larger of the two species, for it became extinct in 1965, before its biology was studied. Its extinction was as needless and mindless as it was tragic. The last known colony survived in the extreme south of New Zealand, on Big South Cape Island. One day in 1962 a fishing boat stopped at the island and some rats as well as fishermen got ashore. Within three years the rats had destroyed every bat on the island, depriving us of the chance of ever knowing anything about the lifestyle of this strange creature.

I have to resort to another rather feeble hypothesis in order to explain why New Zealand's bats did so poorly in competition with the birds. The only reasons I can think of are that they arrived long after the birds had become established and had evolved to fill every conceivable niche. Alternatively, bats may be too specialised in their anatomy to re-evolve into other land-based life forms. If the first hypothesis is correct, perhaps there were, by 30 million years ago, very few vacant ecological niches left to occupy in New Zealand, except those traditionally occupied by bats.

New Zealand's unique soils, climate and history determined that it became a land of birds and frogs, which by a strange evolutionary coincidence include among their ranks nature's greatest vocalists. Joseph Banks, when anchored off Cooks Strait, thrilled to a wonderful morning on 6 February 1770:

This morn I was awaked by the singing of the birds ashore from whence we were distant not a quarter of a mile, the numbers of them were certainly very great who seemed to strain their throats with emulation, perhaps; their voices were certainly the [most] melodious wild musick I have ever heard, almost imitating small bells but with the most tuneable silver sound imaginable.[104]

Even nineteenth century visitors to New Zealand described the dawn chorus of its by then sadly depleted bird fauna as 'deafening'.

I would gladly remain ignorant of the joy of the *Haka*, or even the heart-stopping beauty of Dame Kiri Te Kanawa singing *Songs of the Auvergne*, for the privilege of waking to a symphony of 'the most tuneable silver sound imaginable'. Aotearoa's multitudes of birds performed that symphony each dawn for over 60 million years. It was a glorious riot of sound with its own special meaning, for it was a confirmation of the health of a wondrous and unique ecosystem. To my great regret, I arrived in New Zealand in the late twentieth century only to find most of the orchestra seats empty. Walking through the ancient forest, whose still-living trees were once browsed by moa, I heard nothing but the whisper of leaves blowing in the wind. It was like the rustle of the last curtain fall on an orchestra that will be no more.

CHAPTER 5

MEGANESIAN ENTERPRISES

Meganesia exists as a single landmass only during ice ages when the sea-level is low. At these times Australia, Tasmania, New Guinea and many smaller islands are joined into a single great island (hence the name, meaning 'great island' in Greek) which covers an area of over 10 million square kilometres. The name is not yet in common usage, but biologists are beginning to adopt it because it describes a distinctive biological and geological entity. Meganesian animals and plants share a common biological heritage because their homeland has drifted across the face of the planet as a unit, with a common geological and climatic history. This is reflected in such distinctive characteristics as the abundance of marsupials and monotremes and the lack of native placental mammals, except for murid rodents and bats.

While the lands of Tasmantis (including New Zealand, New Caledonia and Norfolk and Lord Howe Islands) were drifting northwards, Meganesia was to remain attached to Antarctica for at least a further 40 million years. The rift which would finally see it set free was slow to develop at first. It began in the west and over 40 million years or more it gradually widened, an arm of the sea spreading ever eastwards between southern Australia and Antarctica. The old incipient rift between Tasmania and Victoria was now inactive, so the widening channel passed instead to the south of Tasmania. Tasmania's geological history has much more in common with Antarctica than Australia, so this new rift saw it part company with its parent and drift northwards joined to what is geologically an almost foreign land. By about 36 million years ago, Meganesia as we know it today had taken shape and was freed from Antarctica to begin its great journey north.

While this separation was a critical event in the evolution of the

Australian fauna and flora, the fact that Australia had remained part of Gondwana for 40 million years longer than its neighbours was also important, for this allowed relatively recently evolved groups of animals and plants to cross the Antarctic landbridge and enter Australia. Today, Antarctica is more an impenetrable barrier than a landbridge, but in geological terms it became a frozen waste only relatively recently.

The nature and timing of glaciation in Antarctica is still not well understood. Although many people think that glaciers have been present in Antarctica for at least 20 million years, a few researchers suggest that some areas have supported land-based life until much more recently. Indeed, the most extraordinary evidence for the possible late survival of forests in Antarctica has recently been found. A geologist working at an elevation of about 2000 metres in the Transantarctic Mountains found what appeared to be fresh silts which crumbled in his hands. They had been washed out of a glacial moraine (the enormous pile of rocks left behind by a glacier). This was extraordinary enough, for it denoted the presence of free water in what was otherwise a permanently frozen waste. As he examined the sediments, surprise grew to astonishment when he recovered pieces of wood that were still flexible and looked quite fresh, then the impressions of leaves very similar to those of Tasmania's deciduous beech (*Nothofagus gunnii*). When the sediments were combed for microfossils, some tiny freshwater organisms were found which suggested that the plant fossils were no more than three million years old. If this extraordinary discovery is confirmed, then researchers will have to throw away their text books, for according to every model, central Antarctica should have been well and truly glaciated by that time and almost all life banished.[93]

Going back 40 million years and more, however, we find that a continuous belt of cold-adapted forest probably joined southern South America, Antarctica and south-east Australia. Conditions had deteriorated since the age of the dinosaurs and the harsher climate had selected for only the most cold-adapted species. Fossil remains from Antarctica, South America and Australia reveal that forest, similar to fragments of temperate rainforest and woodland that survive in Tasmania and Chile today, extended in a broad band across southern

Australia, Antarctica and southern South America. In these forests southern beeches of the *fuscus* group (to which Tasmania's magnificent deciduous beech belongs) dominated, along with ancient pines, various primitive kinds of flowering plants such as the Winteraceae (their flowers somewhat like magnolias) and Proteaceae.

It is only very recently that we have begun to understand something of the kinds of animals which inhabited those forests. We can reasonably deduce from the nature of the forest and its climate that reptiles would have been greatly disadvantaged, for the perpetual cold and long, dark winter make life for the cold-blooded reptiles virtually impossible. Amphibians have a different metabolism, so it is quite likely that the ancestors of some South American and Australian frogs inhabited that ancient forest. It was the warm-blooded animals, however, that survived best under such conditions.

Those that have left the best fossil record are the mammals, which is very convenient, for their strange distributions have long posed the most difficult questions to zoogeographers. It has long been known that only in Australia and South America have the marsupials undergone extensive evolutionary radiations. Their presence has long been used as evidence for some sort of connection between the southern continents. But some palaeontologists have queried the existence of a continuous forest tract between South America and Australia 70–40 million years ago. If such a forest existed, they asked, why did not monotremes use it to escape confinement in Australia? After all, they had been in existence there for over 120 million years. And why, also, did not placental mammals (a group which includes dogs, cows and ourselves), which had long been present in South America, use it to reach Australia? Furthermore, when did the marsupials cross it and which way did the traffic flow—from Australia to South America— or vice versa?

The answer to the last of these questions began to become apparent in 1979, when Professor Frederick Szalay, Lecturer at the City University of New York, made a sabbatical visit to Australia. Szalay is a gentlemanly Hungarian with a slight accent, dark, soulful eyes, a Sigmund Freud beard—and a foot fetish. For he has long devoted his researches to evolution as it can be interpreted through the bones of the feet.

Most palaeontologists who study mammals concentrate upon the teeth, because they are complex and fossilise well. This has been very limiting, because much good information present in other parts of the body is missed. Until 1979 much of Szalay's work concerned primates.

Szalay came to Australia in 1979 in order to study the foot bones of marsupials. We met because I had independently developed an interest in feet, particularly those of kangaroos. We found we had such common interests that we began collaborating closely—even going so far as to describe a new fossil species of tree-kangaroo, based upon foot bones, which we named after my wife!

After examining the foot bones of every major marsupial group, Fred came up with an answer to a question that had plagued students of marsupial evolution for over a century. The question was this. Had marsupials invaded Australia from South America (perhaps the more popular view), or had they invaded the Americas from Australia and Antarctica? This last view was being championed at the time by some biochemists and particularly nationalistic palaeontologists and zoologists (whoever said that politics does not enter into science was a ninny).

Teeth had been no help at all in answering this conundrum, for the various marsupial groups had become too specialised in their dentitions to yield any clues. The great achievement of Fred's work was his identification of a close relative of the Australian marsupials among the South American species, based solely upon the structure of the foot. He showed that all Australian marsupials, whether now ground-dwelling or arboreal, shared certain structures of the feet which suggested that all had been derived from an arboreally adapted ancestor. Among all of the South American marsupials, fossil and living, only one had a similar foot—the Monito del Monte, or 'small monkey of the mountains' as *Dromiciops australis* is known to Spanish speakers.[128] *Dromiciops* is a tiny marsupial, no larger than an Australian pygmy-possum. It is found only in Chile, where it inhabits ancient *Nothofagus* forests that hardly differ from those that once linked Australia and South America.

When Fred announced his discovery at a scientific meeting in 1981, vigorous debate broke out. Indeed, the exchanges broke down more than once into what could most charitably be called a slanging match, as some Australian scientists, incensed that a cherished theory had been demolished by a New Yorker with an European accent, tackled Fred over his work at the conference barbecue. At first, it seemed that Fred's findings would be rejected, but in the decade following his discovery much remarkable evidence has been forthcoming that supports his hypothesis. Biochemists have now found similarities between *Dromiciops* and Australian marsupials, as have reproductive biologists. Indeed, these latter have produced some striking evidence in the form of studies of marsupial sperm. They have shown that all South American marsupials produce sperm that swims in pairs; all except *Dromiciops* that is, which has sperm like Australian marsupials.

These studies seem to have sealed the debate regarding marsupial origins, for they show that all Australian marsupials form a specialised subset of marsupials, *Dromiciops* being its nearest relative. Many other more ancient marsupial groups are found in the Americas. The inference is clear: Australian marsupials evolved from a South American ancestor which was something like *Dromiciops* and which entered Australia from South America some 60 million or more years ago.

The answers to a second great mystery, concerning the reasons why the monotremes had apparently not used the Antarctic landbridge to escape Australia, began to tumble from the ancient rocks of Australia and South America in 1992. The crucial evidence was found by another American, the palaeontologist Dr Rosendo Pascual, curator at the Museo de La Plata, Argentina.

Rosendo has spent a lifetime trying to understand the fossil record preserved in rocks in the southern part of Argentina. In 1981, in an attempt to understand the context of his fossil sites more fully, he travelled to Australia. There he became familiar with, as well as fascinated by, the unique Australian fauna. Nearly a decade later, while Rosendo was studying 63-million-year-old rocks in central Patagonia, this knowledge was to lead to a great scientific breakthrough. Always a gentleman, Rosendo was accompanying a rather overweight student who had lagged behind the rest of his group. As they sat down to rest, the student felt some discomfort as a pebble bit into his well-padded backside. Reaching around to remove it, he saw that he held in his hand a black, shiny tooth.

Because of its unusual appearance, its worn and rounded state and the fact that there was only one of it, it was a prime candidate to become the kind of fossil that is locked into a bottom drawer of a palaeontological collection to be left until a researcher has more time to grapple with identifying it. It is a sad fact that most researchers never seem to have more time. But something kept nagging Rosendo about this particular tooth, for he was sure that he had seen it before. Finally, he remembered where. It was in Australia nearly a decade before, when he was shown the remains of a fossil platypus that retained its teeth into adulthood (living platypus lose their teeth while still young).

Within a few months Rosendo visited Australia again, where he compared his tooth with the teeth of a 20-million-year-old fossil platypus from northern Australia. To his amazement he found that the South American and Australian teeth were virtually identical. Since then, several more teeth of the now famous Patagonian platypus (named *Monotrematum sudamericanum*) have emerged from the rocks of southern Argentina, confirming Rosendo's identification.[107]

Although these discoveries have stripped Australia of its long-held claim to being the only continent inhabited by monotremes, they have provided clear evidence that monotremes did use the forested landbridge that once connected Australia and South America via Antarctica. As these lands were once one, such similarities should not be unexpected.

The final major unresolved question concerning the nature of Australia's early mammal fauna and its origins centred on the problem of the lack of ancient kinds of placental mammals in Australia. This was a particularly vexing problem, for such animals were well-known from the fossil record of South America and their fossils have recently been found in Antarctica. They are clearly a very successful group, so why were they not found in Australia? Palaeontologists have long postulated that, with the exception of bats and the more recently arrived rats and humans, placental mammals never entered Australia. They argue that if they had done so, the placentals would have outcompeted the lowly marsupials and monotremes as they have done elsewhere.

The curious history of primitive placental mammals in Australia only came to light in 1992. In that year a tiny, 55-million-year-old tooth was found in south-eastern Queensland. At about two milli-metres long, its photograph, when published in the leading science magazine *Nature*, caused a furore.[54] It was found, buried in beautiful green clays below a humble cocky's cottage near Murgon, heart of 'Joh country' (named for infamous conservative Queensland premier Joh Bjelke-Petersen) near peanut-growing Kingaroy. Its discovery was doubly important, for the deposits within which it was preserved are twice as old as the next oldest well-dated Australian fossil site which has produced a fossil mammal fauna.

The Murgon Fauna, as scientists now know it, is intriguing. The sediments that preserve it were laid down in a lake that formed in the quiescent crater of an ancient volcano that was last active some 55 million years ago. Remains of crocodiles and large, soft-shelled turtles (now extinct in Australia but still found in South-East Asia and New Guinea) are common. But occasionally, perhaps once for every tonne of clay sifted, a tiny tooth or bone of another kind of animal is found. Some belonged to mouse-sized and very ancient marsupials, others to primitive bats and yet others to frogs, birds and snakes. Only a single tooth, however, has been identified as belonging to an ancient group of non-flying placental mammals. Palaeontologists speculate, from the arrangement of the cusps at the back of the tooth, that it belonged to an ancient group of placental mammals called condylarths. These were mouse- to tapir-sized, unspecialised placen-tal mammals that are thought to have given rise to many of the living placental herbivores.

Called *Tingamarra porterorum*, this tiny tooth challenges many assumptions. First, it may provide the final piece of evidence that 80–40 million years ago the cool forests of Antarctica allowed ready access between South America and Australia for many kinds of ani-mals. Secondly, it may show that placental mammals are not always superior to marsupials and that sometimes in the competition for survival, marsupials prevail. Thirdly, it opens the way to ask the ques-tion as to what conditions might give marsupials such an advantage.

Today, approximately half of the Australian mammal fauna is composed of rats and bats. Some are doubtless recent invaders, but their success shows that placental mammals can make a living in Australia. This only deepens the mystery of what happened to *Tingamarra*. It left no descendants, and must have become extinct before 25 million years ago, as no traces of the group are found in the abundant fossil record after that time.

Here we have a very strange pattern. It is as if the three major groups of mammals—the monotremes, marsupials and placentals—began a race on two different racetracks. On one (South America) the placentals, those species with the highest metabolisms and energy requirements, finished far ahead, with the marsupials a poor second. The monotremes, with their reptile-like bodies, low metabolisms and low energy needs, did not even finish the race. On racetrack Australia things were very different. The high octane placentals seemed to burn out very early on, while the marsupials, with lower resting metabolic rates and thus lower energy needs, came in clear winners. The monotremes, while taking out a distant second place, at least finished the race there. These two racetracks are still in existence. Innumerable species, including ourselves, continue to race on them.

CHAPTER 6

SPLENDID ISOLATION

For over 40 million years Australia has been physically isolated from the rest of the world's landmasses. This long period of isolation has given rise to a unique flora and fauna which is largely derived from a Gondwanan heritage. Only in Australia can one enjoy the sight of a waratah blooming amid rugged sandstone, or of a kangaroo bounding away between stately eucalypts. Indeed, there are so many species which are unique to Australia that, along with New Guinea and a few nearby islands, it forms one of the world's great zoogeographic realms.

To gain an idea of just how rich the biological heritage of Australia is, consider this: there are more species of ants inhabiting the hill called Black Mountain that overlooks Canberra than there are in all of Britain. I live in a very average suburban house with a small backyard in medium-density suburban Sydney. My tiny backyard is home to seven native species of skinks, while Great Britain is home to just three species of lizard. Furthermore, in Australia's arid deserts there are more species of reptiles than exist in one environment anywhere else on Earth.

Australia supports at least 25 000 species of plants.[50] Europe—if one includes Turkey, the eastern part of the former Soviet Union and the Mediterranean islands—supports only 17 500 species. But the flora of the core of Europe is very much poorer. Great Britain, for example, supports 1600 species of vascular plants compared with the 2000 or more found in the Sydney region. Even if one includes France and the Low Countries with Great Britain, the total is a paltry 6000 species.

These are just a few facts. A few days on the Great Barrier Reef, in Queensland's rainforests, or in the heathlands of Western Australia's south-west are enough to convince even the most phlegmatic of the extraordinary biological richness of Australia. This chapter explores the forces that shaped this biota. As on all of the fragments of old Gondwana, some groups from our great Gondwanan heritage have done much better than others.

Three great factors have worked together to shape the Australian biota: continental drift, regional geology and climate. The delicate and quite extraordinary interplay of these factors has made Australia what it is today. Alter one just slightly and you would have quite a different outcome.

Researchers have recently discovered that as Australia drifted north it carried out an almost miraculous balancing act, for it has moved at a pace roughly commensurate with the pace of changing world climate. Over the past 40 million years the world has chilled considerably. There have, of course, been many oscillations in temperature, but nonetheless the general trend has been towards cooling. This means that climatic conditions that occurred at, for example, 60° south 40 million years ago, occur today at about 40° south.

This extraordinary coincidence means that conditions have remained much more stable in Australia than they might have otherwise. Were Australia moving more slowly, the cooling climate would have overtaken it, driving many species to extinction, as happened in Antarctica. Were it moving faster, it would have outstripped the pace of climate change and the cool-adapted Gondwanan fauna and flora would have been lost as Australia moved into the tropics. This synchronism in the pace of Australia's drift north and deteriorating climate has thus been a great stabilising influence on the continent. Such stability is a necessary pre-condition for diversity, for it takes millions of years for diversity to evolve. It has allowed plant and animal assemblages to persist, evolve and diversify in relatively small areas, as world climate has changed. Europe and North America have not been so fortunate and although tropical habitats once existed there, deteriorating climatic conditions have seen great ice sheets repeatedly wipe all life from the northern parts of these landmasses.

A subsidiary effect of Australia's measured drift north is that it has managed to stay just north of the latitude where glaciers form. This has been made easier by the fact that Australia is the flattest of continents, generally lacking mountains sufficiently high to nurture rivers of ice. Thus, Australia has escaped large-scale glaciation, even during the height of the ice ages. This has had an extremely important effect which is complementary to the next major factor to be discussed—Australia's stable geological history.

No other continent has experienced the degree of geological quiet that Australia has experienced over the past 60 million years. Elsewhere, mountain ranges have risen as a result of continental collision, seas have developed as lands have rifted apart and vulcanism has built great mountains and lava plains. But since Tasmantis broke adrift some 80 million years ago, raising eastern Australia's Great Dividing Range, the Great South Land has slept. To the north, where activity could be expected as Australia rammed into Asia, New Guinea has acted as a buffer, soaking up all the mountain-building impact. As a result, the greatest geological upheavals experienced in Australia were some vulcanism along the east coast, and some gentle upwarping and downwarping, which has seen the sea encroach and recede over parts of the continental margin.

One remarkable result of this is that Australia has preserved some of the oldest rocks on Earth, for they have not been recycled by being carried deep into the Earth's crust and melted, as similar-aged rocks have elsewhere. Not only have the rocks survived, but they have survived largely unaltered. This is extraordinary, for the few other surviving rocks of this age have been pressure cooked by being buried deep inside the Earth. Thus, traces of fossils have either been altered or obliterated. A North American palaeontologist who was searching for remains of the oldest life forms on Earth learned of these extraordinary deposits and began to search in rocks near a particularly hot region of Western Australia known ironically as 'North Pole'. He recently found the oldest evidence of life on Earth, which he thinks dates to about 3.8 billion years ago. This is an almost unbelievable find, for the Earth itself is only about 4.6 billion years old. Thus, these fossils suggest that life has existed on Earth for about five-sixths of the planet's history.[134]

More recent rocks (a mere 900 million years old this time) from the Macdonnell Ranges in the Northern Territory have yielded the perfectly preserved remains of individual cells. The unique geology of Australia, which has allowed the survival of particularly ancient fossils, is of more than academic importance, for much of Australia's mineral wealth has resulted from it. The great iron ore deposits of Australia's north-west are a classic example. They were formed before 2.5 billion years ago, when the amount of oxygen in the atmosphere was increasing. The extra oxygen rusted vast reserves of iron in the ocean water, allowing massive iron-rich deposits to be laid down on the seabed as the rust settled.[134]

But there is more palpable evidence of the remarkable geological coma that Australia has sunk into, for all of our mountain ranges are old and rounded. Until recently, geologists thought that the Australian Alps were a recent development, dating to the last 10 or so million years. Abundant evidence has now proved them to be incorrect and these ranges are now thought to have begun forming some 80 million years ago as Tasmantis began to separate from eastern Australia. As a measure of how quiet things have been since then, geologists have radiometrically dated basalts that fill ancient river valleys. They have found that most of the rivers of the east coast have maintained their positions for tens of millions of years. Indeed, some have cut as little as a few tens of metres deeper into their beds in over 30 million years!

The reasons for Australia's remarkable geological stability are not entirely clear, but many geologists suspect that it is due to the enormous depth of ancient continental crust under Australia. The crust, they argue, is so old and thick that magma (which forms volcanoes) cannot easily penetrate it and, likewise, its thickness and strength prevent it from being easily folded and uplifted into mountain ranges, or torn asunder by rift valleys.

Without geological activity or glaciers, soil cannot be renewed. As a result Australia has by far the poorest soil of any continent. Virtually all of our soils are a fossil resource in the sense that they were made long ago, are now being rapidly used up and cannot be replaced. Many of them are skeletal and extremely badly leached. The worst are virtually nutrient free. Many comparative soil studies have been made, but perhaps the most telling is a study comparing the soils of the world's semi-arid zones. It found Australian soils to contain approximately half the level of nitrates and phosphates (essential nutrients) to equivalent soils found anywhere else. The effect of this lack of nutrients is far-reaching, for not only does it shape life on land, it helps shape life in the sea as well. This is because there are few nutrients flowing into Australian marine environments from our rivers. With a conspicuous lack of cold-water upwellings, this has produced some of the least fertile seas on Earth. In truth, Australia's oceans are a watery mirror of its land—for both are largely infertile deserts.

There are two tiny, yet significant, exceptions to this general rule of geological quietness and great soil poverty. They illustrate magnificently just what kind of impact soil has had on all things Australian. One concerns a scattering of relatively recent volcanoes that have erupted along the east coast of Australia. The second involves the development of the island of New Guinea.

From the Atherton Tablelands area in the north to Tasmania in the south, a number of small volcanoes—none now active but many dating to the last 10 million years—are scattered like an island archipelago. They may result from eastern Australia moving slowly over a 'hot spot' in the Earth's mantle. The Hawaiian and Galapagos Islands, both volcanic in origin, result from the ocean crust moving over a similar 'hot spot'.

These volcanoes have eroded, producing some of the only fertile soil in Australia. Wherever they occur, they support intensive agriculture or prime grazing land. The basalt plain of Victoria's Western District is the largest of these areas. Thus it is no surprise that it has been productive enough to sustain one of Australia's few surviving rural-based aristocracies; former prime minister Malcolm Fraser's family being one of them. The squattocracy, as they are locally known, arrived from Britain in the 1840s, building magnificent manor houses, some complete with family portraits spanning hundreds of years and the skulls of now-extinct Irish Elk in their halls. They often have, of course, magnificent English gardens.

The second exception involves the island of New Guinea. A mere 30 million years ago it did not exist as a separate entity, northern Australia being a flat plain grading imperceptibly to the sea. But things changed about 25 million years ago when Australia, in its slow drift north, came into contact with the Asian Plate. This plate consists of thick continental rock. Australia also consists of thick continental rock, which up until that time had been able to push over the thinner ocean crust of the Pacific. When two resistant masses of rock meet, something has to give. In this case it was the edge of the Australian Plate. It buckled, forming a trough behind the leading edge (in the vicinity of Torres Strait), while the leading edge itself was subject to tremendous force and raised skywards. The result is New Guinea's magnificent Central Cordillera, scattered with peaks reaching over 4000 metres high. The cordillera continues to rise and the very tallest peak—Puncak Jaya at 5030 metres—supports one of the world's few tropical glaciers.

These great, weather-making masses of rock ensure a bountiful rainfall. They include volcanoes as well as mountains formed from the continental crust. Erosion from these peaks has produced wonderful soils. These conditions have created their own unique flora and fauna. They also support a unique culture, for the only agriculturalists ever to develop in Meganesia are found in New Guinea. Their experiments with agriculture began some 9000 years ago. Sugar cane, taro and some varieties of bananas are among the food plants they first brought into cultivation. Today, the highland valleys of Papua New Guinea support some 1614 people per square kilometre—the highest rural population densities supported anywhere on Earth.

The third factor that has been overwhelmingly powerful in shaping Australian ecosystems is climate, which is perhaps the most powerful of all determinants that dictate the way that things are. Over much of the world, climate is either strongly seasonal, or is bountiful year round. These patterns are, of course, vital to life. Northern Europe, for example, may be considered to have a harsh climate in which freezing temperatures halt all biological productivity for half of the year. But because the changes in temperature, rainfall and thus biological productivity are all predictable, Europe is capable of sustaining an enormous biomass, which includes over 660 million people and some of the Earth's great civilisations. This is because no matter how harsh the winter, the predictable, annual cycle means that farmers can store enough grain, straw and meat to live until the next spring. Squirrels can store enough nuts to survive, bears enough fat to last the hibernation and the plants themselves enough energy to sustain the spring growth flush.

Occasionally, however, this predictable cycle is interrupted in small ways. The eruption of a volcano, such as that of Gunung Tambora on the island of Sumbawa in Indonesia in 1815, can deprive Europe of a summer. This is not surprising considering that Tambora ejected 100 cubic kilometres of dust and pumice into the atmosphere. Smaller atmospheric perturbations can bring warm weather in winter, or a freeze in summer. And it is such small alterations to the seasonal pattern, rather than the freeze of winter itself, that is the real threat to life. European farmers have always dreaded these 'unseasonal' happenings, often attributing them to goings-on in the spirit world. Shakespeare described the dread felt by Europeans at these events in *A Midsummer-Night's Dream*:

> *And through this distemperature we see*
> *The seasons alter: hoary-headed frosts*
> *Fall in the fresh lap of the crimson rose,*
> *And on old Heims' thin and icy crown*
> *An odorous chaplet of sweet summer buds*
> *Is, as in mockery, set. The spring, the summer,*
> *The childing autumn, angry winter, change*
> *Their wonted liveries; and the mazed world,*
> *By their increase, now knows not which is which*

All continents, of course, have their deserts where rainfall is rare and not truly predictable, or areas where rainfall and temperature do not vary predictably on an annual basis. But Australia is the only continent on Earth where the overwhelming influence on climate is a non-annual climatic change. The cycle that drives Australia's strange climate is called 'El Niño Southern Oscillation' or ENSO for short. Rural Australians have known about it and have written and sung about it for over a century. Dorothea Mackellar showed an understanding of it when she christened Australia as a land of 'droughts and flooding rains', Lawson saw its effects when a child, having lived through the great drought of the 1890s. His experiences were to prove formative and to provide an underpinning for his writings for decades.

Remarkably, this vast climatic event, which has long affected the lives of millions, was not clearly identified and named by the world's climatologists until the early 1980s. It was in those years that one of the most marked El Niño events ever occurred. No-one living in Australia at that time could forget the dramatic events that El Niño brought to their television screens almost nightly. The images of Melbourne being engulfed in an enormous dust storm, of the precious soil being transported by violent winds to New Zealand where they turned the snowfields red, and the stupendous fires that raced through the fire flume of Australia's south-east are all vividly with me even today. I still clearly remember listening to the radio with horror on the morning of 16 February 1983—later to become known as 'Ash Wednesday'—and hearing that the tiny rural town of Cockatoo, which my wife had left a mere three weeks before, had been entirely engulfed in flames. I later found out that 71 human beings had been burned to death during the conflagration, 2300 houses destroyed, 350 000 domestic animals killed and 350 000 hectares of land scorched, including priceless stands of pine forest which had been ready to harvest.

Such awesome events begin with a seemingly innocuous change far out in the eastern Pacific Ocean.[21] There, off the coast of Peru, the temperature at the surface of the normally cold sea begins to rise. Eventually it can rise to four degrees Celsius higher than normal. Over the course of a year this warm water can spread into a huge tongue at the equator, extending over 120 metres deep and 8000 kilometres eastwards across the Pacific. The warm water comes from the western Pacific, in the vicinity of Australia. The warmer waters of the Pacific are normally kept in this region by the prevailing westerly winds. When the winds weaken, the warm water flows back east.

This rather harmless-sounding event has some very dramatic consequences. Off the coast of South America the warm water sits atop the cold waters of the Humboldt Current. Normally this current brings bountiful nutrients from the ocean depths, but the warm and cold waters do not mix. As a result, the nutrients are kept well below the surface where tiny plants, which need light in order to utilise the nutrients, cannot reach them. The sea surface becomes very infertile and whole ecosystems break down. Tiny fish, called anchovetta, starve. These fish are normally so bountiful that they support one of the world's greatest fisheries. Humans, sea birds and marine mammals, all of which depend upon anchovetta for a living, suffer and vast population crashes occur.

The warm water has equally disastrous effects on land. Normally the western coast of Peru and northern Chile is the driest place on Earth. There are virtually no plants in many places to bind the soil. The parched deserts exist because of the Andean rain shadow and because the sea water off the coast is too cold to allow substantial evaporation to take place. Furthermore, clouds are not created because of the characteristically weak winds of the region. The warm water brought by El Niño changes all of that and cataclysmic storms can batter the landscape, washing millions of tonnes of soils off the exposed hills, swelling normally dry water courses and destroying towns and cities.

In far off Australia, the situation brought about by El Niño is the reverse. There, the coastal water is colder than normal and thus evaporation and cloud formation is decreased. Ghastly droughts, sometimes of years' duration, torture the land. Bushfires reign unchecked, winds strip the land of soil and plant life withers. Effects are also felt as far away as India, where the monsoon is delayed. Brazil, central America and southern Africa can also experience drought. It is only in Australia, however, that virtually an entire continent is held in the grip of El Niño.

Eventually, the westerly Pacific winds re-establish themselves and warm water once again accumulates around Australia. The anchovetta return to the coast of Peru and in Australia the drought is broken—often with floods of such unbridled ferocity that the denuded landscape is turned into a vast inland sea.

The length of the ENSO cycle is remarkably variable, ranging from two to eight years. And it is this variability as much as anything else that makes it so difficult for living things to deal with. The average Australian farmer, who has to repay loans on an unvarying monthly or annual basis, knows this to great cost. It still seems quite extraordinary that such an influential determinant of our climate remained undiscovered by science for so long. Even today, there is still much that we do not understand about ENSO.

From a biological viewpoint, one of the most interesting questions involves how long the cycle has been operating. By examining old weather records, climatologists have established that it has been operating since at least the beginning of the nineteenth century. Judging by the way the Australian flora and fauna have adapted to it, I suspect that it has been in operation for very much longer, perhaps for millions of years.

These features of the soils and climate of Australia have had, as I will show below, a profound effect upon the evolution of life. But, more than affecting the evolution of a single species, they have dictated the way that evolution works in Australia. More than 130 years ago Darwin published his great work *The Origin of Species* and outlined how he thought evolution worked. His ideas were very much based on the concept that nature was 'red in tooth and claw' and that it was the struggle of various species and individuals with each other that drove the evolutionary process.

Biologists in Australia are now beginning to realise that this is not an entirely appropriate way to view evolution in an Australian context. This is because poor soils and ENSO have put a premium upon retaining and rapidly recycling nutrients. This can be done most efficiently by various species developing intimate relationships. Species that belong to an ecosystem that does not have such efficient nutrient recycling are rapidly selected against. Those that cooperate in large, complex systems to maximise the availability of nutrients— such as the corals, fish and other creatures of the Great Barrier Reef, or the plants and animals of Australian rainforests—have a competitive edge. In a sense, it is cooperation rather than competition which has been selected for in many Australian environments.

The highly coevolved ecosystems that have resulted from this evolutionary pressure are extraordinarily good at maximising whatever nutrients are available. They are, however, extremely fragile, for the various species are entirely dependent upon one another to make the system work. Once a few key species have been removed, the entire coevolved structure can collapse.

CHAPTER 7

SWEET ARE THE USES OF ADVERSITY

Australia's infertile soils and the trials of ENSO have forced some unusual adaptations on its plants and animals. These adaptations are varied and sometimes wondrous, but all share a few themes, which are as follows: parsimony born of resource poverty, low rates of reproduction and strict obedience in following and exploiting brief windows of opportunity as they open erratically over the land.

Without doubt the most pervasive and influential of all adaptations in the Australian flora is scleromorphy. The great Australian botanist B.A. Barlow called it 'an expression of uniqueness' and describes it thus: 'Many...major [plant] groups are characterised by relatively small, rigid leaves, by short internodes, and small plant size'.[50] Whether they know it or not, all Australians are familiar with scleromorph plants. Some are those spiky, small-leaved shrubs that cause discomfort to bushwalkers. Others are trees which are recognisable by their hard, relatively thick leaves. Although they may sound unattractive, scleromorph plants are far from ugly, for some possess the most exquisite blooms, while others, such as the banksias of Western Australia, have the most wonderfully intricate leaves and growth habits.

Most eucalypts, banksias, bottlebrushes, ti-trees and a vast number of other non-rainforest Australian plant species exhibit scleromorphy to some extent. It is, according to Barlow 'the most striking aspect of the autocthonous [Gondwanan] element' of Australia's flora.

The fact that scleromorphy has developed in many unrelated plant families suggests that it is a response to some pervasive force in the Australian environment. But what that force might be has until recently remained unclear. Earlier researchers thought that it might be a response to aridity and in particular to the Mediterranean climate that dominates in southern Australia. This view is no longer widely held and it is now believed that scleromorphy is a response to the very low levels of nutrients present in Australian soils. Its manifestations, such as small leaves and small distances between leaves, are the result of limitations on plant growth which probably result from the small number of new cells that can be sustained at the plant's growing tip.[50]

Botanists believe that scleromorphy began to develop at least 50 million years ago in Australia. It certainly dominates in plant communities today; our eucalypts, banksias, heath plants and many others exhibiting it. As I will explain later, I think that scleromorph plants have become particularly abundant and widespread since the arrival of the Aborigines.

Perhaps the most striking effects of scleromorphy are seen among the few mammal species that must feed upon the leaves of scleromorphous trees. The most celebrated of all such species is the koala (*Phascolarctos cinereus*). Its diet consists entirely of eucalypt leaves, but it is extremely selective as to which leaves it eats, preferring those with the fewest tannins and phenolics and the highest levels of nutrients. The koala really lives on the edge, for its food source is so full of dangerous chemicals and so low in nutrients, that it has evolved to restrict its energy needs and thus needs to eat relatively little. Indeed, it is one of the greatest energy misers of all mammals. Its slow movements and low rate of reproduction are obvious results of this, but less well-known is the extraordinary koala brain.

The brain is one of the greatest energy users of all the organs. In humans, for example, the brain weighs a mere two per cent of total body mass, but it uses approximately 17 per cent of the body's energy. It is therefore no surprise that the koala has made some major reductions in brain volume in order to save energy. The strange thing is that the koala brain is much smaller than the cranial vault that houses it. Its hemispheres sit like a pair of shrivelled walnut halves on top of the brain stem, in contact neither with each other nor the bones of the skull. It is the only mammal on Earth with such a strangely reduced brain.

The koala is clearly an extreme, but marsupials in general are not known for their large brains, nor intelligence. The fact that they won out over early placental mammals in Australia suggests that it may indeed pay to be dumb in Australia. You save a lot of energy that way and if, as will be shown below, the major predators are reptiles, great intelligence may not be needed to outwit them.

Wombats, the closest living relatives of the koala, also have an extremely unusual lifestyle which may have been dictated by the limited resources of the Australian environment. The three wombat species are the only large herbivorous mammals in the world that live in burrows. Normally, herbivores need to range over a wide area to meet their energy needs and they spend large amounts of time feeding. This means that they cannot derive much benefit from burrowing. Wombats are able to benefit from living in burrows by keeping their energy requirements extremely low. They require only a third as much energy and nitrogen as a similar-sized kangaroo and spend long hours in their air-conditioned burrow, where they use as little energy as possible. It may be that wombats have been able to evolve such a unique lifestyle by virtue of long evolutionary selection for low energy requirements.[8]

The kangaroos may also have been shaped by selection for low energy requirements, for their most distinctive feature, hopping, is an energy-efficient means of getting about. At low speeds, running and hopping use about the same amount of energy, but at higher speeds, hopping is more efficient. This is because the energy of each bound is recaptured in the tendons of the legs when the kangaroo lands— rather like in a pogo stick—and is used to power the next leap. Likewise, the force of each leap pushes the gut downwards, creating a vacuum which pulls air into the lungs. This saves the kangaroo from having to use the chest muscles to breathe.

Given the efficiency of hopping, and the fact that many Australian marsupials hop, it is remarkable that hopping has not evolved in large animals elsewhere. The kangaroo may well owe its ability to hop in part to its marsupial mode of reproduction. The young are born when about the size of a bean. Therefore the pelvis can be built solidly enough to withstand the enormous pressures that hopping places upon it. It simply may not have been possible for placental mammals—which give birth to much larger young—to evolve such a strong pelvis. For the birth canal must be large and the pelvis flexible if the young is to pass through it at birth.

Another of the effects of scleromorphy on the shaping of the Australian fauna has only recently been discovered. The Australian zoologist Dr Steve Morton was long puzzled by the great diversity of reptiles—particularly lizards—in Central Australia. In some parts of the arid zone, up to 47 lizard species inhabit a single sand dune complex. This is a far higher number of species than is found living together anywhere else on Earth. Morton became even more puzzled when he found that many of the species ate just one food resource: termites. He found that the scleromorphic vegetation of Central Australia is simply too poor a food resource to be utilised by large herbivores and that termites played a particularly important role in breaking down plant matter and returning nutrients to the soil in Australian ecosystems. The abundance of termites in Australia, he argued, has led to unequalled opportunities for termite eaters. Lizards, by virtue of their body form and small size, have been able to specialise variously upon this resource, foraging within subterranean tunnels, in tree limbs, in the bases of grass tussocks and in the open.

Perhaps the most characteristic feature of the Australian fauna is a low rate of reproduction, which is often opportunistic in nature. Our native rats and mice usually have extremely small litters (often one or two) in comparison with rats and mice elsewhere. Indeed, two species of native rat found in New Guinea are unique in that the females possess only a single pair of teats. Many native birds also have small clutches compared with similar species elsewhere. Remarkably, there is also some evidence that the average clutch size of some introduced bird species is declining as they adapt to Australian conditions (C. Dickman, personal communication).

Other reproductive strategies which are rare elsewhere are widespread in Australian species. Many Australian birds have a social structure in which one or more young stay with their parents into adulthood. These individuals forego the chance to raise young themselves, in order to help their parents feed their younger brothers and sisters. Kookaburras, noisy miners and blue wrens all exhibit this behaviour. Indeed, it is extremely widespread, almost characteristic of many Australian birds of Gondwanan origin. Elsewhere, it is an extremely rare strategy and about 85 per cent of all species worldwide which exhibit it are Australian. The strategy is clearly beneficial, for breeding pairs with such helpers at the nest raise more young than those which lack them. The necessity for this strategy illustrates well the difficulty of obtaining enough resources for birds to reproduce in Australia. It is also a good example of how cooperation rather than competition can be fostered by difficult environments and, as will be discussed later, it finds parallels in the adaptations of both Aboriginal and non-Aboriginal people to Australian conditions.

An even more extreme reproductive adaptation, probably related to feeding the young, is seen in the medium–small sized carnivorous marsupials of the genera *Dasyurus, Antechinus* and *Phascogale*. Each year around September, after a frenzied bout of mating, the males die. For a time the population is composed solely of pregnant females or females with young. The lack of males during this period may be critical to the success of reproduction, for the females do not have to compete with adult males for food when they face the great energy demands of lactation. Likewise, when the young are weaned they do not need to compete with males for resources. Remarkably, a small lizard, the mallee dragon (*Amphibolurus fordi*) has developed a similar strategy. The adults live for a less than a year and are all dead by the time their eggs hatch. This means that the young do not face competition for resources from older individuals of the same species.

Yet other small marsupials have developed an opposite, but equally effective strategy—that of extremely long life—for females at least. Researchers have recently discovered that female mountain pygmy-possums (*Burramys parvus*) live up to 11 years in the wild. This makes them the longest-lived small ground mammals in the world. They may need to live that long in order to successfully raise a few litters in Australia's erratic environment.

Where Australian species do produce lots of young, they are often tied to an erratic cycle of reproduction. Perhaps the best example of this is the banded stilt (*Cladorhynchus leucocephalus*). It is a familiar bird to city birdwatchers, for it is often seen feeding on briny coastal lagoons and saltworks. Although familiar and common, its breeding habits have remained largely mysterious, the only occurrence recorded before 1989 being the discovery of breeding colonies on Lake Grace

(Western Australia) and Lake Callabonna (South Australia) in 1930.[108]

In March 1989 the banded stilts disappeared from their usual haunts. They were finally located nesting on three small islands on Lake Torrens (South Australia). Remarkably, they had left the coast within days of rain falling in the inland, thousands of kilometres away, even before the inland lakes had filled. About 100 000 birds were nesting and the whole region was throbbing with life, for rains had filled Lake Eyre and other lakes to a 15 year high.

The reproductive cycle of these birds is incredibly rapid, the three or four young growing quickly on a rich diet of brine shrimp. Soon after hatching the young form into crèches, which are tended by the males. The crèches leave the breeding area to roam over the shallow lakes, in one case swimming 130 kilometres in six days. A mere three weeks after hatching, the young are ready to fly. Even more remarkable, just two weeks after the first clutches of eggs had hatched, the females were nesting again at another location. They presumably mate with the males while the latter are tending the crèches. In all, the reproductive cycle of the banded stilt can be completed in just seven weeks, but even this can be cut down as one cycle overlaps another, when the females mate with the males minding the crèches.[108]

This remarkable reproductive cycle is necessary because the salt lakes upon which the birds depend can support brine shrimp for only a short time. The shorter their breeding cycle, the more young that can be raised. It is important for a pair to raise as many young as possible, for it may be a decade or more before conditions are right for the stilts to breed again. Although the life span of the banded stilt is undocumented, it is certain to be a long-lived species, for adults must often go a decade or more without breeding.

Yet another feature of Australian ecosystems which results directly from its inherent constraints is nomadism. Nearly a third of Australia's bird species are truly nomadic, which is an extraordinarily high percentage in world terms. The red kangaroo (*Macropus rufus*) is also largely nomadic, following the rains that produce the short green pick that it prefers. Of course, Australia's first human settlers remained nomadic.

A possibly related phenomenon is the success of birds in Australia relative to mammals. When the area of Australia is compared with that of other landmasses and the number of bird species calculated on an area basis, we find that Australia has a fauna roughly equivalent to, or a little richer than, that found on the other continents. Its mammal fauna, when viewed on the same basis, is very small. It may be that Australia's very large size, erratic climate and poor resource base makes it beneficial to be able to fly to reach resources as quickly after they become available as possible.

Although few of these features are uniquely Australian, the abundance of adaptations serving the one end is quite remarkable. This is eloquent evidence of the forces shaping the Australian ecosystem. Yet the effects go much deeper. They have determined who should be predators and who prey; and indeed how many species Australia would support.

CHAPTER 8

THE DIVERSITY ENIGMA

I have long been fascinated with the diversity of life on Earth, and the reasons why some environments support a greater number of species than others. An important factor in determining the number of species a given environment can support is its productivity. Every environment has a characteristic rate of formation (through photosynthesis) of new plant material, which fuels the web of life within that environment. If the rate is high, the environment is said to be very productive. There is a widespread belief that such environments are home to a greater number of species than resource-poor environments, where productivity is low. Although this hypothesis makes intuitive sense, ecologists are now questioning it. Indeed, some have pinpointed reasons suggesting that the reverse is (or should be) the norm. This is best understood by examining the relationship between productivity and diversity in three very different Australasian environments: the heathlands of south-western Western Australia, the Great Barrier Reef and the rainforests of New Guinea.

The heathlands of south-western Western Australia support some of the most diverse and spectacular plant communities on Earth. Western Australia supports some 10 000–12 000 species of plants, with the number of species per square kilometre in the extraordinary heathlands of the south-west rivalling those found in the richest rain-forests of Australia's north-east.

Most of the world's banksias, sundews (*Drosera*) and all of its dryandras, to mention only a few, are found in this small region of Australia. Yet it is remarkable that the bulk of these species come from relatively few families, such as the Proteaceae, the family containing species of *Banksia, Dryandra, Grevillia* and others. Thus, many species of rather similar plants coexist. This is all the more surprising as the landscape of the south-west is, superficially at least, rather monotonous. Most of the area is covered by highly infertile sand sheets and there are few mountain ranges or sharp topographic features to provide microhabitats that might speed the evolution of new species. Subtly different soil types do, however, support different plant communities.

Many species in the region possess unusual adaptations to extreme soil infertility. The Western Australian Christmas tree (*Nuytsia floribunda*) is a tree-sized mistletoe that gains nutrients by parasitising the roots of grasses and other plants. The carnivorous plants are particularly abundant. Endemism is high, with two of Australia's seven genera, as well as the majority of species of two others, being found only in the south-west of Western Australia. More than half the world's species of sundews (*Drosera*) are found only there, as is the Western Australian pitcher plant (*Cephalotus follicularis*), which is placed in its own family. Furthermore, the bladderwort genus *Polypompholyx* is entirely restricted to the region, as is one of the only two species of rainbow plants (genus *Byblis*). All of these plants have turned to carnivory to meet their nutrient needs, for they obtain nitrates and phosphates from insects. Their adaptation highlights the harsh constraints that the infertile soils of Western Australia place upon the flora. They add emphasis to the question of how so many different plant species could have evolved and how they can coexist in such an infertile environment. Could the answer lie in the very poverty of the environment itself?

David Tilman of the Department of Ecology, University of Minnesota, suggested in 1982 that in environments where nutrients such as nitrates and phosphates are in plentiful supply, the species that are best at utilising these nutrients can outcompete all similar species.[132] A prime example of such a species (which Tilman calls a 'superspecies') is ourselves. In areas where nitrates, phosphates, water and soil abound we can destroy almost all other species through intensive agriculture, reducing the environment to a monoculture. In less productive areas, such as those used for grazing, our ability to destroy the species competing with us is less, although with modern technology we are improving that ability. In the most unproductive of environments, humans are reduced to just another species among the multitudes, if they can exist there at all.

Incidentally, there is a problem with Tilman's use of the term 'superspecies', for this term is already in established use in taxonomy, where it has quite a different meaning. Perhaps the term 'exterminator species' is more appropriate, for although emotive and used in other contexts, it does adequately describe their effect.

Humans are just one exterminator species among many. Tilman argues that it is only where exterminator species are excluded that many species can coexist. They do this by becoming specialists, exploiting subtle differences in the levels of the critical resources (such as water and nutrients) in different areas. Thus, in the Western Australian heathlands, one species of *Banksia* may be able to survive in runoff areas where more nutrients are available than elsewhere. Another may survive in sand at the foot of dunes where water may accumulate. Yet another may survive in barren interdune areas because it can exist on very few nutrients. Because of the complex interplay of soils, nutrient levels and water availability, many combinations of resource availability are possible in a nutrient-poor landscape, leading to the evolution of many specialist species.

One further aspect of Australia's heathlands has long puzzled me. That is the presence of spectacular heathland flowers, a variety of which bloom throughout the year and most of which drip with nectar. It seems anomalous that such apparent waste of effort should occur in such infertile areas. I now think that I have found a convincing explanation as to why this is so. Because of the diversity of plant species in the heathlands, the nearest potential sexual partner of any individual plant may be a long way off. This means that pollinators, such as insects, mammals and honeyeaters, must be used for effective pollination. Yet in a nutrient-poor environment such organisms—particularly the warm-blooded ones—are likely to be scarce. Thus, competition exists between plants for these few pollinators. It takes fewer nutrients for a plant to make nectar than it does to produce new leaves. This is because only water, air and sunlight are needed to make sugar, while nutrients such as nitrates and phosphates are essential for the production of new plant tissue. Thus, where nutrients are limited, it makes sense for a heathland plant to have blooms that produce lots of nectar to outcompete its fellows for the attentions of the rare pollinators. This is because its ability to produce nectar is not as tightly constrained by low soil fertility.

The presence of abundant nectar alone may not greatly increase the abundance of most pollinators. This is because the abundance of many species may be restricted not by the amount of nectar, but rather by protein, as most species need some protein-rich food such as insects in order to survive. The number of insects present is probably limited by low primary productivity due to poor soils.

The process can perhaps best be thought of as an arms race; a literal war of the roses where flower power wins the day. Some species avoid the arms race by flowering at times of the year when few other blooms are around. This provides a continuous food supply to the pollinators. Thus, in a seeming paradox, these very nutrient-poor soils on which heaths grow, support an abundance of plant species with the most luxurious flowers. Upon these feed specialised pollinators, including the tiny (10 gram) honey-possum (*Tarsipes rostratus*), which is the only non-flying vertebrate to depend totally upon flowers for its food. So rich in nectar and pollen are the heaths of south-western Australia that a honey-possum can feed virtually every day of the year and, for much of the time, find its food requirements from an area of no more than 40 square metres. The fact that ENSO has little or no effect in the south-west doubtless helps ensure the constant nectar supply that such a specialised animal requires.

A somewhat different example of great diversity is provided by the Great Barrier Reef. Coral organisms are animals that support photosynthetic algae inside their bodies. Such a strategy is most likely to have evolved in a very nutrient-poor situation, the ancestors of coral animals taking advantage of a relationship with plants to supplement their food supply. The relationship remains useful, for coral reefs grow only in the most nutrient-poor warm waters which are clear enough for light to reach the algae.

The reefs provide a home for an enormous diversity of fishes and other organisms. It is interesting that the nutrients present in a coral reef environment are cycled through the living animals of the reef very rapidly, with very little loss. Thus, a small amount of nutrient is made to go a long way. Unfortunately, the too-rapid removal of parts of the system (such as large fishes) can have disastrous effects. Today, the relatively small commercial and recreational fisheries on the Great Barrier Reef are having a profound impact as predators, algal grazers and other types of fishes are selectively removed from the environment.

What happens if we enrich such a system? We do not need to guess, for we are already carrying out an immense experiment. Queensland sugar cane growers fertilise their fields with large quantities of nitrates and phosphates. During the wet season these nutrients are washed down rivers and into the sea in vast quantity, eventually finding their way onto the nearer parts of the reef. In consequence, a monoculture of algae takes over from the wonderful reef diversity, much in the way that people and their agriculture take over from heathland with an application of superphosphate.

The effect that phosphates have on many Australian plants is quite intriguing, for not only do they encourage the growth of competitive weeds, they can actually kill native plants. How they do this was only discovered recently. Native plants which are fed fertiliser often produce sickly yellow growth before they slowly wither. This happens because phosphates block the ability of plants to extract iron from the soil. Just why Australian plants should have developed this trait is unknown, but it almost certainly relates to the natural scarcity of phosphate compounds in Australian soils.

To return to Tilman's 'exterminator species' hypothesis: in resource-rich areas, exterminator species can reduce the environment to a virtual monoculture, while resource-poor areas encourage a diversity of highly specialised species. This can be seen in plants such as the few specialist grass species that monopolise western Victoria's basalt plains or in cold-blooded creatures such as tiny, shrimp-like krill in the Antarctic. But what of warm-blooded creatures such as mammals? These animals have high energy requirements and in resource-poor environments there might not be enough resources for them to survive at all.

Biologists from the CSIRO studying tree-dwelling marsupials (mostly various species of possums) in the eucalypt forests around Eden in southern New South Wales have found that 52 per cent of the forests contain no arboreal mammals at all, while 63 per cent of the individuals observed occupy just nine per cent of the forests. They think that much of the forest may be too nutrient-poor to support arboreal marsupials and that only the forests growing on better soils support large populations.[30]

The ability of possums to survive in eucalypt forests is complicated by the eucalypt's propensity to manufacture toxins. Eucalypts growing on more fertile soils respond to the presence of leaf-eating possums simply by growing new leaves as the old ones are eaten. But eucalypts that grow on nutrient-poor soils cannot do this because the lack of nutrients limits their ability to sustain new growth. Instead, they produce carbon-based toxins, especially tannins and phenolics. Like nectar, these compounds do not take large amounts of nutrients to produce. They are thus a cheaper way (in terms of nutrient use) of dealing with leaf loss than growing new leaves. The tannins and phenolics protect leaves by disrupting the acid balance in the digestive system of the leaf-eating possums, thus preventing them from digesting their food.

The effect of soil fertility on possum density in rainforests has yet to be quantified, although there are indications that things may be quite different there. In parts of north Queensland and Irian Jaya, I have found scrubby rainforest growing in white sand or almost bare rock. Yet these forests still manage to support considerable numbers of leaf-eating possums. It seems possible that rainforest plants growing on nutrient-poor soils do not produce tannins and phenolics to the degree that scleromorph trees do.

The reasons for this apparent difference between eucalypt forest and rainforest may lie in the rapidity with which available nutrients are recycled through the ecosystem. In sclerophyll forests the recycling of nutrients is often very slow. Leaves, twigs, bark and possum droppings can lie as litter on the forest floor for months or years until they are consumed by fire and their nutrients released. This is because soil microbes and other recyclers are not efficient at breaking down this material—in part perhaps because of all the chemical defences that the plant has made for it!

In rainforests, nutrient recycling is much faster. During fieldwork in rainforests in New Guinea it was a red-letter day whenever I found a mammal dropping. They were almost invariably surrounded by clouds of bright butterflies and moths, which land upon them and vigorously lap up their liquid. The droppings are always gone in a day or two. This contrasts with eucalypt forests, where droppings are common, laying about intact for months. Because of the rapid breakdown of droppings in rainforests, the nutrients therein rapidly become available to the plants again. Therefore, the loss of leaves to possums by rainforest trees growing on poor soil may not present the same kind of problem that leaf loss does to nutrient-starved eucalypts. Were the rainforest trees to do as the eucalypts and produce toxins, they may incur the cost of slowing the nutrient recycling system, and would thus be self-defeating.

In sclerophyll forest, the nutrient recycling system is naturally slower than in rainforests and is tied to fire frequency. As will be explained below, the involvement of fire often means that many nutrients are destroyed or lost before they can be used by the plant. In such adverse situations, the option of developing chemical defences may be worthwhile, for although it may bear the cost of slowing the recycling system a little, it saves the tree from losing nutrients which may never be replaced.

Some other curious differences between rainforest and sclerophyll forest may be explained by the difference in the rate of nutrient recycling in the two environments. In rainforests, possums use a variety of nesting locations. Some species sleep in tree hollows, but many more shelter under epiphytes such as moss and ferns, make nests (called dreys) or simply sleep exposed on a tree branch, disguised as a lump of moss.

In sclerophyll forests dominated by eucalypts, almost all possum species roost exclusively in tree hollows. This is because there are very few other safe havens for a day-time nap. Large epiphytes are all but absent—perhaps because many of the trees shed their bark and they cannot take hold—fire is frequent and the canopy is very open, allowing exposed animals to be easily spotted. The bark of many species is also smooth and white—thus the strategy of sitting on a branch pretending to be a lump of moss is not a good one. In contrast, tree hollows are abundant, for sclerophyll forest trees (particularly eucalypts), have the peculiarity of developing extensive hollows.

The development of hollows seems to be disadvantageous, as they weaken a tree's structure, making it vulnerable to falling down. It is therefore something of a puzzle as to why hollows occur so commonly among eucalypts. It has been suggested that they may result from the way that nutrients are transferred from the heartwood to the sapwood (a necessary process in nutrient-poor areas), which may encourage decay. But it has also been suggested that hollows actually benefit the trees by acting as gigantic possum-poo traps which retain droppings near the roots until they can be broken down. Because hollows large enough to shelter possums occur primarily in large, old trees nearing the end of their lives, the extra nutrients may give them an important boost just when they need to maximise the energy-expensive business of seed production.

In the case of rainforest as in that of the Barrier Reef, it is clear that not only is the overall level of nutrients important, but the speed at which they cycle through the system is also critical. In systems where nutrient levels are low, but nutrients are recycled rapidly, it is important that nutrients are not lost. Every time we take fish off the reef, or remove a log from rainforest growing on poor soils, we do just that. Although such changes are tiny, their effects need to be monitored to ensure that the nutrient balance is not being disturbed.

The most resource-poor environments, with their long nutrient recycling times, clearly have too few available resources to support many mammals. But what of rainforest environments with quick recycling times? Will Tilman's proposed 'exterminator species' effect mean that areas with less nutrients support more species than those with very abundant resources?

One potential way of examining this issue is to compare the mammals of western and central New Guinea. The Irian Jayan mountains, although less fertile, are home to about 81 species—which is four more than in Papua New Guinea. Although the difference in overall numbers is small, the actual species that inhabit the differing regions is possibly more informative. The more resource-rich central highlands have four fewer smaller marsupials (100 grams–2 kilograms in weight) than the less fertile west. At present it is difficult to be sure what these data mean, but they hint at the possibility that at least for the smaller mammals, Tilman's hypothesis might apply.

Despite what I have said above, variability in productivity alone is insufficient to explain differences in diversity, for time is also an important factor. Australia's extraordinary long-term climatic and geological stability, brought about by continental drift and its thick continental crust, are responsible for the persistence of low-nutrient conditions over tens of millions of years. A long period (I suspect more than a million years in most cases) is needed for new species with long generation times (such as the larger mammals) to evolve. Even if plants and fish can evolve faster to exploit specialised environmental niches, it surely would take many tens of millions of years for the diversity of the Great Barrier Reef and the Western Australian heathlands to develop.

This may seem to conflict with recent research that indicates that the Great Barrier Reef is only 8000 years old. Although it is true that the Barrier Reef has been growing in its present position for only 8000 years, it is certain that somewhere in the vicinity similar reefs have been growing for millions of years. Because of their dependence on shallow waters and the fact that sea-levels have risen and fallen dramatically many times in the past, reefs have become expert migrators.

Similarly, floods, fires and climatic changes have disturbed the south-western heathlands for aeons, but no single shock has been large enough to remove them altogether, for fossil floras indicate that Western Australian heathland plants have been in existence for over 40 million years. Indeed, small scale disturbances such as fire might offer an increased chance for diversity, since some species can make a living as post-fire specialists.

If we could understand more about the determinants of diversity in Australia we might be able to answer some very interesting questions. What, for example, is the significance of the differences between the fossil Riversleigh rainforest animal communities and those surviving at present? How old is our diverse reptile fauna? Why are the plants of the nutrient-poor deserts not as diverse as those of the heathlands?

The arguments developed here also tell us some very interesting things about the future. It has always amazed me that humans (who now utilise about 40 per cent of all the primary productivity of plants on the land surface of our planet) have not caused even greater extinctions than we actually have. It is surely in part at least because we are 'exterminator species', dependent until recently on the richer parts of the Earth and not so much upon the less fertile regions, where most of the world's biological diversity lives. Our depredations of Australia's more fertile regions have certainly been so great that it is difficult to find a very productive region that has not been grossly altered.

Thankfully, we still have those less productive regions. Some humans have dreams though, to make the deserts bloom and to make the depths of the sea and even Antarctica yield their bounty. As each year goes by, we come closer to developing the technologies that will allow us to realise these dreams. Each year we also feel an increasing need to utilise marginal lands in order to feed our growing populations. With our dreams fulfilled we will, I fear, see a wave of extinctions so vast as to dwarf anything that has gone before. For we will have become the exterminator species that broke all the rules. The one that could take not only all the resources of rich lands, but of poor ones as well.

CHAPTER 9

THE DESERT SEA

ustralia's freshwater and marine ecosystems display some unusual characteristics, many of which are related to a lack of nutrients. The freshwater systems show this in an extreme way, for with the exception of Antarctica, Australia is the driest of the continents and thus has the least available habitat. Its freshwater fish fauna is extremely limited, comprising only 200 of the world's 8000 species. Of these, the vast majority have been derived in geologically recent times from marine species.

As mentioned earlier, Australia's infertile seas have produced, in the Great Barrier Reef, one of the world's great regions of biodiversity. Another, much less diverse desert sea exists in the Great Australian Bight. This area has been extensively surveyed, without success, for potential fisheries because it has a wide continental shelf. It seems to be particularly nutrient-poor because there is virtually no river runoff carrying nutrients into it.

Perhaps one of the most striking pieces of information concerning the infertility of Australia's seas concerns the peculiar breeding habits of the Australian sea-lion (*Neophoca cinerea*). One of only two seal species which are unique to Australia, it once inhabited coastal waters from Bass Strait to the Houtman Abrolhos Islands off Western Australia. Sealers drove it to extinction in the nineteenth century everywhere east of Kangaroo Island, South Australia and reduced its numbers elsewhere dramatically. Today 10 000–11 000 individuals survive.

It was only in the 1990s that researchers revealed just how odd the Australian sea-lion is among the world's seals, for it breeds on a non-seasonal cycle of 17.5 months (rather than the usual annual cycle seen in all other seals).[64] What is more, the pupping season lasts for five months instead of two and females suckle their young for an unusually long time—17 months in all.

The unusually long breeding cycle of the Australian sea-lion may well be a response to the few resources available in Australian waters. There may simply be too little food to allow females to fatten the young enough over a 12 month suckling to give them a fair chance at survival.

Because the breeding cycle is not annual, the time of birth varies every year. During the 1970s it occurred in autumn and spring. In the 1980s it was in summer and winter, while the 1990s will again see it again occur in autumn and spring.

Male sea-lions have also been affected by limited resources. Females are few and spread out, so males cannot control a large harem as males of most other seal species do. Instead they have developed a different strategy. An old bull will stake out a female when she first comes ashore, then guard her until she is able to conceive. Unusually fit males can guard four, or occasionally even seven females, but most get by with just one. This seems to leave the males rather cranky and about 20 per cent of pup deaths are due to savaging by males.

Other large mammals seen around the Australian coast overcome the problem presented by our nutrient-deficient waters in other ways. The great whales, for example, that are again beginning to visit Australia's southern coast after decades of overexploitation, have developed a novel strategy. They simply do not feed at all over the months that they spend calving and travelling in Australian waters. They can do this because they spend the summer feeding in the exceptionally rich waters of the subantarctic oceans, where a single whale may take in as much as 2000 kilograms of krill per day. This allows them to put on enough fat to swim to Australia, give birth, produce hundreds of litres of milk per day for seven months, mate, and then swim the thousands of kilometres back to the subantarctic—all without taking a bite to eat.

Although the mammals, with their high energy demands, provide the most striking evidence of the chronic infertility of Australian waters, Australian fish also provide some extraordinary examples.

Australia possesses some nine million square kilometres of fishing zone, an area greater than that of its land. By international standards it is an enormous fishing zone, equivalent in size to that occupied by the USA. Yet the annual catch for 1988–89, was only 201 709 tonnes. This take ranked Australia in fifty-fifth place among world fishing nations. Despite the enormous area available to exploit, surveys indicate that there is very little potential for the catch to expand beyond the 1988–89 figures. Indeed, in 1992–93, the total catch in the south-east fishery declined by 17 per cent and many other fisheries also shrank. This low productivity of Australian seas means that Australia, with a mere 17 million people, imports $500 million worth of seafood each year.

While some might suggest that this situation results from the laziness or ineptness of Australia's fishermen and that an untapped bonanza awaits the adventurous, the experts do not agree. In 1985 Robert Bain, then first Assistant Secretary, Fisheries Division of the Department of Primary Industry, wrote 'Indications are that most of the more readily available and marketable resources of the Australian Fishing Zone are fully exploited or overfished'.[6] Bain was writing before the exploitation of orange roughy was in full swing and before concern at the future of Australian fisheries had turned to alarm.

The present dire state of Australia's fisheries is the result of people realising too late just how limited our marine resources are. Fishermen can hardly be blamed for this, for it is difficult to monitor things under the sea. Furthermore, our fishermen come from the robust and productive fisheries of Europe and Asia. They have no experience of extremely poor fisheries such as those of Australia.

As mentioned above, the paltry 12 million kilograms of fish taken off the Great Barrier Reef each year is causing considerable ecological disruption and damage. In the relatively productive waters off southeastern Australia, one fishery after another has been destroyed by overfishing. For example, before the Second World War a fleet of about 60 boats fished exclusively for flathead in New South Wales waters. Today, because of overfishing, not a single boat remains. Gemfish, one of the mainstays of the south-eastern Australian fishing industry, are in serious trouble, for overfishing has ensured that there are very few young fish in the population.

The Tasmanian whitebait fishery is a fascinating example of an Australian fishery in decline. The Tasmanian whitebait (*Lovettia* sealii) is an unusual fish, for it belongs to a very small family, the Aplochitonidae, the other members of which are restricted to South America. It is almost certainly of Gondwanan origin and has been adapting to Australian conditions for millions of years. It is a small fish, some 60 millimetres long, which is characterised by a high degree of sexual dimorphism. Females have the genital opening in the usual position, but in males the genital opening, urinary pore and anus have all migrated forward and are found just behind the head.

The 1940s were the real heyday of the Tasmanian whitebait industry. In 1947, 480 000 kilograms (about one billion fish) were taken from the estuaries of Tasmania that are their only home. By 1955 the catch had dropped to 1570 kilograms and by 1972 to 1010 kilograms. In 1973 the fishery was closed to allow it to recover.[92] Recent surveys have indicated that whitebait populations are recovering. In 1991 the fishery was reopened to recreational fishermen. Interestingly, Lovettia, the basis of the original fishery, is now present in only small numbers in the take. It its place, two species of Galaxias, a unique southern hemisphere genus of fish, make up the bulk of the small fish (known collectively as whitebait) caught. This suggests that the ecology of small fish in the estuaries of Tasmania continue to suffer major disruption through overfishing more than 40 years ago.

This pattern of overexploitation of marine resources is not a new one. Australia's first industry, seal skins and oil, had overexploited seal stocks to the point of non-profitability and local extinction of some populations by 1810. Likewise, the coastal whale fishery, based primarily upon the southern right whale, had driven stocks to economic extinction by 1840. Robert Bain outlined how the pattern of overexploitation occurs.[6] Typically, a few fishermen will discover a new resource one season and make significantly higher earnings than the rest of the fleet. For the next two seasons a substantial number of vessels convert to exploit the fishery. New vessels under construction in the second season enter the fleet in the third and fourth seasons. Before all of the new vessels have entered the fleet, in the fifth or sixth season, the maximum sustainable yield of the fishery has been reached or exceeded. Meanwhile, fisheries experts begin to collect statistics in the third season of the fishery. By the sixth season enough information has been gathered so that guidelines can be drafted. Two years after overexploitation began, a low quota (but not low enough) is imposed. Many vessels leave the overexploited fishery, which is extremely slow to recover. In the meantime, ecological damage of unknown proportions has been done to Australia's marine environment. Because of the small scale of most Australian fisheries, this pattern seems to be almost universal.

As the more easily accessible fisheries have been destroyed, fishermen have been forced to invest more and more in boats that can fish deeper and deeper. As their mortgages have grown, so has their need to service them from ever less profitable fisheries. Many may have met their Waterloo with the last great unexploited resource, the orange roughy (*Hoplostethus atlanticus*). This small (50-centimetre-long) fish lives in waters of one kilometre depth off the continental shelf of Australia and New Zealand. It spawns around ancient sea mounts where it forms enormous aggregations.

These aggregations were discovered by fishermen in the late 1980s, when they provided an enormous bonanza. At the height of the fishery around 1989–90, trawlers were bringing in 40 000 tonnes of these fish annually. Even today, about 11 000 tonnes are taken in Australia and orange roughy fillets can be bought in Sydney fish shops for as little as $12 per kilogram.

All appeared rosy until fisheries biologists started to look at the annual bands laid down upon the ear stones (otoliths) of these fish. They found to their horror that orange roughy can live for much longer than the humans that consume them, for century-old fish are common. What is worse, they may not breed until they are between 20 and 32 years old.[72] It must be abundantly clear to everyone that 50-centimetre-long fish that live for 150 years and which take longer than a human to reach sexual maturity, must grow very slowly. This is because they live in the most resource-poor part of the world's most resource-poor seas, for the dark waters off our continental shelf get very little nutrient input at all, being far from all sources such as rivers and sunlight.

Everyone knows that the orange roughy fishery is doomed without tight regulation. Indeed, so much damage has already been done that it may be doomed even with it. The sustainable yield is certainly lower than the 11 000 tonnes presently being taken. But it is proving difficult to enforce even this generous quota. Too many fishermen will go broke. All expect a new fishery to emerge and save them, if they can just stave off insolvency that little bit longer.

In the past there *has* always been another virgin fishery just a little bit further out to stave off bankruptcy. But the orange roughy must be close to the end of the line for many operators. If it is overexploited for much longer, the resource may be destroyed, perhaps resulting in irreversible damage to the ecology of the waters off our continental shelf. In return, we might buy just a few more years of economic viability for our struggling fishing fleet.

Typically, another fishery has just been discovered deep off the waters of our continental shelf. It is based upon the king crab (*Pseudocarcinus gigas*). With adults often reaching a weight in excess of 12 kilograms, only the giant spider crab of Japan is larger amongst the Crustacea. They have a spectacular, vice-like right nipper, growing far larger than a human hand. These crabs are unique to the cold and eternally dark waters hundreds of metres deep off south-eastern Australia. In 1990 only a few hundred kilograms of king crab were fished off south-eastern Australia, in all representing a few dozen individuals. But in 1992 over 200 000 kilograms were taken. This represents at least 20 000 individuals. No-one knows how many king crabs there are. We do not know how long it takes a king crab to reach adulthood, or how many young they produce. We have no idea what role the king crab plays in the ecology of the deep waters off Australia's south-east. We do know, however, that we will certainly drive the king crab into economic if not actual extinction before such information and thus knowledge needed to set and implement a sustainable yield, is obtained.

When faced with resource shortages, many people automatically look to the sea as a limitless supplier of food and fertiliser. Unfortunately, this can never be so in Australia, for our oceans are mirror images of our land—they are biological deserts of great fragility.

CHAPTER 10

THE MYSTERY OF THE MEGANESIAN
MEAT-EATERS

I f we return to the land, we find further evidence of the way in
which nutrient shortage and climate has shaped Australia's ecology. For one of the less well-known, although more striking
features of Meganesia is its extraordinary paucity of large mammalian
carnivores. To gain an idea of the number of large mammalian carnivores in the 'natural' Meganesian biota, we need to understand the
fauna of some 60 000 years ago, before the arrival of humans dramatically altered the land. As a result of a century-and-a-half of
palaeontological research we now know this fauna fairly well. Of the
60 or so mammal species then existing in Australia whose weight
exceeded 10 kilograms, there were only two, or possibly three, carnivores: the long extinct marsupial lion (*Thylacoleo carnifex*), the
recently extinct thylacine or Tasmanian tiger (*Thylacinus cynocephalus*) and possibly the giant rat-kangaroo (*Propleopus oscillans*).
Only two other Meganesian carnivores, the still-surviving Tasmanian
devil (*Sarcophilus harrisii*) and the spotted-tailed quoll (*Dasyurus
maculatus*), exceeded five kilograms in weight.

Each of these four or five species fills a somewhat different ecological niche. At about the size of a wolf and very wolf-like in shape,
the thylacine was the only generalised dog-like marsupial carnivore.
The Tasmanian devil is a generalised scavenger and bone cruncher,
perhaps best described as a miniature (five to eight kilograms in
weight) marsupial hyena. At up to five kilograms in weight the spotted-tailed quoll, although marsupial, is stoat or civet-like in its
behaviour and appearance.

The marsupial lion was one of the few carnivores to have arisen
from herbivorous ancestors. Its teeth are highly unusual, for one pre-

molar in each jaw developed into an enormously long slicing blade, while the rest of the cheekteeth atrophied. These large, slicing premolars led some nineteenth century scientists to speculate that it may have fed upon melons, but discovery of an articulated hand has revealed a large and hooded claw, which was doubtless used in killing. It is now universally considered as a specialised carnivore. Although bearing the common name 'lion' it was only 40–60 kilograms in weight, about the size of a leopard. It may have been a marsupial equivalent of the medium-sized cats of other continents.

The ecology of the extinct giant rat-kangaroo (*Propleopus oscillans*) puzzled me ever since I first studied it while completing my doctoral study on kangaroo evolution. At about 40 kilograms in weight (the size of a female grey kangaroo), it had a dentition similar to that of much smaller insectivores. It existed throughout eastern Australia during the last ice age, even in areas of grass steppe. I wondered how such a large species, whose teeth seemed suited to eating insects or nuts and fruit, could have survived in such a harsh environment. Finally, I examined one of the slicing teeth from the front of the jaw under an electron microscope. I found that it was covered with long grooves, very similar in shape to those seen on the teeth of marsupial lions. This suggested that it may well have been partially carnivorous, eating plant matter but scavenging as well as opportunistically taking bird eggs and small vertebrates. If this interpretation is correct, this primitive kangaroo may have filled a similar ecological niche to some of the smaller bears.

Australia is unique in that its two largest mammalian carnivores (the marsupial lion and the giant rat-kangaroo) evolved from herbivorous ancestors. This may be further evidence of the constraints placed upon carnivores in Australian ecosystems.

In the entire Australasian region, therefore, there was only one mammal species filling each of the broad ecological niches of dog-like, cat-like, civet-like, scavenging and possibly bear-like species. Just how unusual this situation is in a worldwide context is best shown by comparison with the mammalian fauna of the United States of America. Even today the USA supports many carnivores that exceed five kilograms in weight. These include three bear species ranging in weight from 90–675 kilograms, five members of the dog family, six

cats, six members of the weasel family and three species of the rac-
coon family. Yet even this plethora of carnivores pales in comparison
with what was present during the Pleistocene and before humans
arrived some 11 000 years ago. Then, in addition to the above list,
there was the dire wolf (*Canis dirus*, which was larger than living
canids), the dhole (*Cuon*, which still exists in Asia), an extinct bush
dog (*Protocyon*, the size of a wolf), two species of enormous and
carnivorous short-faced bears (*Arctodus*), the spectacled bear
(*Tremarctos*), a cheetah, the sabre-tooth cats, the lion and a hyena.

This diversity of mammalian carnivores is by no means exceptional. Europe, Asia, Africa and South America either did, or still do,
support similarly diverse carnivore guilds (in biological terms a guild
means a group of species doing roughly the same job, for example,
dog-like). In all these regions the cat-like and dog-like niches are sub-
divided according to size, prey type and habitat, allowing many species
to coexist. The environment is also productive enough to support
large numbers of specialised or unusual carnivorous mammal species.

Biologists have long speculated on the cause of this great imbalance in the Meganesian mammal fauna. One idea is that marsupials
for some reason have found it difficult to evolve into truly predatory
species, perhaps because of their relatively small brains. However, a
quick look at the fossil record of South America disproves this
hypothesis, since many species of dog-like marsupials of the subfamily Borhyaeninae evolved there during the Tertiary (65–2 million
years ago). These ranged in size from that of a bear to that of a civet.
A second, even more remarkable subfamily of South American carnivorous marsupials, the Thylacosmylinae, evolved into animals
resembling sabre-tooth cats and were probably capable of killing the
largest of prey. A third subfamily, the Sparassocyninae, related to the
living didelphids (such as the American opossum), also included large
flesh-eaters. All of these carnivorous marsupials became extinct upon
the arrival of placental carnivores in South America over the past five
million years, but what is more remarkable is that they thrived for
many millions of years, preying mainly upon large herbivorous placental mammals!

There being no intrinsic bar to carnivory in marsupials, we
should look to the environment for an explanation of Meganesia's

paucity of large carnivores. Large, warm-blooded carnivores sit at the apex of a broad-based food pyramid and are thus the most vulnerable of life forms to disturbance in the food chain. As an example, an area of grassland that supports billions of individual grasses may be able to support a few thousand large herbivores. These in turn may be able to support less than 100 large carnivores. If the environment is poor, large herbivores will be rare and spread thinly. A critical point is reached where the density of prey is so low that a self-sustaining population of large carnivores cannot be supported. Likewise, if during a period of environmental change, 90 per cent of the grasses of an area are destroyed, then almost certainly only a few hundred herbivores would survive. This would be insufficient to support *any* large carnivores.

One of the most important limitations affecting carnivores is simply the size of the landmass that they inhabit. The New Guinea harpy eagle (*Harpyopsis novaeguineae*), the largest New Guinean bird of prey, is a good example. It is not found on a single island, even though many were connected with the mainland during the last ice age. This is because it is only on a very large island such as New Guinea that the resource base is sufficient to support a viable population of such large and energy-hungry carnivores. But this constraint does not fully explain the mystery of the Meganesian meat-eaters. This is because Meganesia (or even Australia alone) is such a large landmass.

Although Australia is the smallest of the continents, it is still a large land—the same size as the contiguous 48 states of the USA. Meganesia was much larger. There are no long-isolated landmasses of similar size to which Meganesia can be compared, for the isolated landmass next closest in size—Madagascar—is only one-twentieth the size of Meganesia. There, two species of cat-like fossa (genus *Cryptoprocta*) existed, the living one at 7–12 kilograms in weight and a puma-sized species that became extinct about 1000 years ago. This compares with three similar-sized species in prehistoric Australia. Furthermore, at least seven species belonging to the mongoose family—which weigh 0.5–4 kilograms—inhabit Madagascar. In contrast, there are only five carnivorous mammal species in the same size range in all of Meganesia.

While land area alone does not adequately explain the mystery of the Meganesian meat-eaters, low productivity, as explained above, almost certainly does. Meganesian soils are extremely poor and ENSO makes for great variability, with long periods of low productivity. When highly variable productivity is superimposed on an inherently infertile environment, top-order carnivores become particularly vulnerable.

Although these arguments are readily accepted for the drier parts of Meganesia, what of the rainforests, where the effect of the ENSO cycle may be less severe and where biomass productivity may be higher? The largest rainforest block in the region is found in New Guinea, which is, if anything, more remarkable for its lack of mammalian predators than Australia. Here we have no evidence of indigenous large cat-like, scavenging or civet-like predators. The only large carnivore present before humans came was the thylacine. This was in a fauna of about 200 species of mammalian herbivores and insectivores! One possible explanation may lie in the size distribution of mammal species in pre-human New Guinea. Unlike Australia, there were few large herbivores. Indeed, only nine herbivorous species that weighed over 10 kilograms are presently known (although one or two more may remain to be discovered). This again is strikingly different from rainforests elsewhere, which harbour mammals such as elephants, rhinos, tapirs, okapis and many others. These observations hint that Meganesia's rainforests may also be environments of relatively low productivity compared with rainforests elsewhere. It may well be that the inherently nutrient-poor soils and/or climatic variability of Meganesia have an impact even in these most favourable areas.

If large mammalian carnivores are disadvantaged in a low-productivity system such as Meganesia, what would the consequences be for other kinds of carnivores? Carnivores that, on other continents, are less 'fit' may be advantaged by Meganesia's unique ecology. There is in fact evidence that this has occurred, for Australia has produced a remarkable assemblage of carnivorous reptiles. Not since the age of dinosaurs had such a bunch of cold-blooded killers stalked the landscape.

Before the arrival of humans, a gigantic goanna (*Megalania prisca*) and a land crocodile (*Quinkana fortirostrum*) were the largest carnivores in the region. Both weighed more than 200 kilograms and would have been able to kill the largest of marsupials. There was also a very large (over 100 kilograms) python-like snake (*Wonambi narracoortensis*) that inhabited southern Australia, as well as several now-extinct kinds of crocodile.

The giant goanna must have been one of the most formidable carnivores of all time, an echo of the time when dinosaurs ruled the Earth. Reaching seven metres long and over a tonne in weight, it was about the size of a medium-sized *Allosaurus* and, if the agility, intelligence and rapacity of living goannas is any guide, must have been a terrifying predator. Its body was bulky, with a relatively short, thick tail and its teeth were stout, finely serrated on one side and curved, rather like a steak knife. There is no doubt that it was the top carnivore in the Australian ecosystem, for it would have been capable of killing even the largest diprotodon.

At three metres in length, the poorly known land crocodile *Quinkana* was also a formidable carnivore. Its snout was deep and box-like, quite unlike that of any living crocodile. Its teeth were rather like those of the giant goanna, except that they were serrated along both edges. It probably had hoof-like feet rather than the webbed feet of living crocodiles. Its tail may have been encased in circular arrangements of scales. We know that it was largely terrestrial because its remains have been found in cave deposits far from water, associated with an entirely terrestrial fauna.

The giant snake *Wonambi* was an entirely different kind of predator. Its remains have been found in cave deposits in the far south of South Australia, where they are associated with faunas which suggest that the climate then was much as it is today. Large snakes do not presently live under such cold conditions, so *Wonambi* must have had some special metabolic adaptations. Perhaps it inhabited rocky areas, which trap and store heat and offer shelter from the wind. Reaching six metres long and 30 centimetres in diameter, it was an enormously bulky snake. Its head was the size of a shovel, and was similarly flattened and broad. Its mouth was filled with hundreds of tiny teeth. This suggests that it might have specialised in taking wallaby-sized prey. It was probably an ambush predator, waiting by waterholes or along runways in rocky areas.

The extinct crocodiles, although aquatic, are also worthy of mention here. The largest was *Pallimnarchus*, an inhabitant of Australia's inland waters. It was enormous, resembling the largest living salt-water crocodiles, except that its snout was short and wide. Its snout shape suggests that it may have fed upon large mammals, ambushing them at the water's edge.

Not all of Meganesia's cold-blooded killers are extinct, for among the surviving goannas, there are at least five species that exceed five kilograms in weight. A further five species of pythons are in the same size range. There are also two surviving crocodile species, the larger of which (the salt-water crocodile *Crocodylus porosus*) is the world's largest surviving reptile. This remarkable assemblage of large, predatory and predominantly land-based reptiles has no parallel outside Meganesia.

Why should large reptilian land carnivores have been so successful in Meganesia? An obvious difference between reptiles and mammals is that reptiles are cold-blooded. Because they do not need to burn energy to create heat, they need to eat much less frequently than warm-blooded species. This means they can survive long periods of food shortage and can exist at higher densities relative to their prey. These factors may well have given them an advantage over mammal predators in resource-poor and unpredictable Meganesia.

Armed with this knowledge, we can look back over the Australian fossil record to investigate if there were periods when Australia supported a greater number of mammalian carnivores. This might indicate a time when Australia was more fertile than it has been over the past few million years. Many of the fossils necessary to answer this question remain unstudied, but it appears possible that at least two species of dog-like carnivores (thylacinids) coexisted at Riversleigh (north-western Queensland) during the Miocene, about 20 million years ago. It is also known that both a large and small species of marsupial lion coexisted in eastern New South Wales in the Pliocene period, five to two million years ago. Even given this slight increase in carnivore diversity in earlier times, the number does not compare with those that have existed on every other continent (except Antarctica) throughout the age of mammals. Thus it seems possible that Meganesia has, at least for the past 20 million years, been a resource-poor land which has suffered a further decrease in fertility over the past two million years.

A final question needs to be raised concerning the success of the larger carnivores which have been recently introduced to Australia. Three species are most important: humans, dogs and foxes. All three have thrived since their arrival in Australia, suggesting that mammalian carnivores can be successful here. However, a closer analysis of the situation reveals an interesting story.

Humans doubtless present the least understood and most controversial case. I will argue here that, when they arrived in Australia some 40 000 years ago, they were spectacularly successful predators, for they appeared to have killed off all of the Australian land-based creatures which were larger than themselves. This included all of the land carnivores larger than the thylacine. Thus, in part, humans can be seen as a replacement for—not an addition to—the carnivore assemblage of Australia. Also, by world standards, the pre-European human population of Australia has always been small (probably between 300 000 and 600 000). Humans are omnivorous as well as being capable of exploiting marine resources. This ability to fall back on marine or plant resources may have given humans an advantage during periods when land-based protein resources were sparse.

The dingo has also been successful, but when it was introduced about 3500 years ago it led to a decrease in carnivore diversity rather than an increase. This is because it apparently drove both the thylacine and Tasmanian devil to extinction on mainland Australia. Therefore it too is a replacement, not an addition.

The case of the fox is intriguing, for it has been extraordinarily successful since its introduction in the 1850s. Its diet, like that of humans, is remarkably broad. It can survive on insects and berries if no vertebrate prey are available. It can also kill animals as large as young wallaroos (weighing about 20 kilograms). A part of the fox's success may be due to the abundance of rabbits in Australia. The rabbit itself may only be so successful because of the prior extinction of many species, which left abundant niches vacant in the Australian landscape. Thus the fox and rabbit may be utilising productivity that otherwise would have fuelled a whole megafauna. Furthermore, the smaller marsupial carnivores (quolls) have declined wherever foxes are present. Overall, I feel that the success of these introduced carnivores does not detract from the idea that Meganesia is generally a hostile environment for mammalian carnivores because, by and large, they have been replacements for the few existing species.

The hypothesis developed here is a useful one in that it allows us to examine many other questions. Could it be, for example, that our herbivores are (and were) especially susceptible to predation by placental carnivores because, before people arrived, they only had to deal with 'thick-witted' reptiles? Additionally, does the hypothesis tell us anything about the relative abundance of species of megafauna? Are there other resource-limited places, such as islands, where reptiles have become the dominant predators? Komodo and its dragons and New Caledonia with its newly discovered extinct large varanid come to mind. Whatever the case, the Meganesian meat-eaters remind us of just how unusual the biota of the Australian region is. Because of its unusual conditions and long isolation it truly is a separate experiment in evolution.

CHAPTER 11

A BESTIARY OF GENTLE GIANTS

The great carnivore assemblages of Australia's past depended on a large array of herbivores and insectivores. Although not as strikingly different as the carnivores, they are of interest because study of their fossil remains has revealed much about the nature of Australian ecosystems. Before examining the bestiary of Australia's past, it is worth making a few points about the assemblage as a whole.

Before humans came, the large herbivore and insectivore fauna of Australia consisted of four major groups: echidnas, marsupials, flightless birds and tortoises. Their remains have been found in virtually every habitat, from what are now the driest deserts, through rainforest, to the alpine tundra of New Guinea. The exact species present varied from place to place, but in most habitats a large number of species coexisted.

One curious feature of the assemblage is that it did not contain any extremely large species. The largest mammals ever to exist in Australia were the species of diprotodon. At about 2000 kilograms in weight, the biggest weighed roughly a third as much as elephants. Before 10 000 years ago, more than a dozen species of elephants occupied every habitat, from the arctic tundra to rainforests, on the continents other than Australia.

It seems to be a common occurrence that Australian herbivores weigh a third or less as much as their ecological counterparts overseas. The many extinct and living kangaroos are a good example. Their closest counterparts elsewhere are the African antelopes. The largest kangaroos (now extinct) may have weighed up to 200 kilograms, while the largest antelopes weigh 1000 kilograms or more. The smallest kangaroos weigh 500 grams, while the smallest antelopes are a few kilograms in weight. The same pattern holds true for the insectivores. At 20–30 kilograms in weight our giant extinct echidnas weighed only a third as much as the termite-eating aardvark of Africa or the giant anteater of South America, while our tiny termite-eating numbat weighs only one-twentieth as much as its African counterpart the aardwolf.

Many other instances of this trend exist and indeed it seems to be a general rule for the Australian warm-blooded animals. Again, low levels of nutrients may be responsible for this, as most Australian environments may have insufficient food resources to select for the 'normal' large sizes of many species overseas.

Another important point about this vanished assemblage is that there was considerable overlap in size between the smallest extinct species and the largest surviving ones. The largest living marsupials—the red and grey kangaroos—can weigh up to 90 kilograms. The smallest of the vanished mammals may have weighed as little as 20 kilograms. All of the smaller extinct species, however, were short-limbed and slow-moving, while the largest living species were the very fleetest of all Australian mammals. Among the flightless birds, some quite small species (weighing only a kilogram or two) vanished.

Having outlined the few characteristics that seem to unite this assemblage of extinct species, it is time to examine them group by group and try to assess what role they played in Australia's pre-human ecosystems.

The horned turtles of the family Meiolanidae must have been among the most striking of all Australia's animals. At 200 kilograms or more in weight, they were enormous. Their elephant-like feet were rather similar to those of the surviving Galapagos giant tortoise, but in almost everything else they differed. Their carapaces were lower and broader, more like those of the living aquatic side-necked tortoises of Australia. Their tail was unlike that of any living tortoise, for it was heavily armoured with spiked, bony rings and it terminated in a heavy, bony club which was bristling with blunt spikes. The head, which could not be retracted under the carapace for protection, was likewise heavily armoured. Its most striking feature was two horns resembling those of cattle. In some species the horns were straight and stuck out at the sides, but in others they curved out and up rather like the horns of Texas longhorns.

Shadow-drawings of Australia's lost and dwarfed fauna. The human hunter at left gives some idea of their size. (*Courtesy Peter Murray*)

Row 1   Tree-feller (*Palorchestes*), marsupial rhino (*Zygomaturus*), diprotodons, both large and small species (*Diprotodon* spp.), *Euwenia*

Row 2   Marsupial lion (*Thylacoleo*), five species of extinct wombats (*Ramsayia, Phascolonus, Phascolomys, Vombatus*), giant rat-kangaroo (*Propleopus*)

Row 3   Seven species of giant short-faced kangaroos (*Procoptodon, Simothenurus*)

Row 4   Eight species of extinct kangaroos (*Simosthenurus, Sthenurus, Troposodon, Wallabia*)

Row 5   Seven species of extinct or dwarfed kangaroos (*Protemnodon, Macropus*)

Row 6   Extinct or dwarfed marsupials and monotremes (*Macropus, Sarcophilus, Zaglossus, Megalibwiliba*)

Row 7   Extinct birds and reptiles (*Progura, Genyornis, Megalania, Wonambi*)

Among the Pleistocene birds which once existed in Australia can be counted both a large and small species of the logrunner genus (*Orthonyx*). These are medium-sized perching birds that dig in the leaf litter of forest for worms, spiders and crickets. The extinct species may have been poor fliers. There was also a giant, flightless or nearly flightless, member of the cuckoo family (*Centropus colossus*) and, most curiously, a species of pilot bird (genus *Pycnoptilus*). The only surviving species of pilot bird is a small, obscure, brownish, ground-dwelling bird of Australia's eastern forests. The extinct species was even smaller than the surviving one. It might seem very strange that such a tiny bird should become extinct along with all of the giant fauna. But the pilot bird has some curious habits that may have made it vulnerable. The living pilot bird gets its name from the fact that it is often found in association with the much larger lyre bird (*Menura novaehollandiae*). Indeed, the calls of the pilot bird often reveal the presence of the more secretive lyre bird to humans. The pilot bird seeks out lyre birds because it feeds upon tiny insects missed by the lyre bird as it turns over leaf litter with its large feet. It seems possible that the extinct species of pilot bird had a similar relationship with one of the other extinct Australian birds, perhaps the extinct logrunner, whose remains occur in the same deposits. The extinction of its large, poorly flighted companion may well have spelled doom for the small pilot bird.

Among other vanished birds are one or possibly two gigantic and now extinct species of eagle, a small inland pelican, and two species of flamingos. Many Australians are surprised to learn that flamingos once counted amongst their continent's diverse fauna. But the two species that survived to the Pleistocene (2 million–10 000 years ago) appear to have been merely the last survivors of a great evolutionary radiation of flamingos that had lived around Australia's inland lakes for over 25 million years. Their extinction may have been due to hunting and/or climate change.

The remains of two other large species of extinct birds have been found in Pleistocene deposits in Australia. The first was a gigantic species of mallee fowl known as *Progura gallinacea*. The sexes appear to have differed greatly in size, males weighing as much as seven kilograms and females up to four. They almost certainly built mounds, within which they laid their eggs, much as the living mallee fowl does today. They were probably poorer fliers than surviving mallee fowl.

The second extinct species was the largest of all Australian birds of the last two million years. At up to 200 kilograms in weight, *Genyornis newtoni* was an enormous and stockily built flightless bird. It resembled an emu, only its legs were much thicker, its head larger and its beak deep and blade-like. Surprisingly, it was not related to emus at all, but to chickens or ducks. It seems to have been that old Australian curse incarnate 'may your chooks turn into emus and kick your dunny door in!'.

Despite these moderately diverse assemblages of large flightless birds and herbivorous reptiles, it was the mammals that dominated the scene. Among the smaller of the now-extinct species were the giant echidnas. The most common species appears to have been about the size of the living long-beaked echidna of New Guinea. At up to a metre long and 16 kilograms in weight these are substantial animals. They were widespread in the forests of eastern Australia. Because of their long, stout beaks with ridges on the palate, scientists have speculated that they fed upon the larvae of scarab beetles, the best known of which are the ubiquitous Australian Christmas beetles. Their larvae can be particularly common and in pasture land in the New England area of New South Wales, the combined weight of Christmas beetle larvae can exceed the combined weight of grazing stock on the land. Both, unfortunately, compete for the same re-source—grass—the beetle larvae eating the plant from below, the sheep from above.

The very largest of the extinct echidnas was found only in the south-west of Western Australia. At about 20–30 kilograms in weight, it appears to have stood more erect than other species. Unfortunately, no skull has ever been found, so its dietary preferences remain unknown.

Today, members of the kangaroo family tend to dominate the large herbivore niche in Australia. This was also the case in the past, for at least 20 additional species of kangaroos and their relatives existed 40 000 years ago. A few were large relatives of the larger living kangaroos, but most were very different. Three species were members of the genus *Protemnodon*. With their large forearms, short

feet and stout tail, they resembled gigantic New Guinea forest wallabies. The largest weighed over 100 kilograms (more than the largest
living kangaroo). They appear to have been quite common throughout Australia.

By far the greatest number of now-extinct species (some 14; their
classification still being uncertain) belonged to the subfamily of
short-faced kangaroos known as sthenurines. Today, only a single
member of this subfamily survives. It is the banded hare-wallaby
(*Lagostrophus fasciatus*) of Western Australia. It maintains a precarious
foothold on existence, the only known colonies surviving on Bernier
and Dorre Islands in Shark Bay. It is a very pretty little wallaby, with
chocolate and silver bands alternating on the rump, tiny forepaws
and a short tail with a graceful crest of black hairs running along it.
François Péron, the nineteenth century naturalist who discovered it
there, gives a sympathetic account of its behaviour:

> *their whole care was lavished on their young...If wounded, the
> mother fled carrying the young in her pocket [pouch] and did not
> abandon it until, overcome with fatigue or exhausted through loss
> of blood, she could carry it no longer. The mother would then stop,
> help the young one out of the bag, hide it in a safe spot and con-
> tinue her flight with what speed her remaining strength would
> permit.*[136]

Unfortunately, very little is known of the habits of the 14 extinct
species of sthenurines. The smallest was about the size of a swamp
wallaby (*Wallabia bicolor*), while the largest were the largest kangaroos that ever lived. All had some striking features in common. As in
horses, the hindfoot had a single toe. In a further remarkable parallel
to the horses, the claw was flattened and broad, resembling a miniature hoof, while the digit was retracted by a pair of cruciate
ligaments—the same ligaments that bend a horse's hoof backwards if
it is lifted off the ground. It seems probable that they could hop, but
most species would have been rather slow, for it is only the very
largest that have hindlimb ratios suggesting that they could travel at
similar speeds to the living large kangaroos.

Despite these similarities in the foot, the rest of the animal was very un-horse-like, for the tail was short and exceedingly thick. They probably had a pronounced pot belly in order to accommodate the large fermenting vat of a stomach, for most species appear to have fed on low quality browse; a diet that must be taken in considerable quantity if the animal is to survive. The forelimbs were also peculiar in shape. Two of the fingers were very long, with exceedingly long, flat claws, while the rest of the fingers were reduced. The arms themselves could, as in humans, reach high above the shoulders. This is something that no living kangaroo can do. It seems likely that these peculiar arms allowed the sthenurines to reach into the canopy of bushes and bring leaves in reach of their mouth. They needed to use their arms to do this because the neck was very short—necessarily so—for the head was massive and may have given trouble when the animal hopped if it were balanced at the end of a long neck.

The head was shaped unlike that of any living kangaroo, for the face was very short and broad—almost human-like in proportion in the largest species. Their jaws were massive and fused together just as in humans and their relatives. Some even had molars with the same thickened enamel that characterise hominid teeth. The nasal region of the skull of the short-faced kangaroos was unique, for it was short, but often massively domed. Just what these kangaroos did with such expansive nasal cavities is not known, but it is possible that air was cooled or filtered, or that sound was amplified there.

There were three distinct kinds of short-faced kangaroo. The most primitive kinds, which are placed in the genus *Sthenurus*, had slightly longer faces than the others and simpler molars. They appear to have been common throughout Australia, but were perhaps particularly abundant in the drier regions. A second kind (genus *Simosthenurus*), had shorter faces and more complex molars. The third kind, which included the very largest species (which may have weighed about 200 kilograms) were the species of *Procoptodon*. These appear to have been restricted to the drier regions of south-eastern Australia. They had very short faces, massive jaws, reduced incisors and very large and heavily enamelled molars. Overall, the teeth and jaws of these kangaroos present a remarkable parallel with that of our distant, now extinct relatives, the robust australopithecines such as *Australopithecus boisei* of southern Africa. It seems likely that these strange kangaroos and the robust australopithecines shared a similar diet.

The wombat family is also well-represented among Australia's now-vanished fauna. The largest wombat of all time was the aptly named *Phascolonus gigas*. At nearly two metres long and up to 200 kilograms in weight it dwarfed living species. Over 100 000 years ago some individuals became mired in the muddy bed of Lake Callabonna, where they perished. Their beautifully preserved skeletons have revealed much about their lifestyle. They show, for example, that this mammoth wombat was capable of digging burrows.

Their burrows must have been enormous and their backdirt piles must have been prominent features of the flat inland Australian landscape of the time. The burrows doubtless provided other animals with shelter in the harsh land and the churning and loosening of the earth must have encouraged the growth of plants on the enormous piles of excavated soil. These great wombats were thus almost certainly important to other life forms in the area. With a diameter of up to a metre, their burrows could have been entered by humans at a stoop. This, perhaps, was their downfall.

Their teeth reveal some interesting adaptations. As in all wombats, the teeth are formed into columns that grow throughout life. For a wombat, a chipped or broken tooth is no problem, as a new section soon grows as the damaged part is abraded away by the intense wear that the teeth endure. This wear is brought about by the wombat's diet, which is high in very abrasive foods, such as grasses that grow crystals of silicon in their cells. Silicon is harder than enamel and can wear teeth appallingly quickly. If an animal consumes silicon with its food its teeth must either be extraordinarily tough or ever-growing in order to cope with the wear.

A further strange feature of the teeth of the giant wombat is that its upper incisors are much broader than the lowers. The upper incisors are also very thin from front to back, making them very fragile. It is very difficult to imagine how they functioned. Perhaps they had some role in display, for their large white surfaces would have been very striking when the animal opened its mouth.

Among the smaller but now-extinct wombats were several species that weighed between 50 and 100 kilograms. Most are known only from their skulls, which resemble those of living wombats. There was also a pygmy species, which may have weighed as little as 10 kilograms. The larger species all appear to have been burrowers, broadly similar to the living wombats, but the pygmy wombat was very different. Named *Warendja wakefieldi* it was made known to the scientific community as late as 1981. An inhabitant of the south-east corner of Australia, it was evidently rare even before people arrived. It was a very primitive wombat. A partial skull and skeleton have recently been found which suggest that it may not have been capable of burrowing. Without the shelter of burrows, it was probably vulnerable to overhunting.

Other large marsupials became extinct without leaving any relatives. Perhaps the most interesting are the tree-fellers of the genus *Palorchestes*. The larger of the two Pleistocene species weighed up to a tonne and was a most imposing creature. Both the large and small marsupial tree-feller resembled the now-extinct ground sloths of North and South America in general shape, but their head bore a moderately long trunk. Their barrel-shaped body, short tail and massive limbs gave them a very compact appearance. The hands and feet bore extraordinarily sharp, deep claws up to 10 centimetres long. When I first examined the fossilised remains of these creatures, I was fascinated by their forearms. The humerus (bone of the upper arm) is broad and short, with extremely powerful muscle attachments. It resembles the humerus of a mole or some other creature that exerts enormous pressure upon its forearms. The main bone of the lower part of the forearm is, however, even more extraordinary. Short, massively deep and marked with numerous pits and crests denoting muscle attachments, it is composed almost entirely of solid bone. Most bones possess a marrow cavity, but the ulna (lower arm bone) of the marsupial tree-feller has the narrowest marrow cavity of any limb bone that I have seen. As impressed as I was by the structure of these bones, I was confounded when I discovered that the elbow joint could not move and the forearms seem to have been locked into a semi-flexed position!

Whatever the marsupial tree-fellers were doing with their forelimbs, they clearly required enormous strength. They do not appear to have been burrowers or diggers, for this would have quickly blunted their sharp claws, which in any case are the wrong shape for digging. Furthermore, the elbow must be able to flex to dig efficiently. There are very few other lifestyle options for an animal with such forearms. Perhaps the most plausible explanation is that they ravaged trees for their leaves and bark. They presumably used their trunk to strip branches within their reach and then used their incredibly strong forearms to either fell smaller trees, or to break branches or strip bark from larger ones. All would have been fed through their battery of tough, high-crowned molars.

The remains of both the large (500–1000 kilograms in weight) and small (200 kilograms) species are found only in eastern Australia, where they are usually rare. A large number of bones have been found at one site only—Foul Air Cave at Buchan in eastern Victoria. It is possible that the marsupial tree-fellers entered the cave to hibernate and, becoming lost in the dark, fell into a hidden shaft or simply died in a torpid state. The general rarity of fossil remains of the tree-fellers probably reflects a rarity in life, for it is likely that unproductive Australia could afford to support only very few such destructive tree-wreckers.

The final extinct family of giant Australian herbivores included the largest marsupials of all time. Known as diprotodontids, it included species that varied in weight from between 100 kilograms and two tonnes. All of the Australian Pleistocene species were large, but a number of smaller species lived in New Guinea. They appear to have filled a variety of ecological niches in the broad browsing and grazing categories. Some were doubtless the equivalents of rhinoceros, bison or elephants.

The largest member of the family were the diprotodons (*Diprotodon optatum* and *D. minor*). Resembling an enormous wombat in shape, *D. optatum* would have stood a little under two metres high at the shoulder, was three-and-a-half metres long and would have approached two tonnes in weight. The remains of diprotodons have been found throughout most of Australia. Entire skeletons are known, while at Lake Callabonna in South Australia, trackways (showing the form of the foot pads), remains of the foot pads themselves and the impressions of the fur of the legs have been preserved. Unfortunately, the fur impressions are poorly preserved, but they do indicate that diprotodons were fur-covered, not naked like rhinos and elephants.

Their skeletons reveal that, like elephants, they bore their weight on pillar-like limbs and were probably slow-moving. Their tails were short, their bodies stout and the head relatively enormous. I know the skull quite well, for while still a student I was given the delightful if challenging task of preparing (freeing from the surrounding rock) some skulls that had been found in a clay pit near Bacchus Marsh in Victoria. I remember being impressed by an enormous and stout bony strut that projected upwards from the front of the skull, roughly where the nose would have been. I am still puzzled by this structure and can only suggest that it supported a large rhinarium or nose pad, much as in the koala but different in shape and more massive.

It was the back of the skull, however, that gave me the most food for thought, as well as heartburn. I soon learned that the bones in this area were almost wafer-thin. One day, I accidentally broke through one and, expecting to be able to peer into the brain cavity through the hole, was amazed to find myself staring into a large and empty space above the brain. I slowly discovered that most of the head of the diprotodon was made up of air-filled sinuses, the brain itself taking up a ridiculously small amount of room at the very rear of the skull. This indicates that the diprotodons were not terribly bright creatures. The reason for their air-filled sinuses is still not fully apparent. They would certainly have made the head very light, and it may be that this lightness saved the animal energy. The thin bone enclosing them may have provided anchor points for the muscles that worked the jaws. As we have seen previously, energy savings would have been an important consideration for a large, warm-blooded Australian animal. There were drawbacks to the design, however, for if they suffered from sinusitis it must have been amongst the most painful of all of their complaints.

A study of plant fossils preserved along with the diprotodon skeletons at Lake Callabonna has provided some interesting insights into its habitat. The plants growing around Lake Callabonna at the time the diprotodons lived were eucalypts and *Callitris* pines, as well as many shrubs. They are typical of parts of Australia's inland that receive about 250 millimetres of rain a year. Callabonna is in the driest part of Australia today and receives only 125 millimetres of rainfall. Yet today there are vast areas of semi-arid Australia which, on the evidence of Lake Callabonna fossil plants, would have been capable of supporting diprotodons.

Several other curious members of the diprotodon family survived into the Pleistocene in Australia. Some of the more common were the species of *Zygomaturus*. Dubbed 'marsupial rhinoceros' by some scientists, they were, at a metre high at the shoulder and two-and-a-half metres long, about 500 kilograms in weight. Their fossils are most common around the margins of the continent, indicating that they preferred moister habitats. Although the body was shaped like that of the diprotodons, the skull was very different. The muzzle was pinched-in in front of the eyes and the forehead rose steeply behind the snout. The eyes themselves faced forward, while the snout broadened out towards the front, terminating in a roughened patch of bone on either side of the nose. It is likely that these areas supported either horns like those of a rhinoceros, or a thickened pad of some kind. As with diprotodons, the marsupial rhinos were slow-moving creatures. Their molar teeth are lower-crowned than those of the diprotodons, suggesting that they fed upon softer plant matter.

The only other Australian species of this extraordinary family to survive until the Pleistocene was a very strange creature called *Euryzygoma*. Found in Queensland, it was about the size and shape of a marsupial rhinoceros, but it differed principally in that the males at least had enormous cheek bones. So enlarged are these bones that the skull is actually broader than long. No-one has any clear idea what the function of the cheek bones were, although it has been suggested, rather unconvincingly, that they may have supported cheek pouches which were used to store food. Alternatively, they may have been used in displays relating to reproduction.

This bestiary of cold-blooded killers and gentle giants were the cogs which kept the great machine of Australia's ecosystem functioning. Millions of years of coevolution had ensured that their gears and teeth intermeshed to form an intricate yet perfectly functioning ecosystem. Remove one or two species, and unless they happened to be critical cogs in the mechanism, the machine would continue operating. But remove all of the cogs and it must stop.

The last 60 000 years have seen all of the great cogs described above removed, and a different sort of machine reassembled from the small, left-over pieces. Two hundred years ago the machine was disrupted again when new cogs and gears were thrown in. But these great herbivores and carnivores were derived from a very different machine—the one that drives the ecology of Europe. They have once again brought the machine which drives Australian ecology to its knees.

CHAPTER 12

LOST MARSUPIAL GIANTS OF NEW GUINEA

The fascinating fauna of New Guinea has thus far played a relatively minor role in our story. This is because New Guinea is the newest of the 'new' lands, for it has developed only over the last 25 million years, when it rose out of the sea as the result of the collision of Australia with Asia. Before this, it was just a low-lying northern peninsula of Australia. The development of the mountains of New Guinea has had some important effects. It has created a moist, cool environment which has sheltered a part of the Gondwanan heritage of Meganesia that would otherwise have perished as the continent dried out. Indeed, the drying out of Australia and mountain building in New Guinea may be related, for New Guinea casts a rain shadow in its wake to the south.

At about 900 000 square kilometres the island of New Guinea is one-ninth the size of Australia and, after Greenland, is the world's largest island. Although it was colonised by relatively few groups of animals when it began its separate evolution 25 million years ago, these groups evolved and diversified rapidly. Today, it contains one of the most diverse floras and faunas on Earth. An idea of its extraordinary richness can be gained from the following comparisons: it supports 210 species of mammals, only 35 less than the much larger Australia; the Papuasian region supports 840 bird species, against 750 from Australia and its nearby islands. Furthermore, it supports about 300 reptiles and 220 frogs, against Australia's 750 reptiles and 250 frogs. At present, the counts for New Guinea's reptiles and frogs are probably severe underestimates, for the classification of these creatures is poorly understood. The numbers should probably be doubled to get a realistic estimate of diversity.

The reasons for this great diversity of life on a relatively recent and comparatively small landmass lie in New Guinea's tropical climate and topography. Because it lies in the tropics, it supports fantastically diverse jungles, which vary in composition with elevation. If one stands upon the peak of Puncak Jaya—at 5030 metres the highest mountain in the entire south-west Pacific—a vast and marvellous panorama is revealed. Near at hand and stretching to the south are a snow field and small glacier. Below that stretches an alpine tundra, complete with boulder-strewn moraines, then a rich alpine herbfield, covered in the flowers of dwarf rhododendrons, heaths, daisies and blueberries. Beyond that there is forest; at first a mossy and short elfin woodland redolent of the dwarfed, mossy forests of northern Europe so well described by the author of the Anglo Saxon epic *Beowulf* over a millennium ago:

the king rode ahead
to search out the way, till suddenly he came
upon stunted firs, gnarled mountain pines
leaning over stones, cold and gray.

Below and beyond, the forest becomes taller and southern beeches raise their heads high above the pines. Lower down still grow stands of great oak trees, bearing bunches of acorns, each nut five centimetres across. Below them the columnar klinki pines (*Araucaria hunsteinii*) grace the ridges and below them again the riot of the lowland jungle can be seen. At the coast, savanna, then mangrove give way to tropical sandy beaches and finally coral reefs. This marvellous place, which houses perhaps more diversity than is to be found in an equivalent area anywhere in our hemisphere, is part of the proposed Lorentz National Park, named but not yet gazetted by the Indonesian government.

With such topographic diversity it is no wonder there is such a diversity of life; for each vegetation type, from tundra to mangrove, has its own unique species. But this diversity is multiplied because often each of the separate mountain ranges that make up New Guinea have been sufficiently isolated to have given rise to whole assemblages of species that are unique to that small area.

The marsupial fauna of New Guinea is entirely derived from that of Australia—it supports no unique families. Recently discovered fossils indicate that marsupial giants were once part of the New Guinean fauna. So far, the remains of seven species of large, extinct, herbivorous marsupials have been found. They all belong to either the giant wallaby genus *Protemnodon*, or the diprotodon family.

The three species of giant wallabies thus far found in Pleistocene sediments in New Guinea were small in comparison with their Australian relatives, being about the size of a female grey kangaroo. They were browsers of the forest and subalpine grassland.

The three extinct species of the diprotodon family were more interesting. The largest may have weighed about 200 kilograms and was similar in form to the marsupial rhinoceros of Australia. It was a browser of the lowland jungle. The mid-montane forest was inhabited by another species, *Hulitherium tomassetii*. Its name means 'Huli beast', because its remains were first found in the land of the Huli people in Papua New Guinea's Southern Highlands. At about 100 kilograms in weight the Huli beast was shaped much like a panda, with the same high forehead and forward-facing eyes. Its hindlimbs show that it could stand upon its hind legs like a panda. Indeed, the Huli beast and the panda may have had similar diets, for bamboos are abundant in parts of New Guinea's mountain forests.

The third species, *Maokopia ronaldi* was the smallest of all the Pleistocene diprotodontids. At 50 to 100 kilograms in weight it lived in the alpine grassland and glacial tundra of the high mountains of Irian Jaya. Today, about 5000 square kilometres of such habitat exists in New Guinea, but during the ice ages it expanded to cover over 50 000 square kilometres, so during those times *Maokopia* was particularly widespread and abundant. Very little is known about *Maokopia*, for I only described its remains in 1993, having found its fossilised bones in caves and rivers of the remote Kwiyawagi area of Irian Jaya a few years before. Like *Hulitherium*, it may have resembled a panda, but scaled down to near-wombat size. Its skull suggests that it had a tiny, short snout, large, forward-pointing eyes and a high forehead. Its teeth indicate that its diet consisted of tougher plants than *Hulitherium* fed upon. It may have browsed the silica-rich ferns and grasses of its icy homeland.

I would love to have seen a living *Maokopia*, for both it and its environment are very special to me. I can imagine how it must have been all of those thousands of years ago, with small groups of teddy-bear cute marsupials browsing daintily among the flame orange flowers of the ground orchids and the brilliant red blooms of the pin-cushion rhododendrons. Was their dense, luxuriant fur black and white, or brown perhaps, with dark eye-rings and a pale belly? As now, the air was chill and carried well the odd tinkling sound of the wingbeat of Macgregor's bird of paradise. The great clumps of orange, bronze and olive moss covered rock and scree slope alike, and there, standing above all with its glistening glaciers, was the great Carstenz Peak. A dense fog, advancing faster than a person can walk, descends down the valley—and all is gone.

Although few in number, the extinct marsupial giants of New Guinea have much to tell us. It is now fairly clear that the fauna of giant marsupials of New Guinea was very limited. In all, New Guinea has lost only 10 species of large animals over the past 50 000 years. This is far fewer that any other of the 'new' lands and adds up to less than five per cent of the total mammal fauna found there. Outside Africa and Eurasia, no other landmass in the world has lost so little. Thus New Guinea is a wonderful place for research, for its fauna and ecosystems are still largely intact. In this it is unique in our region, a gem of inestimable value, the study of which will help us to understand the workings of the 'new' lands that support us all.

The reason that New Guinea's fauna has survived the human invasion so well may lie in its forested landscape. Over millions of years it has selected predominantly for smaller kinds of animals. Even before humans arrived there were very few large (and thus vulnerable) species. Additionally, the jungles of New Guinea are difficult places for humans to make a living. So dense is the undergrowth in the mountain forests that people are virtual prisoners of the paths they have trodden for generations. At lower elevation most life is found thinly dispersed in the treetops, where it is difficult for a person to reach. Thus, before people acquired agriculture and dogs, New Guinea must have been one of the most difficult of lands for its new settlers.

Having outlined the situation of the 'new' lands as they were before they were peopled, it is now time to examine what happened on the day that the first human footprint marked the shore and what it all led to. But in order to do that, we need to understand the nature and ecology of humans themselves.

Part Two

ARRIVAL OF THE FUTURE EATERS

*To appreciate their general
powers of mind is difficult. Ignorance,
prejudice, the force of habit, continually
interfere to prevent dispassionate judgement.
I have heard men so unreasonable, as to
exclaim at the stupidity of these people, for
not comprehending what a small share of
reflection would have taught them they aught
to have expected. And others again I have
heard so sanguine in their admiration, as to
extol for proofs of elevated genius what the
commonest abilities were
capable of executing.*

Watkin Tench, commenting upon European evaluations of the
Aborigines of Port Jackson, 1792.

CHAPTER 13

WHAT A PIECE OF WORK IS A MAN

Humans have long lived in ignorance of their origins, which is a terrible thing, for those who do not know their history are destined to repeat it. It was only in 1872, with the publication of Darwin's *The Descent of Man* that we got our first inklings of where we came from. It was not until 1890 however, that our first non-human fossil ancestor, *Homo erectus* was discovered in Java, and it was only in the 1970s that the partial skeleton of one of our earliest ancestors, *Australopithecus afarensis*, or Lucy, as she is now affectionately known, was found. Her skull remained unknown until 1994. Presently, almost every year brings new discoveries made in Africa and Asia which illuminate our origins. Just as often new hypotheses are formulated and books are written which retell the story of human evolution with the newly unearthed fragments as their centrepieces.

Much of this vast field of human palaeontology is of little relevance to an understanding of the history of Australasia. This is because it deals with events that occurred in Africa and Asia long before people discovered the 'new' lands. The history of humanity in its Afro–Eurasian homeland cannot be entirely ignored, however, because of the importance of the changing ecological niche that humans made for themselves over the million or more years that they were restricted to Afro–Eurasia.

This story is vitally important because it tells of the long process of coevolution that occurred between humans and their ancestral environments in Africa and Asia. A similar coadaptation could not occur in the 'new' lands, for people arrived there as fully modern humans, landing among a fauna that had never seen a primate of any kind before. The history of people in Africa and Asia reads like the story of a kitten and a bird that grow up together and who, through their early association, came to an understanding that allowed them to coexist. In contrast, the story of people in the 'new' lands is that of a meeting between a cat and a bird, but in this case the cat grew up amongst cats, while the bird had never seen a cat before.

Until around a million years ago, the entire panorama of hominid evolution had been played out on the wide plains and rift valleys of Africa. It was our immediate ancestor, *Homo erectus*, that became the first member of our family to leave the ancestral African home. Possessed of the most basic of stone tools, but bereft of anything else that we might recognise as technology, *H. erectus* was perhaps the most rudimentary being that we would grace with the name 'human'.

What gave these human ancestors the ability to migrate out of Africa and expand into the vast, well-watered regions of Asia—from the Levant to Java—no-one knows. It is unlikely to have been a technological advance and was probably a change in the environment, such as a shift in climate that opened a fertile corridor across the normally arid Middle East. Such changes have allowed animals to sporadically migrate from Africa to Asia and vice versa for tens of millions of years, so it is hardly surprising that our ancestors should take advantage of such opportunities.

For more than half-a-million years after the great migration, *Homo erectus* made a living preying upon smaller animals and foraging upon plants, before our own species, *Homo sapiens*, began to spread across the landscape. As Jared Diamond, Professor of Physiology at the University of California, has pointed out, we would like to think that the arrival of our own species heralded a great change, but this was simply not so. Even the great technological advance of the age, the taming of fire, seems to have belonged to *H. erectus* rather than our immediate ancestor *H. sapiens*.[36]

It has been rather difficult to determine the ecological niche occupied by early hominids. Archaeologists have long known of sites where the bones of large mammals are found in association with those of our ancestors. Perhaps the most important of these sites are a series of deposits which accumulated in caves and fissures in southern Africa, which were studied by the Australian-born archaeologist Dr Raymond Dart in the 1920s. He, along with many contemporaries, interpreted bones with sharp points as tools and human remains as evidence of ancient cannibal feasts. He clearly envisaged early hominids as big-game hunters and war-like people.

Recently, closer examination of some of the more important of these sites has shown that this interpretation was almost certainly wrong and that rather than being the hunters, our ancestors, whose bones are represented in these sites, were actually victims of carnivores such as leopards. One famous skull with two round holes in the top—long thought to be evidence of an ancient murder resulting from a blow on the head with a sharp object—is now known to have resulted from a fatal encounter with a leopard, for the two holes match exactly in size and position the lower canines of this fearsome carnivore.

In order to more fully understand the circumstances of these ancient bone accumulations, researchers have recently analysed many of the bones in great detail. They have found that in many cases the cut marks left by the stone tools of our ancestors are superimposed upon the tooth marks of other predators. This is clear evidence that our ancestors were scavenging from the kills of other carnivores. A few bones have the tooth marks of predators superimposed upon cut marks. This indicates that other scavengers were consuming the leftovers of our ancestors' meals. But these meals themselves may have been salvaged from the meals of larger carnivores.

As a result of these and other studies researchers have begun to re-evaluate the lifestyle of our early ancestors. Many now see stone tools not as hunting-related implements, but as adaptations to rapidly cut off hunks of flesh from carcasses of animals killed by other carnivores, so that they could be carried away and eaten in safety.[20]

In short, the most recent research suggests that the earliest members of our genus, along with our more distant relatives, probably made a living as generalised and rare savanna apes which took a little small game, but did a lot of scavenging and plant harvesting. This basic ecological niche had been occupied by hominids for a million or more years by the time that the South African sites were deposited. It was to remain important to human ecology for a million or more years after. Indeed, except for the importance of scavenging, it is the same ecological niche occupied by our nearest living relatives, the two species of chimpanzees, in the rainforests of Africa today.

We do not know when humans first began to expand this basic ecological niche, but it is clear that the change was gradual. By around 100 000 years ago a definite shift in ecology had occurred, for remarkable new evidence indicates that our ancestors had been raised above the serried ranks of the scavenging jackal, hyena and vulture, into the exclusive club of the big-time killers.

Five remarkable sites, all in South Africa, have provided by far the best evidence of the ecological niche occupied by our ancestors at around 100 000 years ago.[20] Of these, the spectacular cave site at beautiful Klasies River Mouth near the southern tip of Africa has provided us with perhaps the most important details. At this time relatively archaic types of humans, such as Neanderthals, occupied most of the world outside Africa. So these African sites are doubly important, for not only do they document our ancestors' transformation from scavenger to top carnivore, but they also reveal something of the lifestyles of the first of our ancestors who were virtually identical in their anatomy to us modern humans.

Some biochemists have developed a theory that all living people are descended from a single woman, nicknamed Eve, who lived in Africa some 130 000 years ago. This is because they can trace all variations in human mitochondrial DNA back to a single type that existed at that time. The mitochondria are the 'power packs' of the cells. There are no mitochondria present in the sperm head, so males do not get to pass on their mitochondrial DNA to their children. Mitochondria are present in the egg, so they get passed on exclusively through the female line. This is why biochemists have identified an Eve rather than an Adam. The proponents of the Eve theory postulate that Africa was the Garden of Eden for fully modern humans because the greatest genetic diversity among people exists there today.

Although useful in some ways, this hypothesis may be rather misleading. This is because it is almost certainly through statistical chance alone that all living human mitochondrial DNA descends from a female who lived 130 000 years ago. Nonetheless, there does seem to be some validity in the idea that a new kind of human, who was ancestral to all living people, spread out of Africa, possibly some 130 000 years ago. The people of Klasies River Mouth may have been early members of this group.

The Klasies River Mouth site is—as its name suggests—by the sea. Not surprisingly, the cave sediments contain the remains of marine animals such as seals, penguins and shellfish. Yet intriguingly, there are few bones of fish and seabirds. Perhaps the people who occupied the site lacked the technology to catch them.

Among the bones of land animals, those from medium-sized species are most abundant. Curiously, the bones of the relatively rare eland (an african antelope weighing some 400–1000 kilograms) are extraordinarily well-represented in the sites. Today, eland have been exterminated over much of their range because they are relatively easy targets for human hunters. Archaeologists think that people living 100 000 years ago were able to kill eland relatively easily because they are rather tame, not dangerous, and easy to drive in herds. The bones represent animals of all ages, suggesting that the hunters may have made mass kills, perhaps by driving whole herds into traps. In contrast, the bones of the decidedly more dangerous Cape buffalo (*Syncerus caffer*) are either from very young or very old individuals, while bones from the most dangerous game (elephants, rhinos and wild pigs) are virtually absent from the sites.[36] These finds indicate that while the ability of the Klasies River Mouth people as hunters was greater than anything achieved earlier, it was still a long way from the hunting ability of modern hunter–gatherers.

The fossil record indicates that by 50 000 years ago our ancestors had progressed even further as hunters. South African cave deposits dating to around this time contain the abundant remains of such vicious prey as pigs and Cape buffalo, while elsewhere people were killing rhino and elephant, which are the very largest of land-based prey.

By 40 000 years ago people living in northern Africa were making stone points with tangs (a finely worked area at the base to which a handle could be attached). Although spears, perhaps made entirely of wood, had probably been in existence well before this, the tanged spear tips from north Africa are important, for they represent the development of a sophisticated stone-tipped weapon. Such a weapon makes it possible to bring down large animals from a considerable distance.[20]

By 20 000 years ago people the world over were regularly hunting big game (where it still existed) and had developed sophisticated and diverse stone armouries. Even the largest and most powerful of land mammals, such as mammoths, were a substantial part of the diet of some human groups.

This gradual shift in the ecological niche occupied by humans had seen them develop from scavengers and hunters of small game, to the principal predators of the most formidable of all land animals. The shift meant that in little less than 100 000 years, humans had developed an ecological niche that overlapped with virtually all other carnivores. In addition, the human ecological niche had become broader than that of any other mammalian species. Despite the dramatic nature of these changes, it is important to remember that, on the scale of human lifetimes, they happened imperceptibly slowly. Furthermore, they were part of a continuum of change in the shifting ecological niches of the large mammals of Africa and Asia.

The relationship between predator and prey in such a situation can be seen as a kind of arms race. Two groups of people who are in continuous conflict will gradually hone their weapons. Both sides do their utmost to keep up with advances made by the other, for they know that to allow the other side to gain too great an advantage is fatal. It has taken this process less than 10 000 years to transform the armouries of western Europe from caches of stones to nuclear weapons. Yet even with the possession of such horrifying weapons, no one group in Europe has been able to gain a complete victory. This is because change has been relatively slow and everyone carefully monitors everyone else and will sacrifice virtually anything in order to catch up with a technological advance.

Similar arms races also occur between species. Natural selection acts slowly upon predators to increase their efficiency. Thus, over time cheetahs have become faster relative to their ancestors. This has placed selective pressure upon their prey, because only the fastest individuals survive. Thus, the average speed of the prey species of cheetahs—principally gazelles—has increased commensurately with that of cheetahs.

Humans and their prey species have undergone a similar kind of coevolution in Africa and Asia, but instead of speed, it is probably various behavioural adaptations that have been selected for. Thus, each time that human hunters adopted a technological or behavioural change, the animals adapted to it by developing behaviours that granted some immunity to the new technique. It is likely that some species, by virtue of their ecology, could not adapt and that these were hunted into extinction. Such species may have included some of the African and Asian elephants and the giant African buffalo, whose skull approached two metres in length.

The greatest danger that a species faces in a rapidly coevolving ecosystem is the loss of contact with its competitors. For if one party in a multi-sided conflict is somehow isolated, but then contact is re-established, combatants bearing stone tools can find themselves squaring off against opponents bearing nuclear weapons.

While the very broad ecological niche that humans had opened for themselves by 50 000 years ago was an impressive achievement, it was just the beginning. For since then humans have, through the use of technology, broadened their ecological niche fantastically. Indeed, if one feature sets humans apart from other animals, it is the breadth of the ecological niche that we presently occupy.

The development of traps, nets and spears had, by 50 000 years ago, allowed people to become direct competitors with virtually all land-based carnivores, including foxes, lions and tigers. The development of fishing technology would soon see them competing with barracuda and killer whales. The development of agriculture brought competition with plants and entire ecosystems, while modern technology allows us to compete—and win—against virtually any organism or group of organisms on Earth.

The speed at which a species evolves is clearly critical in determining the impact that it has on the species with which it coexists. For much of our history, humans have evolved at the relatively slow pace that is dictated by physical evolution. The pace is slow because it depends upon natural selection acting upon successive generations. Because humans are long-lived and reproduce slowly, their physical evolution cannot progress as rapidly as it can in species with shorter generation times. It may take thousands of generations to bring about significant change in a species. For humans that means tens, or even hundreds of thousands of years. This may explain why humans have not been able to overcome certain diseases, for a bacterium with a generation time of an hour or less will always adapt more quickly than a species with a generation time of 20 or more years.

The development of technology allowed people to speed up evolutionary change. Humans no longer had to wait for the evolution of a sharp tooth before they could tackle large prey, for they could simply fashion a spear instead. Likewise, modern people can don an aqualung or a space suit, travel in an aeroplane or car, think with computers and fight diseases with manufactured drugs. All of these adaptations, from the most simple to the most complex, have the effect of speeding the rate of human adaptation relative to other species from a snail's pace to that of the speed of light. It is this very fundamental shift in evolutionary speed that has upset the ecological balance throughout the world today.

Sixty thousand or more years ago human technology was developing at what we would consider to be an imperceptible pace. Yet it was fast enough to give the first Australasians complete mastery over the 'new' lands. Freed from the ecological constraints of their homeland and armed with weapons honed in the relentless arms race of Eurasia, the colonisers of the 'new' lands were poised to become the world's first future eaters.

CHAPTER 14

GLORIOUSLY DECEITFUL, AND A VIRGIN

How wonderfully apt a description of the 'new' lands is Horace's thumbnail sketch of the mythological Greek maiden Hypermestra. Her deceit lay in her unique refusal, among 50 sisters, to slay her new husband at her father's request. Ironically, her virginity was lost at the moment of her honourable deception. Like Hypermestra, none of the 'new' lands had 'known man' until very late in their histories—and each has been, in its own way, gloriously deceitful to its new colonists. As each of the 'new' lands has lost its virginity, it has suffered frightful ecological disruption. Over time, some sort of ecological balance has usually been restored.

The deception experienced by each wave of human immigrants into the 'new' lands is one of the great constants of human experience in the region. To the earliest Aborigines, it must have seemed as if the herds of diprotodons stretched on forever. To the Maori, the moa must have appeared a limitless resource. European agriculturalists saw what they imagined were endless expanses of agricultural land of the finest quality. Early Chinese immigrants to Australia saw *san gum shaan* (new gold mountain), and now each new immigrant sees an opportunity to prosper in the land of plenty. In short, all have seen a cornucopia where there is in fact very little.

Worse, the new settlers have usually assumed that their new homeland is a virgin, often in the face of the most blatant contrary evidence. *Terra nullius* is not solely a British delusion. Each new wave of people, arriving from the resource-rich lands to the north, sees in the unoccupied regions of Australia and her neighbours room for development and space in which to flourish. In part perhaps because of this sheer sense of space, each new wave of settlers can identify some virgin resource—some field untilled, sea unfished, or forest unfelled—with which they can make their future.

Yet these unoccupied spaces and apparent opportunities in fact represent something very different, for they are the necessary accommodation that each group makes to life in a hard land. For the Aborigines, that accommodation meant foregoing agriculture and hence leaving a very different kind of mark on the land. For the Australians of European origin, it has meant leaving the centre and north largely empty and the creation of vast national parks on what appears to be useable land. It has meant the imposition of what are—by European standards—extraordinarily low stocking rates on rangelands and a low level of utilisation in other areas.

These necessary accommodations have created a sense of paranoia in living Australians. Perhaps because of their own recent use of the concept of *terra nullius*, many fear that people from Asia will perceive in Australia, if not an empty land, then an under-utilised one.

In an ecological sense, the history of all of the colonists of the 'new' lands has followed the same trajectory. Their histories look so different to us because we see human groups at different moments: Europeans 200 years, Maori about 800–1000 years and Aborigines 60 000 years, after the time of colonisation. The trajectory, or pattern of development, experienced by each group is as follows. The initial deception leads to a sense of unbounded optimism. But this soon turns to bitter disillusion as resources are exhausted. Finally, there comes a long and hard period of conciliation, during which the land increasingly shapes its new inhabitants.

In this chapter, we will examine the common trajectory followed by the Australasians by looking at a variety of people who have entered the new lands at different times and thus have variously progressed in the same great game of adaptation.

The pre-European colonisation of the 'new' lands was undertaken at different times by two very different peoples. Meganesia was colonised in its entirety at least 45 000 and probably 60 000 years ago, by the ancestors of the Aborigines and New Guineans. The colonisation of Tasmantis, by contrast, did not begin until 3500 years ago, when the seafaring Lapita people arrived in the Pacific. Their colonisation was not complete until some 800 years ago. Even then they missed a couple of tiny, isolated fragments of land which were to remain true virgins until the European colonisations in the late eighteenth and nineteenth centuries.

Even on the world stage, these human expansions in the Pacific are events of enormous importance. The settlement of Meganesia resulted from the very first human foray into areas outside the vast Afro–Eurasian homeland that they and their ancestors had occupied for a million years or more. The much later colonisation of Tasmantis marked the development of the first truly ocean-going human culture of traders and deep-sea fishermen. Together, the people who undertook these invasions were I believe, through the adaptations they acquired as new colonists, to transform the world.

No-one knows precisely when humans first arrived in Australasia, for we are yet to find clear evidence of the very first Australians. But we do know that the archaeological record of Australasia must hold evidence for the first-ever colonisation of a continent since *Homo erectus* crossed the Red Sea from Africa to the promising land of Eurasia, some million or more years before. For this reason, Australasian archaeology holds a particularly important place in world archaeology. But in part because it is a relatively new field, very few archaeological sites of extreme antiquity have been discovered. Thus it holds the promise of many more great discoveries which may well transform our view of the past.

The big questions in Australasian prehistory: the when, from where, how and why of the peopling of Australasia, are gradually being answered, at least in part. Answers to the easiest questions have long been evident, if sometimes a little puzzling. Perhaps the easiest of all to answer has been the 'from where'.

Most researchers (with the notable exception of Dr Alan Thorne of the Australian National University, who argues for a hybrid South-East Asian/Chinese origin) agree on a South-East Asian genesis. Of the great island chain that stretches eastwards from South-East Asia, the very closest to Australia is Timor. It is presently separated from the north-west coast of Australia by a distance of over 500 kilometres, but at times of lowered sea-levels during the height of the last ice age, Timor was then only 90 kilometres away.

To the north, the islands of the Sula Group were separated from Sulawesi only by a narrow strait during the ice age. They, in turn, are separated from Grand Obi (a piece of old Gondwana that is joined to New Guinea by a series of stepping-stone islands) by a distance of around 100 kilometres, ice age or no (the distance does not change because the islands slope off steeply into deep water).

It is reasonable to assume that humans first entered Meganesia across one or both of these routes. But really, their crossing was just the final step in a long journey. For the area between mainland South-East Asia and continental Australia is a maze of islands. Some were connected with each other and with one or the other mainland during the ice ages, while others have always been isolated.

To return to the question of South-East Asian origins for the Aborigines and New Guineans, a curious but immediately obvious fact arises. This is that there are presently no people resembling the Aborigines and New Guineans in South-East Asia west of the Moluccas and Timor. One has to go as far afield as the Philippines, peninsular Malaysia and the Andaman Islands to find a few tiny groups of pygmy negrito people who look anything like Australian Aborigines. Researchers have therefore long assumed that the ancestors of the Australian Aborigines were displaced from their South-East Asian home by invading Mongoloid peoples.

There is now a considerable body of evidence which supports this view, including the discovery that the Mongoloid peoples arose in the dry, near-polar deserts of north Asia (fatty cheeks, narrowed eyes and the Mongoloid eye fold are all adaptations to dry, cold conditions) who only recently invaded the tropics. It also includes the discovery of archaeological sites which document the presence of Australoid people in areas today occupied by Mongoloids. Then there is of course the long-standing evidence of relict groups (such as the Andaman Islanders) who exist scattered within the distribution of the Asiatic peoples.

The displacement of Australoid peoples from South-East Asia by Mongoloid invaders from the north appears to be a very recent event.[15] It began only 8000 years ago, when Mongoloid people living along the lower Yangtze River developed agriculture based upon the cultivation of rice. They also developed pottery and had domestic dogs, pigs, chickens and possibly cattle and water buffalo. These developments allowed the growth of great human population densities. By 7000 years ago towns like Hemudu in what is today Zhejiang Province, China, were supporting vast and wealthy populations.[15] Conditions further south, in South-East Asia, were ideally suited to the development of similar communities based upon rice agriculture, for the subtropical climate suited rice, while the vast, fertile, swampy and alluvial environments created by river floodplains were ideal for expansion into rice paddy.

The movement of agriculturalists into South-East Asia appears to have been rather slow, for it was not until 5000 years ago that the entire southern coast of China had been taken up by rice-growers. By 2000 years ago, large-scale land clearance had begun on Java and Sumatra, heralding the arrival of the agriculturalists there. Just why some of the Australoid peoples who were indigenous to the region did not adopt agriculture and avoid extinction or assimilation is not clear. After all, the Australoid peoples of the mountains of New Guinea have been agriculturalists for 10 000 years. Instead, the Australoids of South-East Asia seem to have vanished in the face of the Mongoloid invasion. It may simply be that the agriculturalists, who were present in far greater numbers, absorbed the relatively tiny non-agricultural human populations. Whatever the case, it is clear that the development of rice agriculture, with its ability to sustain large populations, gave the Mongoloids the ability to make South-East Asia their own.

East of Wallace's Line the replacement was virtually absolute, for today, on islands such as Java, there remains no physical evidence at all among the more than 80 million inhabitants of Australoid ancestry. It was only in areas with very poor soils, or in isolated situations such as the Andaman Islands, that some Australoids have persisted.

The question as to why the Asian agriculturalists did not pursue their conquest across Wallace's Line and into New Guinea and Australia is a fascinating one. The answer may lie in part in the unique nature of Meganesian environments. Even on the margins of Meganesia, on the Wallacean islands of Halmahera and Seram, agriculture is a risky business. I have visited a *Transmigrasi* settlement in northern Halmahera. There, in the middle of a vast clearing made in what was once a densely forested plain, some few hundred transmigrants from Java were attempting to coax a crop of rice from the infertile soil. It had not rained for some time and the soil, composed of fine, whitish volcanic ash, blew between the rough wooden houses, coating everything in a fine siliceous grit. In infertile and ENSO-dominated Australia, Javanese style *padi* is often not a viable option. Under such conditions, people adapted to living with scarcity will perhaps ultimately prevail.

But there was another reason that the invasion stopped at the invisible Wallace's Line. And this is that those few areas of Meganesia which are suitable for agriculture had already been occupied and cultivated for millennia by the time the rice-growers arrived in the region. Indeed, some Mongoloid groups soon gave up rice agriculture and began growing taro, sugar cane, bananas and sago, varieties of all of which had first been domesticated by the Melanesians. They may have found—as Australian would-be plantation owners who arrived in Papua New Guinea in the 1950s did—that there is simply no room in New Guinea's fertile regions for any more agriculturally based people.

If the question as to *where* the first Australians came from is relatively easy to answer, then perhaps the most vexing question of all concerns *when* people first arrived in Australasia. Without knowing how long people have inhabited Australia, it is extremely difficult to interpret evidence concerning their impact on the land, or even how the land has shaped the people themselves. Ten, or even five years ago, many scientists gave much more confident answers about when the first Aborigines arrived than they do today. Their misplaced confidence grew through the 1970s and 1980s as a series of discoveries of ancient human remains were made in Australia which were initially dated at 40 000–30 000 years old. At the time, these discoveries were considered to be of world importance, for they suggested that Australia had been inhabited for much longer than anyone had ever suspected.

What are arguably the most important of these discoveries were made near Lake Mungo in western New South Wales in 1970. There, the cremated remains of a woman were found buried in a sand dune. She had been cremated while lying on her left side, then the bones had been gathered up, broken into small pieces, and placed in a pit. Radiocarbon dating undertaken at the time suggested that the woman, later to become known as Mungo 1, last walked the earth some 38 000 years ago. It has since been found that this date was incorrect, and that she lived more recently.

On the 26 February 1974 an uncremated, fully articulated skeleton of an adult man was discovered in the same area. Pink ochre powder had been sprinkled over the upper part of his body. Radiocarbon dating revealed that he lived some 32 000 years ago.

Since that time, a number of other discoveries have been made which, by the 1980s, convinced archaeologists that humans had occupied Australia for about 40 000 years. First there was the excavation of cave sediments from Devil's Lair, Western Australia, which demonstrated a human presence from 30 000 years ago. Then a remarkable series of discoveries were made in Tasmania, showing that humans had been occupying caves in the south-western part of the island for 35 000 years. Additional, less well-documented evidence of a human presence at this remote time, was published to back up the claims of a human entry into Australiasia 40 000 years ago.

These finds convinced many people who were not intimately familiar with the technique of radiocarbon dating (including myself), that humans must have arrived in Australia around 40 000 years ago. For it makes sense that the cluster of sites dating to this time was evidence of the first human presence, and that the absence of firm evidence for older sites indicated that humans were not present prior to this.

But there were always a few researchers who held out against this interpretation. Chief among them was Professor Rhys Jones of the Australian National University, one of the most illustrious and respected of all Australian prehistorians. His argument, although long unheeded by many, is a sound one, and is based upon the mechanics of the radiocarbon dating technique itself.

The radiocarbon dating technique is based upon measuring the ratio of a radioactive isotope of Carbon (called C^{14}) to normal Carbon (C^{12}). C^{14} is being formed and decaying in the atmosphere at approximately the same rate. Therefore there is always about the same amount of it in the atmosphere and oceans. All life is based upon carbon, and while an organism is alive, the ratio of C^{14} to normal carbon within it stays the same because carbon is constantly being exchanged between the organism and the atmosphere. When an organism dies, however, no new carbon is taken in or exchanged and the C^{14} gradually decays.

After 5700 years, approximately half of the C^{14} which accumulated during the life of the organism has decayed. After 38 000 years, only around one per cent of the C^{14} that was present in the living organism remains.

Although a wonderful basis for dating, the rate of decay of C^{14} limits the usefulness of the technique to a rather short time span (approximately the last 40 000 years). Dates of older than about 35 000 years are particularly problematic. This is because it is very difficult to detect the small amount of decaying C^{14} emitted by the sample (in standard radiocarbon dating it is the decay rate, not the C^{14} itself that is counted).

A recently developed technique called AMS actually counts the amount of C^{14} remaining in a sample. This allows for older materials to be dated more accurately.

Contamination with younger carbon is the great bug-bear of all C^{14} dating. If a single tiny rootlet of a modern plant has penetrated a sample and is not detected, it may give off enough decay of C^{14} to make the sample look more modern than it really is. Likewise, if a researcher handles a sample carelessly, tiny flakes of skin, sweat or grease from the hands can cause contamination. These obvious contamination problems have long been known, but what is not often realised is that organic matter carried in ground water can contaminate samples quite easily. This kind of contamination is much more difficult to detect.

Unless the amount of contamination is massive, it is not such a problem for samples less than 20 000 years old, for there is still lots of original C^{14} and the small amount of contaminant does not greatly affect the outcome. It is a bit like being one dollar out when counting a thousand. But contamination can be a very severe problem for very old samples, when so few atoms are being counted. The contamination may then be like miscounting by a dollar when there are only two dollars! At the extreme, a tiny amount of contaminant could make a million-year-old sample appear to be 38 000 years old. This is because the only C^{14} counted comes from contamination. Indeed, there are documented examples of coal from the Carboniferous period (some 300 million years ago) giving a date of 38 000 years before present, despite the best efforts of researchers to eliminate contamination.

Jones' familiarity with these limitations of the radiocarbon dating technique made him examine the evidence for a human arrival in Australia some 40 000 years ago in a new light. He has suggested that the cluster of dates on human bones and occupation sites around 35 000 years ago is partially the result of contamination of materials that probably vary widely in age. Some samples may be much older. He argues that the hypothesis that humans arrived in Australia only 40 000 years ago is based upon very little evidence, as almost all dates are based upon the radiocarbon technique.

One of the most important pieces of evidence not based on radiocarbon dating was found on the Huon Peninsula of New Guinea. Today, the northern slopes of the Huon Peninsula look like a giant's staircase extending more than a kilometre above the level of the sea. Each step in the staircase is a raised coral reef, the one above being older than the one below. The staircase has been formed in geologically recent times because the mountains of the Huon Peninsula are rising at a faster rate than mountains almost anywhere on earth. The 1993 earthquakes that killed tens of people in the Finisterre Ranges are symptoms of this rapid uplift.

In the 1980s archaeologists working at the University of Papua New Guinea found great stone axes buried in sediments that lay behind a reef that was raised above the sea some 45 000–40 000 years ago.[59] Intriguingly, the sediments associated with reefs which were raised more than 50 000 years ago lack human artefacts. Because researchers know how rapidly the mountains of the Huon Peninsula are rising, the estimate of the age of the axes is not based entirely upon radiocarbon dating, but upon the height of the enclosing sediments above the sea and upon another dating technique called thermoluminescence.

Jones, along with other researchers, is currently pioneering new dating techniques which should shed new light on the early history of humans in Australia. One of the most promising of these is optically stimulated thermoluminescence dating. It is a complex technique, and currently is poorly understood. Preliminary results, however, suggest that some human occupation sites in northern Australia may be between 60 000 and 53 000 years old.

Until the new dating techniques are further developed, prehistorians must remain satisfied with our present imperfect understanding of the early history of humans in Australasia. We currently know that people have been present in Australia and New Guinea at least 40 000 years, but no-one knows for sure how much earlier they arrived. They may have been there for 60 000, or even 120 000 years. It is unlikely that they have been present for much longer than this, however, as abundant fossil deposits which are older than this contain no evidence of a human presence, and fully modern humans—as the first Aborigines were—had not evolved much before this time.

The question as to how people arrived is, at one level, easy to answer, for the first Australasians must have travelled across water to reach their new home. Most people therefore assume that the first Australians had watercraft.

Dr Jonathon Kingdon of Oxford University has recently developed a hypothesis about the nature of the ancestors of the Australians that seems to fit the few known facts well.[77] He suggests that when modern humans finally reached South-East Asia some 100 000 years ago, they rapidly evolved to suit the unique environment in which they found themselves. He notes that South-East Asia is particularly rich in near-shore marine environments which offer extraordinarily rich pickings for large omnivorous primates such as humans. The endless reefs, mudflats and seagrass beds of the Malay Archipelago can supply tonnes of dugong and turtle meat, fish, molluscs, prawns, crabs and much more. Indeed, they probably offer more food, which can be gathered in greater safety, than exists in the jungles that cover the land.

If people were to exploit such resources, they would have had to develop a number of physical and cultural adaptations to help them deal with their new environment. Previously, humans had been children of the sun, their activity patterns being driven by the inexorable rhythms of day and night. They foraged in the morning and evening, avoiding both the dangerous night and the deadly rays of the midday sun.

The rhythm of the littoral is, in contrast, driven by the moon, for it is the tides that determine when resources are available there. If people were to gather food on the littoral, they would have had to spend long hours exposed to the glare of the midday sun while foraging on beaches, mudflats and reefs. The risk of sun stroke and skin cancer would have made this extremely dangerous.

Their defence consisted of the development of very black skin. Kingdon points out that the primitive skin colour for humans is an all-purpose brown, and that very black skin is as extraordinary a deviation from this as is white skin. This is because truly black skin represents a radical departure in structure from that of other skin types.

The pigment in skin is called melanin. In all skin types but very black skin, the melanin is held in special cells called melanocytes. These lie clustered together in the epidermis, but in truly black skin the melanocytes are widely dispersed and are embedded in the epidermis in a unique way. It is as if the bags containing the melanocytes had been punctured, allowing the cells to spread everywhere. Kingdon calls these first truly black people the Banda, after the islands of the Inner and Outer Banda Arcs north of Australia which, he postulates, were their ancestral home.

Along with this unique physical adaptation would have gone a number of cultural adaptations. It seems likely that the Banda were the very first people in the world to develop seaworthy watercraft. Such watercraft would have been a tremendous advantage to people who live on the littoral. They may have started out simply as a floating platform upon which to pile molluscs and fish. They would soon have developed into platforms large enough to sit upon and spear fish and turtles from. At some stage they became seaworthy enough to carry people to nearby reefs and islands that they had not previously been able to exploit. The opportunities that this opened up would have been a tremendous boost to the development of maritime technology.

Kingdon suggests that the Banda people possessed considerable abilities in the use of fibres for making string, rope and netting. This makes a great deal of sense considering their ecological niche.

Admirably adapted for life on the littoral (an environment which links continents) Kingdon hypothesises far migrations for the Banda. The move east into Australia/New Guinea would have been the shortest, and perhaps the first major migration that they undertook. But they also appear to have migrated far to the west, along the littoral of the Indian Ocean, out to the Andaman Islands (where their descendants survive today) and back into the African homeland of all humanity. There, they displaced the original honey-coloured Africans from many habitats in coastal and equatorial Africa.

I believe that the peopling of Australasia was an event of major importance for all of humanity. This is because I think that it altered the course of evolution for our species. The first Australasians were, after all, the very first humans ever to escape the straightjacket of coevolution. As such, they were the first of the future eaters.

Professor Jared Diamond is one of the world's greatest living scientists. He postulates that at the time humans were colonising Australia, people around the world underwent an event which he calls the 'great leap forward'.[36] He suggests that this change transformed our ancestors from just another species of large mammals into world conquerors.

He suggests that if extraterrestrials had visited the Earth as little as 100 000 years ago, and had tried to decide which species of animal would inherit this wonderful planet, they would have given the guernsey to the bower birds of New Guinea and Australia for their wonderful constructions, artistic accomplishments and complex social life. Our ancestors, on the other hand, had little to recommend them. All but lost among the great herds of mammals of their Afro–Eurasian homeland, only their use of rude stone tools and possession of fire would have given any indication of future greatness. Yet by 40 000 years ago they were already well underway in their conquest of the planet. Although at that time their architectural achievements still lagged far behind that of the best bower-building birds, their art, social life and communication far exceeded anything that had come before.

The cause of this transformation of humanity—from one in a crowd to the dominant species—is still hotly debated. Diamond has his own ideas on what powered it, and considers that only the acquisition of something as fundamental as language could be responsible. Language, of course, makes and defines us. Without language we cannot have a complex social life, pass on ideas, nor investigate our place in nature. Without these fundamentals we cannot have art, science or religion.

One can easily imagine then how the acquisition of language could transform our archaic ancestors into fully modern humans. But did the development of language occur approximately 60 000–40 000 years ago? Two telling pieces of information have been unearthed that suggest the contrary. The most important concerns the mechanics of speech.

Humans have undergone the most remarkable physical changes to make speech possible. Mammals have a fundamental problem in this area. This is because the channels that carry food and air into the body follow a common course at the back of the throat. In order to prevent the dangerous possibility of food 'going down the wrong way', they have a structure called a larynx that locks into another called the nasopharynx. When interlocked, these structures effectively isolate the two channels, allowing most mammals to breathe and eat or drink at the same time.[79]

Speech demands a large space between the larynx and nasopharynx so that noises made by the vocal cords can be modified into more complex sounds. This space, called the pharynx, is formed in adult humans by migration of the larynx down the neck. This migration occurs at around the time we are about two years old. While allowing for speech, it creates a potentially dangerous condition, for a common canal is formed down which, for a short way, both food and air must pass.

We have an epiglottis to guide material into the right channel, but as we all know food or drink can still 'go down the wrong way', causing inconvenience or even death through blockage of the air passages. Only strong evolutionary pressure could have led to the development of such a potentially dangerous structure. Perhaps only something as valuable as speech could compensate for it.

Although changes in soft tissue such as the larynx and pharynx are not preserved in the fossil record, they can be traced through associated changes to the bones forming the base of the skull. From these, palaeontologists know that a pharynx capable of producing speech had developed by 400 000–300 000 years ago—long before the great leap forward.

A second piece of evidence comes from the distribution of world languages. If, as Diamond suggests, language arose as recently as 60 000 years ago, one might expect that some very primitive languages may be preserved among some long-isolated groups of humans. After all, many isolated groups do (or did) preserve the technological equivalents in their tool kits. Prime candidates are the peoples of Australia, Tasmania and New Guinea, who have been largely isolated for at least 40 000 and possibly 60 000 years. But these people speak languages every bit as complex and advanced as those found in Europe, Africa or the Americas. Indeed, so complex are some New Guinean languages, such as Miyanmin (a non-Austronesian language of the Mountain Ok group of Papua New Guinea) that children are not expected to master them until they are eight or nine years of age. People just laughed when I expressed an interest in learning Miyanmin as an adult.

The Miyanmin and their language have remained relatively isolated in the mountains of New Guinea for at least 40 000 years. If language did arise as recently as 60 000 years ago, it seems that it must have become universally adopted almost instantaneously in order for it to have achieved its present universal level of complexity. Alternatively, perhaps language has been evolving slowly, just as technology has, over a very long period.

If the development of language did predate the great leap forward by hundreds of thousands of years, are there any other developments which could have powered it? Indeed, instead of leaping, could our ancestors have been pushed? I think that very few evolutionary shifts occur because of the development of a new faculty or organ. Rather, changes in ecological circumstance place organisms under intense and quite different selective pressure, to which they respond by evolving new structures or behaviours. Taking this approach, we can ask whether any changes in the human environment occurred 60 000–40 000 years ago, which might have influenced the great leap.

To investigate this we must understand a little of the situation of humanity at that time, and what it is like to be an omnivore in a co-evolved ecosystem. Sixty thousand years ago humans occupied the same land area that they and their ancestors had occupied for over a million years. They existed as small bands of hunter–gatherers which were scattered throughout suitable parts of Asia, Africa and Europe. They were just one uncommon omnivorous species among a plethora of other large mammals. Admittedly, they already had a broader ecological niche than most comparable species, but still, they were far from being the world conquerors that they were to become by 30 000 years ago.

Because of the arms race effect described above, fundamental shifts in ecology (and thus in culture, technology and outlook) are very difficult to achieve in complex, coevolved ecosystems. We can imagine, for example, that humans had been hunting horses, deer and other prey for over a million years. The fossil record suggests that they had gradually become better predators, for over tens of thousands of years they had increasingly begun to take larger and more formidable prey. But just as gradually the prey species had adjusted to the new predator, so the balance of power between them usually did not shift.

Likewise, other carnivores preyed upon people, keeping them within strict bounds. The fossil record, as well as present experience, shows just how effective big cats are as predators on humans. Where tigers are present in India, it is a foolish and short-lived person indeed who enters dense thickets of vegetation. Likewise, wandering out onto the African savanna after dark, even today, is a short-cut to becoming a lion's dinner. Sixty thousand years ago, entering dense thickets or venturing out at night were not options for people, because such large carnivores were plentiful and occurred everywhere that people occurred. The restrictions that this placed on human activity left refuges in time and place for prey species that humans might otherwise have destroyed.

Living in a tightly coevolved ecosystem makes taking the great leap forward terribly difficult. It is hard, for example, to gain the resources necessary to develop sufficient population density to sustain complex social structures, art, or the flow of ideas that leads to improved technology. It is also difficult to utilise new resources because there is always another, more specialised species already harvesting each resource efficiently. In brief, living in a complex, coevolved ecosystem is a little like living in a prison cell on short rations. The constraints of ecology are tight and it is difficult to take a step or two forward, much less a leap. Worse, if an organism does make an advance, the other species which are affected by it in the system—whether they be predator, prey or parasites—take a step also. As a result, there is often a zero sum gain.

I suggest that the great leap forward could not have taken place under such circumstances. In order to have achieved the leap, humanity must have experienced conditions which allowed population to build up substantially, which encouraged experimentation and a 'managerial' mentality, and where competition between various human groups was intense. Until 60 000 years ago, humans did not inhabit—and had never inhabited—such an environment. By 40 000 years ago, however, they were well established in over 10 million square kilometres of such habitat, for they had invaded Australasia.

We do not know where the beachhead for the invasion was, but it is a fair guess that the narrow strait between Bali and Lombok was the first and most fundamental barrier to be breached. For a million years humans and their ancestors had lived on Bali, from where they had probably looked out daily upon the uninhabited lands to the east. Then, sometime before 60 000 years ago, someone crossed that barrier and laid the first human footfall on a naive land.

The arguments put forward by Kingdon suggest how people gradually came to use more and more marine resources. Their use of rudimentary watercraft may have moved them into a situation where it was only a matter of time before some people were lost at sea— only to make landfall at Lombok. Alternatively, people may have arrived at Lombok on a deliberate voyage of discovery, seeking new areas of littoral to exploit. Whatever the case, it would have become immediately apparent to that first voyager and his companions (if indeed he had any) that they were in a very different place from that which they had just left.

For a start, the small islands to the east of Bali lack large, warm-blooded predators. Therefore, despite the probable presence of relatives of the Komodo dragon, it was a much safer place than Bali with its tigers and leopards. This alone released a major constraint upon human activity. But in addition to this, there would have been a major difference in the availability of human food between Lombok and Bali, for the mudflats and reefs of Lombok would have been strewn with molluscs and crabs that had never been harvested, while the seas themselves were filled with fish that had never been fished. Although there were relatively few kinds of land animals, pygmy elephants of the extinct genus *Stegodon*, birds, bats and large rats would have abounded. None had ever seen a mammalian predator before. Their first reaction to the presence of a carnivorous ape, whose skills had been perfected in a million-year-long arms race with Asia's biggest and best, was probably simple curiosity. We suspect from their fossil remains that the larger species such as the pygmy elephants were rarely given the chance to experience a second reaction.

The castaway, if indeed the first settler was one, would first have grown plump, then lonely. He would have known that his people lived within view across that small strait. Perhaps one day he summoned the courage to recross it. Then perhaps he, a little like a prehistoric Moses, led his people across the waters to the land of milk and honey. If people had arrived upon a deliberate voyage of discovery, then the process of colonisation would have happened that much more quickly.

The situation that these new settlers found themselves in must have had a profound psychological impact. This is because, for the first time ever, humans had been released from the rigid grip coevolution. They were therefore able to have a profound impact on an entire biota. It is probable that for thousands of years their ancestors had been practising rituals to give them success in the hunt. But in the new land, every hunt would have been successful. Without predators and surrounded by naive prey, people would have become, in a sense, gods. For they were now all-powerful beings in a land of plenty.

The myriad of new ecological opportunities and challenges that this situation opened up to them would have encouraged innovations in technology and thought. People would have been well-fed and have had abundant leisure time. Moreover, in this land without apparent ecological limits, their offspring would have gone forth and multiplied. The carrying capacity of an island such as Lombok in such circumstances may have been quite extraordinary, for humans could coopt virtually every resource available to them. In a place like Bali or Java, these same resources would have gone towards supporting thousands of other species, including such big energy users as tigers, leopards, elephants and rhinos.

Perhaps the best analogy to their situation is that of rabbits, following their introduction to Australia. There, rabbit densities reached hundreds of times that typical of populations in their homeland, because they were able to monopolise the resources previously used by such diverse species as diprotodons, wombats and rat-kangaroos.

Because the first human invaders of places such as Lombok could monopolise so much of the land's resources, it is possible that the human population expanded enormously. People may have existed at hundreds of times the population density that existed elsewhere on earth. Such a dense human population would have greatly facilitated human ingenuity, borne both from competition and the enhanced flow of ideas. With a large and dense population, much leisure time and a surfeit of resources, one can hardly imagine a situation more conducive to the development of the great leap forward.

But the resources of small islands such as Lombok could not last indefinitely and finally the crunch would have come. But perhaps even this acted to propel people towards taking the great leap. As the last of the pygmy elephants was eaten—which in turn had controlled the build up of plant matter—fire was probably already ravaging the vegetation of the mountain slopes. If the human impact on Pacific islands is any guide, erosion would have washed soil into the fringing lagoons and swamps of the lowlands, smothering rich marine resources but creating a fertile coastal plain. Forced away from their animal resources, people may have concentrated on plants. Many human food plants are pioneer species which by definition do best in newly created habitats and disturbed areas. The rich soils of the newly created coastal plain would have been quickly populated with wild bananas, taro, yams, sugar cane and many other such species. With a large human population dependent upon plant resources, it seems likely that any previously rudimentary knowledge of agriculture and plant selection would have developed rapidly.

Along with these inducements to experiment with agriculture, the resource crash would have sharpened competition between humans for the few remaining resources. They may have developed social structures which enhanced their ability to defend and take resources from others. They may also have been prompted into developing ingenious technologies or behaviours in order to gain access to as yet untapped resources.

The golden age of Lombok could not have lasted long, nor could the subsequent period of resource decline, for Lombok is a small island and equilibrium would soon have been re-established. Nonetheless, the general conditions conducive to development of the great leap forward surely endured elsewhere. For those who had experienced the golden age resulting from colonising a new land, or who had an oral tradition of it, the desire to undertake the risky crossing from Lombok to Sumba must have been irresistible. Human abilities in developing seaworthy craft may have improved rapidly at this time.

If the great leap forward did take place in island South-East Asia, then the ideas and technologies that it gave rise to, must have been carried back into the great Afro–Eurasian homeland of humanity. It is possible that the increased understanding and utilisation of certain food plants, or enhanced military skills, or even improved technologies, allowed them to reinvade the Afro–Eurasian homeland and find an ecological niche for themselves.

It is also possible that ideas and technologies were passed from one area to the other, but that there was no human migration. If so, the situation may have been rather like that experienced between North America and Europe in the nineteenth and twentieth centuries. Although America was settled by people from Europe, the subsequent flow of technologies and ideas, from mass production to fast food, has been almost entirely the other way—from America to Europe. As a result, the economy and lifestyles of the Europeans have been transformed. I suspect that it is no accident that the traffic has been largely one-way, for the challenges and opportunities of the American frontier, with its vast resources, have encouraged the North Americans to reorganise their economy upon new and more efficient lines. Perhaps the resources and challenges of the first humans to leave their Afro–Eurasian cradle provided a similar stimulus.

This story is necessarily a speculative one, for the changes that encouraged the great leap forward have not been fossilised; or if they have, they have not yet been recognised. Yet evidence concerning the consequences of the great leap surrounds us everywhere.

Despite the paucity of direct evidence, our current knowledge of the fossil record is generally congruent with the scenario developed here, except that agriculture is thought to have developed at a later date. But then, how much do we know about the very earliest period of agricultural experimentation in the lowland tropics? It is quite possible that low-intensity agriculture was practiced there for millennia, yet has not left a clear fossil record.

A final curious fact seems to support the idea that it was colonisation of the 'new' lands that permitted the great leap forward. As might be expected from the history of large mammal extinction around the world, almost all of our animal domesticates arose from species inhabiting the great Afro–Eurasian homeland. This is because it was the only region in the world to retain a diverse megafauna, including cows, horses and elephants, until historic times. Many of our plant domesticates, on the other hand, are from the invaded lands of the Americas, and island Australasia. This is just the pattern expected if faunal extinction and its consequent environmental degradation led to agricultural intensification.

Finally, it is worth considering humanity's second great leap forward, which occurred in the nineteenth and twentieth centuries. This leap resulted in the transformation of small, largely agrarian human populations into the world economic and technological complex of the present. What was the cause of this transformation? I would argue that it was largely fuelled by agricultural exploitation of the previously inaccessible but immensely rich soils of the Americas, as well as the riches and opportunities afforded by other European colonies. Others may argue that it was technology driven. But this, I think, is equally as short-sighted as suggesting that the first 'great leap forward' was powered solely by cultural change. Surely technological change must follow some shift in ecology that either necessitates it or makes it possible.

Whatever the case, it is clear that an understanding of what powered the first great leap forward is critical to humans living today. For we are in the middle—or perhaps at the end—of a third leap. If I am right, the first great leap forward was only achieved at the cost of the destruction of whole biotas, while the second leap resulted in more destruction. We cannot afford to leap at such expense again.

CHAPTER 15

PEOPLING THE LOST ISLANDS OF TASMANTIS

While evidence concerning the colonisation of Meganesia is currently shrouded by the mists of time, evidence concerning the peopling of Tasmantis is much better. This is because humans colonised Tasmantis so relatively recently and because the first settlers left such unmistakable evidence of their presence. New Caledonia was the first part of Tasmantis to be reached by people, who arrived there some 3500 years ago. Radiocarbon dating works particularly well for the time interval of 25 000–2000 years ago, and so there are no significant doubts regarding the timing of this colonisation event. New Zealand, in contrast, was one of the very last of the world's larger landmasses to be settled. The dating of early settlement there has run into problems that occur when archaeologists are working with a time scale of hundreds of years rather than thousands or tens of thousands of years.

Until recently archaeologists thought that New Zealand had been settled around 1200 years ago. The chronology was based upon hundreds of radiocarbon dates from many sites. Although only a few samples gave dates older than 1000 years, researchers were generally happy with the scenario. That was until an archaeologist, Professor Atholl Anderson, who can proudly claim Maori ancestry and is now head of Australia's prestigious Prehistory Department in the Research School of Pacific and Asian Studies at the Australian National University, began looking at the dates a little more critically.[2] He discovered a quite different story.

The early Maori built their houses from timber. In the colder parts of New Zealand trees grow very slowly and logs such as those from some species of southern pines can lay around undecayed for centuries. Anderson showed, quite conclusively, that many of the older dates associated with Maori settlements had come from 'old' wood. That is, wood from trees which had died long before they were used by the Maori.

This could have happened in a number of ways. Maoris could have used timber from old trees (the heartwood of which had been dead for centuries), or they may have burned old driftwood and debris in their fires, leaving old charcoal (a favourite material for C^{14} dating). As a result of his studies, Anderson has now produced convincing evidence that the ancestors of the Maori arrived in New Zealand between 1000 and 800 years ago. The more recent date is only 450 years before the first European, Abel Tasman, sighted Aotearoa in 1642.

A reduction of Maori history by 200–400 years may seem trivial, but because their history is so short, 200 years represents more than 20 per cent of total Maori history in New Zealand. Anderson's work is a critical advance in our understanding of New Zealand history. Indeed, many of its aspects, from the extinction of the moa to the transformation of the New Zealand landscape by fire, would look quite different were Maori history to stretch back those additional 200 years. It is also clearly a political issue and so this major scientific breakthrough has not been universally lauded.

The colonisation of both New Caledonia and New Zealand occurred as part of a much broader saga of colonisation which resulted from one of the great expansionary phases of human history. This began some 3500 years ago. Yet this massive human expansion was preceded by an earlier one. The first wave of colonisation, begun by the ancestors of the Aborigines and New Guineans, had spread to Australia, New Guinea and Tasmania, but it extended into the Pacific only as far as the Solomon Islands. Extensive as it was, it left an ocean full of islands whose immense biological wealth was yet to be plundered by humanity.

Interestingly, almost all of the islands in the long chain between South-East Asia and the Solomon Islands are within sight of each other, or at least the next can be seen before one loses sight of the first if one paddles a canoe far out to sea. During the ice age, when sea-level was up to 160 metres lower than at present, the intervisibility of islands would have been much improved and it is possible that, with the exception of the gap between New Ireland and the

Solomons, and New Ireland and Manus, every island was within sight of another. This intervisibility may have been crucial to the people who first settled these lands, for it is likely that their water-craft and navigational skills were not highly sophisticated. Thus, it may well have been exceedingly dangerous for them to have lost sight of land. The sea gap between the Solomon Islands and Vanuatu has always been broad enough to prevent the islands on one side being seen by a near-shore paddler from the other. It thus remained a barrier to human migration for at least 25 000 years and was not breached until a new people, with a new technology and enhanced maritime confidence, arrived in the Pacific.

The evidence for the arrival of these new people is both clear and abundant throughout the Pacific, for they were the first makers of pottery in the region and broken shards of pottery form some of the most indestructible evidence of human occupation that there is. The Lapita people, as they are now known, made particularly intricately decorated pots, the shards of which were first found on Watom Island, just off New Britain, in 1909. Just why they subsequently became known as Lapita ware, after a place on the west coast of New Caledonia, no-one seems to know. The recognition that Lapita pottery is evidence of a distinctive and ancient culture has gradually been growing. Now, shards of Lapita ware have been found throughout the islands of Melanesia and Oceania.

Curiously, the quality of the pots made by the descendants of the Lapita people declined over the centuries. They first became less ornate and well crafted, until finally the art of pottery making was lost altogether on most islands. The classic Lapita period, when people were making the distinctive Lapita ware pots, was brief, lasting only from 3300–2000 years ago.

There is still much vigorous scientific debate concerning the origins of the people who drove this second great wave of invasion into Australasia. No less a personage than Captain James Cook himself began the speculation when he wrote in 1799:

The inhabitants of the Sandwich Islands are undoubtedly of the same race with those of New Zealand, the Society and Friendly Islands, Easter Island and the Marquesas; a race that possesses,

*without any intermixture, all the known islands between the
latitudes of 47° South and 20° North, and between the longitudes
of 184° and 260° East... From what continent they originally
emigrated, and by what steps they have spread through so vast a
space, those who are curious in disquisitions of this nature, may
perhaps not find it very difficult to conjecture... they bear strong
marks of affinity to some of the Indian tribes, that inhabit the
Ladrones [Mariana] and Caroline Islands; and the same affinity
may again be traced amongst the Battas [Bataks of northern
Sumatra] and the Malays.*[27]

While most researchers agree with Cook that the Lapita people
must have come from South-East Asia, this long-held view has not
gone entirely unchallenged. Several researchers, including the eminent Professor Peter White of Sydney University, argue that the
Lapita people evolved in Melanesia from ancestors of Australoid
stock.

Various lines of evidence support differing conclusions, and it is
worth examining some of the evidence here. White's view is not supported, for example, by the study of the skeletons of Lapita people,
for they reveal that the Lapita people closely resembled the living
Polynesians who are descended from them. These people in turn
appear to be Asiatic in origin. But White would argue that Melanesians are very variable and that some naturally resemble
Polynesians. If the first canoe loads of Lapita people had, by chance,
contained a good proportion of such people, their genes may have
greatly influenced their descendants, resulting in a shift towards
lighter skin and wavy hair.

White also has answers to other features that seem to point to an
Asian origin for the Lapita people. He suggests that their pottery was
a home-grown invention and that shards found in South-East Asia
derive from pots which were traded there from Melanesia. His case
was strengthened recently when blades made from obsidian (volcanic
glass) which had been mined on New Britain in the Bismarck
Archipelago were found in Borneo. They had been carried there some
3000 years ago.

Despite these arguments the balance of evidence is gradually shifting in favour of an Asian origin for the Lapita people. A recent genetic study of many biochemical systems by a great number of workers concluded:

> [the view that] Polynesians evolved within Melanesia from a population resident there for some 30 000 years, is untenable in the light of the genetic evidence. It seems quite implausible that a group evolving within Melanesia could have acquired, by chance, so many non-Melanesian genes![121]

The food animals and camp followers of the Lapita people provide more clues as to their origins and lifestyles. Their main livestock consisted of pigs, dogs and chickens, all of which are Asian in origin, while their pests included the Pacific rat (*Rattus exulans*), which is a native of South-East Asia. The pest species are particularly interesting, for the study of ancient rat bones in archaeological sites is beginning to reveal much about the movements of ancient people throughout the Pacific.

I became interested in this research when Peter White approached me to help identify some bones from archaeological sites on New Ireland. I soon found that the rat bones, which were abundant, fell into four categories. The very largest bones, which have not been found anywhere else but in archaeological sites on New Ireland, must have belonged to an enormous member of the 'spiny rat' group which is native to New Guinea. They were unlike the bones of any known species, so we named them as a new kind of rat, *Rattus sanila*, in honour of Sanila Televat, a wonderful old man who owned the caves where Peter dug.[49]

At around half a kilogram in weight, Sanila's rat must have been an impressive creature, but it was marked for extinction. Very abruptly, at around 3000 years ago, its bones cease to be deposited in the cave sediments and instead the bones of two smaller *Rattus* species are found. These are the large spiny rat (*Rattus praetor*) and the Pacific rat (*Rattus exulans*). The Pacific rat is a small and dainty species, half way in size between a mouse and black rat. It has a long association with people, and while it is a garden pest, it was important to some Polynesians as a food source. The presence of the Pacific rat is clear evidence that Lapita people had arrived in New Ireland, for it is very much out of place there. Originally it was an inhabitant of disturbed areas and villages on mainland South-East Asia, but it spread with the Lapita people and then the Polynesians, throughout the Pacific. Today it occurs on virtually every island from lonely eastern Kure Atoll in the extreme east to small islands off Australia in the west.

The presence of the large spiny rat is a bit more of a mystery. Originally restricted to the northern lowlands of New Guinea, it is a large (300 grams), foul-smelling and spiny garden and house pest without any redeeming features at all. In the native languages of New Guinea its name usually translates as something like 'stinking mouth' and although it is sometimes eaten by protein-starved women and children, it is invariably the first bit of bush tucker to be given up when store-bought food becomes available, probably because of the disgusting taste of its flesh and the large number of worms that emerge from the mouth and anus of freshly dead specimens.

I am certain that no self-respecting Lapita person would have voluntarily allowed one to board their canoe, yet it is a very large rat to remain undetected aboard a canoe on the lengthy voyage from New Guinea to New Ireland. I think that its presence tells us two things. The first is that Lapita vessels were very large, with plenty of hiding space for stowaways. The second is that Lapita people must have spent some time on New Guinea before travelling to New Ireland in order to have picked up their unwelcome guest. Upon arrival in New Ireland it probably played a role in the extinction of the larger *Rattus sanila*.

Rats are currently providing further insights into the history of Lapita travel and colonisation in the Pacific. For example, Peter and I have just discovered that the Pacific rat is very much a latecomer to Micronesia. It appears to have arrived with the Spanish in the sixteenth century, probably aboard a Manila galleon. The early Micronesians were not entirely rat-free, however, for their houses and gardens were infested with the larger and less pleasant *Rattus tanezumi*, the Asian version of the plague or black rat. The prehistoric absence of the Pacific rat in Micronesia suggests that there was little contact between Polynesians and Micronesians until European times. This newly discovered rat-borne evidence contradicts traditional wisdom on this point, for it was previously believed that eastern Micronesia was settled from Polynesia.

An even more curious story of prehistoric human journeying has emerged from the study of the distribution and ecology of an obscure sucking louse. Dr Gordon Corbett, a parasitologist with CSIRO, was amazed when he found that the louse *Heterodoxus spiniger* is widespread on dogs in Asia. The reason for his surprise is that this particular sucking louse and its relatives evolved in Australia, where they parasitise members of the kangaroo family. The fact that this one species has adapted to life on dogs and become so widespread in Asia suggests that it has been there for some considerable time. Because it is entirely dependent upon its host, the only way that the louse could have migrated from Australia to Asia was aboard a dog or a kangaroo. But how could such an animal have travelled from Australia to Asia in prehistoric times?

Scientists have long known that dogs travelled from Asia to Australia prehistorically. This is because the ancestors of the dingo (*Canis familiaris*) arrived in Australia some 4000–3500 years ago, probably aboard a Lapita canoe which landed in northern Australia during one of the Lapita people's great voyages of discovery. The only way that dogs could have been re-exported to Asia is if there was more substantial contact between prehistoric Australians and Asians than anyone has previously imagined. It may have been that Lapita voyagers often stopped in Australia to trade and acquire dogs owned by Aborigines. Alternatively, similar trade may have been undertaken more recently by Buginese or other Asiatic peoples.

Dogs were probably carried by Lapita people and others as a source of meat. They are still considered a delicacy in places such as Ambon in the Moluccas, where black ones are particularly esteemed for their culinary properties. Perhaps lousy young dingos were acquired as a tasty snack by Lapita voyagers for their return journey. Those carried aboard canoes powered by favourable winds evidently reached an Asian landfall before dinner time and survived long enough to pass on lice to their Asian brethren.

Studies of the languages of the descendants of the Lapita people and of other people who adopted and developed a variant of their tongue, are also revealing some fascinating facts about Lapita lifestyles. Linguists suggest that the Lapita people spoke a language resembling Bahasa Indonesia, Motu and Maori. All of these languages form part of a language family known as Austronesian, which has been traced back to an origin in southern China or Taiwan some 6000 years ago. This vast language group is one of the most widespread in the world, for it is spoken today by people spread over half of the Earth, from Madagascar, to parts of Vietnam, Malaya, Indonesia, the Philippines, non-Chinese Taiwan and the whole of the Pacific except for places like Australia and New Guinea, where the more ancient languages of the Banda are retained.[14]

Even today, a remarkable number of words are shared between the dispersed Austronesian languages of the Pacific. *Lima*, or some version of it, has the near-universal meaning of five in Austronesian languages, as has *wanita* for woman, *mat mat* for cemetery or death, and *tumbuna* for ancestors. Thus, if you speak Bahasa Indonesia you can understand a little Maori, Madagascan and Motu.

Dr Peter Bellwood of the Australian National University has devoted his research career to understanding the colonisation of Asia and the Pacific by various people. He has reported some fascinating research in reconstruction of the original Austronesian language. By tracing the origins of words, various researchers have been able to determine which words (and thus which materials) were in the original Lapita vocabulary and which were subsequently borrowed. By determining which words formed part of the basal language, one can gain an idea of which crops and activities were important to a vanished people. As Bellwood puts it, his research:

> indicates [that the first Austronesian speaking people were] a
> Neolithic, pottery-using society with rice and millet agriculture,
> domesticated pigs and dogs, and a well-developed technology for
> activities such as sailing-canoes and house construction.[14]

The Austronesian-speaking Lapita people were clearly great traders, for regardless of where they originated, their arrival in the Pacific opened communication between the peoples of Melanesia and South-East Asia on a scale never experienced before or since until modern times. Indeed, perhaps like the contemporary Buginese of

Sulawesi, trade was their *raison d'être* resulting in many of their great journeys. Whatever the case, the distances over which goods were traded is truly remarkable.

The Lapita must also have included among their number fine artisans, expert potters and canoe makers. They were a social and hygienic people who raised their houses on poles and placed them over shallow water, either in lagoons or over reefs. Expert fishermen, they caught even the true blue-water fish such as the tuna and their kin. No-one before them had been able to utilise this resource. Their canoes were substantial, being built up at the sides with planks. They carried great sails and were stabilised with outriggers. These canoes were capable of carrying considerable loads: perhaps tens of people, along with their mats, pots, pigs, taro and coconuts and had room enough left over to hide stowaways the size of the large spiny rat, which hitched rides as far as remote Tikopia Island, near Vanuatu.

All of these things indicate a central fact about the Lapita people: they were the world's first true blue-water mariners. If the Banda were the first people of the seashore and shallows, then the Lapita called the vast Pacific itself home. In this, they represent a major breakthrough in human adaptation, for they developed a mobility previously undreamed of, gained access to resources previously unknown and developed a way of life in which trade was an integral part. In all of these things the Lapita people closely resemble ourselves—the humans of the late twentieth century.

The Lapita appear to have been a remarkably peaceful people, for there is no evidence of fortified Lapita villages in Melanesia and no evidence of massacres in either Australia or New Guinea. Instead, there is much evidence of trade, intermarriage and cohabitation between these diverse people. Perhaps it was their unique maritime orientation which allowed the Lapita people to avoid conflict with their neighbours, the land-based Melanesians and Aborigines.

For Melanesians and Aborigines, one thing comes before everything else—that is their land. But for the Lapita, the sound and smell of the sea may have meant contentment. Their preference for settling on tiny islands, their fearlessness in repeatedly crossing the trackless waste of the Pacific, and their blue-water fishing skills, shows that they were truly at home on the open ocean. Happy with the *pied-à-terre* provided by a reef flat, they did not seek to dispossess the landwards-looking indigenes of Meganesia. Great traders, they probably provided for their needs from the soil at least partly through peaceful trade.

Although at one level the ocean-oriented existence of the Lapita was a great advantage, in another way it limited them, for wherever the Lapita people went they shunned the larger landmasses. Indeed, they probably lacked the skills necessary to make a living in difficult situations away from the sea. We know that they touched land on New Guinea and other large Melanesian islands. But evidence of their settlements are never found far inland. Indeed, their settlements are most common on tiny islets off the coast. Today, languages descended from theirs are mostly spoken in coastal areas in New Guinea. They tend to penetrate inland only in a few drier areas, such as the south-east of Papua New Guinea and the Markham Valley. This suggests that they paralleled the Europeans in their colonisation of New Guinea, for both people found the seasonally dry tropical savanna preferable to the jungle. In the case of the Europeans, the Port Moresby and Jayapura areas formed the beachheads. Just why these areas were favoured is not entirely clear to me, but it may have been that the incidence of malaria was initially lower in these seasonally dry regions.

We know, from the study of sucking lice, that the Lapita people probably visited northern Australia repeatedly. This is hardly surprising for a people whose descendants sailed as far as South America. What is more surprising is that they failed entirely to establish a colony anywhere along Australia's vast coast. This may be because their agricultural technology was singularly inappropriate for Australian conditions. Their main crops were root crops such as taro, which need enormously rich soils and abundant water in order to thrive. Australia's thin soils and dry, erratic climate mean that such staples cannot be grown consistently and economically.

I personally think it quite likely that the Lapita people and their descendants, the Polynesians, saw quite a lot of Australia. That they knew the north coast is unquestioned, but they may well have discovered the east coast as well. The Maori have long known of a land which they called *Ulimaroa*. It was believed to be a vast land lying one month's trip by canoe to the north-east. *Ulimaroa* sounds remarkably like Australia. Archaeological evidence indicates that Polynesians discovered tiny Norfolk Island, which lies some 1600 kilometres north-east of Sydney and they settled virtually every island east of Australia. Indeed, Austronesian speakers even settled Madagascar, off Africa, in prehistoric times. Given these great voyages of discovery, it seems highly likely that eastern Australia was visited, probably more than once. Some tentative evidence suggesting this concerns the Aborigines of the New South Wales coast. They, unique among their neighbours, began to make fish hooks around 800 years ago. Did they adopt the idea from visiting Polynesians? No-one knows, but it remains an intriguing possibility.

By 3500 years ago the Lapita people had swept past New Guinea and out into the central Pacific, reaching Vanuatu and New Caledonia. As is evidenced by the people surviving there today, they carried with them large amounts of Melanesian genes. A fascinating study of mitochondrial DNA among the peoples of Fiji has suggested how these Melanesian genes may have been transported into the Pacific.[121] Mitochondrial DNA is passed on only through the mother, the father making no contribution. This means that female ancestry can be investigated in detail. The Fijian study reveals that Lapita women comprised approximately 80 per cent of all females among the early Fijians, even though Lapita people as a whole comprised only 20 per cent of the population. Thus, it seems likely that women were underrepresented on the voyages of discovery that led to the colonisation of Fiji. One possible explanation for this is that Melanesian men instigated those voyages, perhaps aboard canoes that they had learned to build from Lapita men. The fact that they carried many more Lapita than Melanesian women is also intriguing. While it is difficult to be sure exactly what the data means, it is tempting to imagine Melanesian men acquiring both Lapita canoes and women as a result of some conflict, then setting sail into the unknown to escape retaliation.

By 3000 years ago Fiji, Tonga and Samoa had been settled and it was from these last two island groups—and at this time—that the ancestors of the Polynesians would spring. By 2100 years ago they had spread throughout most of Polynesia, including the Cook

Islands, the Marquesas and Tahiti. By 1600 years ago they had travelled north to Hawaii, and 1100 years ago they had even reached remote Easter Island. They were then ready for their greatest sailing ventures of all time. For they would cross the entire Pacific, reaching the coast of what is now Chile and Peru, but typically, they established no settlement there. Although they left no trace on the South American mainland, we know that they visited because of the traded goods that they brought back with them from the new world, chief amongst which was the sweet potato (*Ipomoea batatas*). Because it produces so bountifully, and because its tubers can be stored or used to fatten pigs, the sweet potato is among the most highly regarded of Pacific Island foods. It will also thrive in much less fertile and drier soil than the indigenous taro and it is much quicker to yield.

By 1000 years ago the sweet potato had travelled as far west as Mangaia in the Cook Islands. By 400 years ago it had reached the heart of New Guinea. There it was to cause a revolution, of which more will be said later. Meanwhile, back in the Pacific, around 1000–800 years ago the Polynesians had set out upon a great journey to the south, in search of the last great uninhabited landmasses of the Pacific. Although they had never seen it, the Polynesians almost certainly knew that New Zealand existed and in which direction it lay. For they would have seen vast flocks of birds which, before human settlement, radiated out from their headquarters in New Zealand by their tens of millions. Some were migratory species which, each year, disappeared to the south, returning in the autumn. Others made up vast halos of seabirds that sallied forth from their breeding grounds to gather food from the far reaches of the Pacific before returning, crops full, hundreds or even thousands of kilometres to land.

The final trigger that pushed the ancestors of the Maori to venture south is not known. Perhaps it was a famine, or a war, or perhaps a clan just decided that life would be better elsewhere. Whatever the case, it was probably only a matter of time before someone followed the birds on their long journey southwards. It was to be an epic voyage, for never before had Polynesians ventured so deep into temperate latitudes and rarely had they gone so far from home.

The golden age of Lapita, which lasted from 3300–2000 years ago, seems to have been a time of great resource abundance. For on all of the uninhabited islands that the Lapita people touched, the food supply must at first have seemed limitless. Every mudflat had shells of enormous size that had never before been harvested, every reef teemed with huge fish that had never seen a hook. Almost every island had great assemblages of flightless rails, huge pigeons and other species and many beaches were turtle nesting sites which, in season, must have swarmed with the huge reptiles.

Such great bounty probably lasted several hundred years on each individual island. Over this time people kept trading, sailing and making their wonderful pots. The archaeological record shows that eventually, however, all of the large birds were eaten and the shells and fish began to shrink in size due to overharvesting. The turtles vanished, their nesting sites banished to ever more remote coral islets and atolls. As the resource base shrank, so did the material culture and perhaps the numbers of people themselves. Traders stuck closer to home and people gave up making pottery. In general, the physical and cultural world of the Lapita shrank as their exhausted resource base collapsed.

We know a little about the world that the Lapita voyagers encountered in the virginal Pacific and about how easily it was destroyed, all because they missed one tiny island during their wanderings, leaving it just as it was before humans arrived. At about 14.6 square kilometres, Lord Howe Island is minuscule. But it is a gem of a place, quite rightly nominated for world heritage listing. Lying some 500 kilometres off the coast of northern New South Wales, it is well west of New Zealand and New Caledonia, its next-nearest neighbours.

In 1788 the French navigator La Pérouse became aware of Lord Howe's existence using the same method that the Lapita people probably used to locate new lands.[141] He records than on 17 January, his ship was surrounded by innumerable seabirds. Even though no land was in sight, he knew that an island must lie nearby. Exactly a month later Lord Howe's isolation came to an end when it was sighted by the crew of HMS *Supply*, a vessel of the First Fleet which was sent from Port Jackson to found a settlement on Norfolk Island. David Blackburn, a crew member, records the historic first contact in a remarkably prosaic and gastronomically oriented account:

*[it] abounds with turtle much superior to any I have seen, on the
shore we caught several sorts of birds, Particularly a land fowl of a
dusky brown about the size of a small pullet, a bill 4 inches long
& feet like a chicken. Remarkably fat & good, plenty of pigeons, a
white fowl—something like the Guinea Hen, with a very strong,
thick & sharp pointed bill of a red colour—stout legs & claws...
Some of them have blue feathers on the wing. Here is also a web
footed fowl in general of a deep blue. Its bill 2 inches long—
straight but suddenly bent downwards at the end... We took them
in burrows in holes like rabbits—the bay abounds with a variety
of excellent fish.*[141]

All of the birds were so tame that they could be approached on
the beach and simply knocked down. Many had lost the power of
flight. Three months after the *Supply* landed Arthur Bowes Smyth,
aboard another First Fleet vessel, gave a more emotive and perhaps
truer account of the island. He wrote:

*When I was in the woods amongst the birds I cd. not help pictur-
ing to myself the Golden Age as described by Ovid to see the Fowls
or Coots some white, some blue & white, others all blue wt. large
red bills & a patch of red on the top of their heads, & the Boobies
in thousands, together wt. a curious brown bird abt. the size of a
Landrail in England walking totally fearless & unconcern'd in all
part around us, so we had nothing more to do than to stand still a
minute or two & knock down as many as we pleas'd wt. a short
stick—if you throwed at them and missed them, or even hit them
with out killing them, they never made the least attempt to fly
away & indeed they wd. only run a few yards from you & be as
quiet & unconcerned as if nothing had happened—.
The Pidgeons were also as tame as those already described & wd.
sit upon the branches of trees till you might go & take them off
with your hand or if the branch was so high on wh. they sat, they
wd. at all times sit till you might knock them down with a short
stick—many hundreds of all the sorts mention'd above, together
wt. many Parotts, Parroquettes, Magpies & other birds were
caught & carried on board our ship & the Charlotte.*[68]

Tragically, within a very short time this last naive land was utterly transformed. Thomas Gilbert, who visited the island in May 1788, records how easily it was often done:

Partridges [are] likewise in great plenty... Several of those I knocked down, and their legs being broken, I placed them near me as I sat under a tree. The pain they suffered caused them to make a doleful cry, which brought five or six dozen of the same kind to them, and by that means I was able to take nearly the whole of them.[141]

Such hunting, initially by the crew of passing whaling vessels, and after 1850 increasingly by settlers, had caused the remarkable white fowl to become extinct by 1844, followed by the Lord Howe pigeon (1870) and the Lord Howe parakeet (1870). The Lord Howe Island woodhen was also dramatically reduced, and the vast cloud of seabirds noted by La Pérouse was no more. Although the island had thereby lost its largest and most spectacular species, a unique fauna of 12 endemic birds survived this early hunting.[141]

Then, in 1918, the supply ship SS *Makambo* ran aground, and black rats (*Rattus rattus*) got ashore. Within a few years the Lord Howe boobook owl, the vinous-tinted thrush, Lord Howe warbler, Lord Howe fantail, robust silvereye, and Lord Howe starling were all extinct. Today, only six of the original 15 endemic land birds of Lord Howe Island survive.

The birds were not the only victims. The turtles were decimated, as were the fish of the lagoon. MacGillivray, who visited the island in 1853, recorded that a fishing party had met with indifferent success, taking only three or four small fish over a day's effort.[141] Clearly, the small area of the lagoon had been badly overexploited even by this time.

Were Lord Howe still as it was in early 1788, it would be one of the wonders of the world. Despite the terrible destruction, the island is still an extraordinary place—the last South Pacific virgin to be deflowered—and the only one whose fate was recorded. She tells us how the islands of the Pacific came to be as they are today, and what extraordinary places they once were.

Were it not for Lord Howe we may never have known in our hearts what a remarkable world the first Australasians stumbled upon. Today, a vast research effort is aimed at understanding the nature of that world and the impact that humans had on it. This research is in some ways the most important being undertaken in our region, for it will provide guidelines for the management of the 'new' lands in future.

CHAPTER 16

THE GREAT MEGAFAUNA EXTINCTION DEBATE

In a sense, the lands inherited by the living Australasians have been made by their ancestors. We do not, as yet, understand all of the processes involved. Indeed, our understanding of even the basic forces is probably far from complete. But we know enough to discern three major factors—extinction, dwarfing and fire—which were important in shaping the Australasia of today. I would argue that all three result from human actions. But this is difficult to prove, as they occurred against the dramatic backdrop of climate change over the past 50 000 years.

I would also argue that extinction, dwarfing and fire are intimately interrelated phenomena and that they are partially interdependent. Just how they are related and what the past dynamics between them have been, are some of the most hotly debated topics in Australia's prehistory today. My guess is that the primary—indeed catalytic—of the three phenomena is extinction. And so it is here that we will begin our explorations of how humans created the ecosystems that the living Australasians occupy.

As knowledge of the history of colonisation of the 'new' lands has grown, researchers have argued over the impact that the first people had on their new-found homes. Debate has thrived because planet Earth has undergone enormous changes in geologically recent times—perhaps the most extreme and rapid changes to have occurred since the extinction of the dinosaurs nearly 65 million years ago. Among the most important and obvious of these changes concerns the extinction of most of the world's larger animals. The great nineteenth century explorer and co-founder of the theory of evolution, Alfred Russell Wallace, summarised the phenomenon splendidly as long ago as 1876:

We live in a zoologically impoverished world, from which all the hugest, and fiercest, and strangest forms have recently disappeared; and it is, no doubt, a much better world for us now that they have gone. Yet it is surely a marvellous fact, and one that has hardly been sufficiently dwelt upon, this sudden dying out of so many large Mammalia, not in one place only but over half the land surface of the globe.[135]

As early as last century, the views of those involved in the 'great megafauna extinction debate', as the scientific jousting concerning the causes of this great 'dying out' has been called, crystallised into two opposing camps: those who blamed climate change for the loss and those who blamed human hunters. Until recently, those who believed that humans had little impact and that the extinctions were the result of climate change, have had the upper hand. Their argument is that the giant animals of Australasia and the Americas in particular, became extinct as a result of the dramatic change in climate associated with the last ice age.

Opposing them is a growing band of researchers, led by Professor Paul Martin of the University of Arizona, USA, who argue that humans alone have caused virtually all recent extinctions of large animals worldwide. He has developed a theory called the 'blitzkrieg hypothesis', the central tenet of which is that the extinctions occurred almost immediately (certainly within a few hundred years) of the first arrival of humans in any one place. Proponents of the theory suggest that the large animals and flightless birds were easy prey to early humans because they did not recognise humans as predators. An alternative (which I consider to be much less probable) is that, in Australia particularly, the fires that humans lit somehow caused the extinctions.

Recently, some researchers have shifted to a middle position, suggesting that while humans may have delivered the *coup de grâce* to the giants, climate had deteriorated in such a way as to make them especially vulnerable to human hunting or other activities. Opponents of this theory criticise it because they feel that it does not differ greatly from the human-caused extinction theory. It is, nonetheless, a testable hypothesis and so will be examined below.

The hypothesis that climate change caused the great extinctions depends upon the nature of the change brought about by the ice ages. The last ice age is the best understood of all ice ages and was, in broad outline, probably typical of the others. Before its peak between 25 000–15 000 years ago, the world was cooler than at present, but nothing like as cold as during the ice age. The onset of the ice age was rapid, perhaps occurring over a century or two; or even within a few decades. World temperatures fell by an average of eight degrees Celsius. As a result, the polar ice caps grew until they covered much of the world's landmasses which lay north of the latitude of Wales. At the southern edge of the northern ice cap, in places like southern England and the Netherlands, a vast and lifeless polar desert developed.

As a result of so much of the world's water being locked up as ice, the sea-level fell by as much as 160 metres. Large landmasses suddenly appeared in the world's oceans, as shallow seas such as those around the Norfolk Rise, became dry land. At the same time the continents and islands enlarged—sometimes unifying into a single landmass—as the continental shelves became exposed. As the ocean cooled, less water vapour was carried by the wind to form clouds and rain, thus the world's deserts generally expanded, although in a few places, such as the south-west of the USA, dry areas experienced more effective precipitation than previously.

The tropics were not as dramatically affected as the higher latitudes, but even on the equator glaciers formed on the higher mountains. Thus, all of the 'new' lands, with the exception of New Caledonia, were dramatically affected by the last ice age.

In New Zealand, the sea receded so that all of the islands joined to form a single landmass. Most of the land was either covered by glaciers or became a treeless tundra—the South Island in particular became virtually deforested.

In New Guinea the seas also receded from the continental shelf, joining it with Australia. Glaciers appeared on all of the higher peaks. Below them, subalpine grassland, which today covers some 5000 square kilometres, expanded to cover 50 000. The tree-line was lowered from its present 3900 metres to 2100 metres, and below that the various forest types that typify particular elevations in New Guinea today may have been mixed. But despite these changes, vast areas of rainforest remained.

Australia was particularly dramatically affected, for the continent became cold and extremely arid. The centre was turned into a vast dustbowl of swirling sand dunes where vegetation could not survive. The deserts were so expanded that the desert-dwelling hare-wallabies (*Lagorchestes*) occurred as far south as north-western Tasmania. The aridity was accompanied by intense cold. Red kangaroos grazed among plants typical of the alpine herbfield, near where Melbourne stands today. Because of the cold and dryness, trees vanished from much of the land to the west of the Great Dividing Range and many moisture-dependent species were doubtless restricted to small refuges in favourable locations along the east and south coasts. Even in places like Tasmania, rainforest disappeared almost entirely. The plant species that compose the rainforests today probably survived as scrubs in the river valleys.

Suddenly, about 15 000 years ago, in perhaps as little as three to five years, the grip of the ice age loosened. Ice-caps rapidly melted, ocean basins filled and plant communities began to migrate. Change was so rapid that by 9000 years ago the world resembled the one we know today.

The proponents of the climate-caused extinction hypothesis argue that under ice age conditions, or conditions when the ice age is either waxing or waning, the large animals, which they suggest were dependent upon ample supplies of water, could not survive and all perished. Alternative or supplementary theories suggest that climate changed too rapidly either at the onset or demise of the ice age, for the large animals to keep pace, or that disturbed seasonality affected their breeding seasons, or that somehow plant biomass was reduced, or that plants became less palatable. As dramatic as ice age climate changes were, the theory runs up against a number of problems when the fossil evidence is examined in detail.

Perhaps the most immediate problem with attributing the extinctions of large mammals to changes associated with the last ice age is that it is only the most recent of 17 ice ages that have gripped the Earth over the past two million years. Some, including the so-called Anglian Glaciation of 400 000 years ago, have been more severe, resulting in more extensive ice sheets at least in Europe, than the last ice age. Sixteen of the world's 17 ice ages passed without causing dramatic extinctions. Why, researchers ask, should the last one have had such a dramatic effect? Unfortunately, no-one who supports the climate-caused extinction theory can find firm evidence for anything that might have been special about the last ice age. Clearly, as dramatic events as the ice ages were, the world's fauna was used to living with and adapting to such change, probably through migration or survival in relict suitable areas.

The second and perhaps most critical problem concerns timing. Because the last ice age occurred roughly contemporaneously over the entire globe, peaking at 25 000–15 000 years ago, one would have expected that its effects on the large fauna would also have been felt more or less contemporaneously everywhere. But an examination of the fossil record shows that the extinction of large animals did not occur synchronously around the world. At the extremes in this respect stand the neighbouring lands of Australia and New Zealand. The earliest extinctions appear to have been those of the large marsupials, reptiles and birds of Australia and New Guinea. They probably occurred before 35 000 years ago, which is well before the height of the last ice age. In New Zealand, in contrast, the extinction of the moa occurred as late as the twelfth to sixteenth centuries AD.

This question of timing also has implications for the hybrid human–climate hypothesis, for its proponents must argue that the effect of climate on fauna was sufficiently strong to seriously disadvantage the fauna. This might make sense if the extinctions all occurred during one, critical period during the ice age. If, on the other hand, they are shown to have occurred at times when it was much colder than at present, as well as at times when it was almost as warm, the effect of climate becomes so negligible that proponents of the theory end up arguing that climate almost always disadvantages megafauna. Thus, the theory becomes virtually indistinguishable from that suggesting that humans alone caused the extinction.

Great advances in understanding the timing of megafaunal extinction have recently been made worldwide. A growing body of evidence (to be discussed below) suggests that the extinctions may have occurred by 35 000 years ago in Australia. In North and South

America there is unequivocal evidence that a diverse large mammal fauna (at least eight now-extinct species) survived up until 11 000 years ago (several thousand years after the ice age finished). Furthermore, there is good evidence that many other North and South American species survived to near 11 000 years ago. In northern Europe, there is good evidence that the extinctions occurred a few thousand years earlier, with most megafauna becoming extinct by 14 000–12 000 years ago. On various Mediterranean islands the extinctions occurred between 10 000 and 4000 years ago, while on islands in the Arctic Sea off northern Russia, pygmy mammoths may have vanished as little as 4000 years ago. On New Caledonia the extinctions occurred 3500 years ago, while on New Zealand they occurred a mere 800–500 years ago.

The great disparity of timing of megafaunal extinction worldwide presents almost insurmountable problems to those who support the climate-caused extinction theory, for they must explain how conditions at widely different times on different landmasses conspired to cause extinction. Their case becomes doubly difficult to argue when it is realised that the timing of extinction coincides closely with the initial arrival of people in many regions.

A further problem for the climate theorists is that some species were affected, that the climate theory might predict should not have been, while others, which one would predict might have been vulnerable to climate change, survived. As Wallace put it over a century ago, it was the 'largest, fiercest and strangest' species which became extinct. The largest species covered a wide range of ecological niches —from elephant to moa and giant tortoise. The fiercest were all the large carnivores, while the strangest were often the ecologically most specialised. These species share little in common ecologically. They are all similar, however, in being either close competitors of humans in their expanding ecological niche, or their prey.

Among the groups that, on the basis of their biology one would predict should not have been affected, but which were, are giant marsupials that inhabited the wet subalpine grasslands and montane rainforests of New Guinea. All seven species became extinct despite the fact that they should have been protected from climate change, as their habitat never vanished, but merely shifted a thousand metres or so up and down New Guinea's mountains in response to cooling and warming climates. The survival of many drought-sensitive organisms in New Guinean rainforests today testifies to the fact that their habitat has never been subject to a critical water shortage.

Among the species that one would expect to have been affected by extreme aridity, and yet which survived, are the hairy-nosed wombats and some larger kangaroos. The hairy-nosed wombats are larger than some species that did become extinct. They live in burrow complexes around the margins of the arid zone and cannot readily migrate in response to climate change. Furthermore, they depend upon good quality short grass for survival. They seem like prime candidates for extinction under the climate hypothesis, yet they survived. Likewise, the large red and grey kangaroos inhabit areas that were greatly affected by the ice age climate shift. They also are larger than many extinct species. Why did they survive?

Another interesting case concerns diprotodon. Entire skeletons of these largest of all marsupials have been found in Lake Callabonna, South Australia. There, they became mired in the sticky mud of the lake when they wandered out, probably in search of water. At the time that the diprotodons lived around Lake Callabonna, climatic conditions were probably not much different from what they are near Broken Hill today. Diprotodons were clearly arid-adapted. Even during the height of the ice age there would have been vast areas of Australia which supported habitat suitable for them. After all, several million square kilometres of continental shelf had been exposed by the drop in sea-level and this would have been, by and large, more moist than the continent's centre.

These still unanswered objections to the climate theory are the primary reasons that I cannot embrace it with enthusiasm. Other reasons concern the vast and growing body of evidence relating to the impact of humans upon their environment. I have personally seen and recorded humans exterminating mammal species through over-hunting in the mountains of New Guinea. In the next few chapters I will try to relate my experiences with extinction and will give additional evidence to support the view that the overkill hypothesis is the correct one.

CHAPTER 17

MAKING THE SAVAGE BEAST

Some circumstances are observed so regularly and unvaryingly in everyday life that we assume that they are part of the natural order of things and that they have always been so. One such circumstance concerns the behaviour of wild animals. No-one imagines that they might have behaved in any other way than the way in which they do today. This assumption is a great stumbling block to an understanding of the question of megafaunal extinction.

We observe, on a daily basis, the dichotomy in behaviour between 'wild' and 'tame' animals. Tame or domestic animals show no fear of us, while wild animals are often extraordinarily wary. Indeed, so skilful are many at avoiding us that we can live side by side with them yet never see them. How often, for example, do we see the two possum species that are ubiquitous in Australian suburbs, or the suburban foxes, bandicoots and gliding possums that manage to survive in our urban parks and backyards?

Even more remarkable are the experiences of professional biologists who try to locate elusive species. Dr Roger Martin, who is one of the most experienced field zoologists that I know, recently completed a study of Bennett's tree-kangaroo (*Dendrolagus bennettianus*), a large-dog-sized member of the kangaroo family that inhabits rainforests in the Cooktown area of north Queensland. He told me in despair that he had searched for six months and still had not had more than a fleeting glimpse of one. Yet after years of patient effort he found that he was working in the midst of an abundance of tree-kangaroos. In fact, half a dozen or more individuals had lived all the time within a few hundred metres of his camp.

My experiences in New Guinea are similar. I still find it hard to accept, for example, that after 13 years working in the most remote parts of New Guinea I am yet to see a wild cassowary. I have found a footprint so fresh that water was still oozing into it, droppings still steaming and everywhere older signs of cassowaries. Yet so wary are these great birds that to this very day they have remained as invisible to me as ghosts.

If this had always been the state of affairs, then we could quite rightly ask how it was ever possible for humans to exterminate any animal species, especially over an area as vast as Australia. For if animals had always been so wary, the effort involved in hunting them to the last must have been enormous—far too great indeed for a small human population ever to have accomplished.

As counter-intuitive as it seems, I am now convinced that wild animals have not always behaved as wild animals behave today. This is because our belief that animals must be tamed in order to show a lack of fear of humans is wrong, for there was a time when all wild animals behaved like tame ones. Long ago we humans made the savage beast—and ever since we have been attempting to unmake it, in order to provide ourselves with animal companions.

Except at the very grossest level, the behaviour of animals usually leaves no trace in the fossil record. How then, can we know how animals behaved in the past? The answer is that we know what the pre-human behaviour of various species is because there were, until recently, a few tiny parts of the Earth where humans had not penetrated. These regions were inhabited by a wide variety of animals. Thankfully, a few were observed by some of the greatest biologists of all time before human hunters either exterminated them or made the animals 'wild'. The most extraordinary and telling of these special areas is doubtless the Galapagos Islands. Charles Darwin visited the archipelago in 1835 and, as is widely known, his experiences there were seminal to his development of the theory of evolution. What is less well-known, however, are Darwin's extensive observations regarding the behaviour of the Galapagos fauna. His own words tell it more elegantly than I ever could:

> *September 17th 1835, Chatham Island. The day was glowing hot, and the scrambling over the rough surface and through the intricate thickets, was very fatiguing; but I was well repaid by the strange cyclopean scene. As I was walking*

> *along I met two large tortoises, each of which must have*
> *weighed at least 200 pounds: one was eating a piece of cac-*
> *tus, and as I approached, it stared at me and slowly walked*
> *away; the other gave a deep hiss and drew in its head. These*
> *huge reptiles, surrounded by the black lava, the leafless*
> *shrubs, and large cacti, seemed to my fancy like some ante-*
> *diluvian animals.*[33]

The resemblance of these great testudines to 'antediluvian' animals was, I suggest, more than passing, for the giant tortoises of the Galapagos Islands are among the last survivors of a great assemblage of giant herbivorous tortoises which once inhabited every continent except Antarctica. They are also the megaherbivores—the ecological counterparts of mammoths and diprotodons—of the Galapagos Islands. Their behaviour was, I think, typical of the megaherbivores of lands where people are newly arrived. Australia's diprotodons probably greeted the ancestors of the Aborigines with an equally bucolic stare and huff.

As delightful as Darwin's observations on tortoises are, his records on the Galapagos birds are even more revealing. He recounts that extreme tameness was a feature of the disposition of all the birds—the mocking-thrushes, finches, wrens, tyrant-flycatchers, dove and carrion-buzzard. He wrote that:

> *A gun is here almost superfluous; for with the muzzle I*
> *pushed a hawk off the branch of a tree. One day, whilst lying*
> *down, a mocking-thrush alighted on the edge of a pitcher,*
> *made of the shell of a tortoise, which I held in my hand, and*
> *began very quietly to sip the water; it allowed me to lift it*
> *from the ground while seated on the vessel.*[33]

Remarkably, this degree of tameness had persisted after over 150 years of sporadic human visits and finally colonisation.

Cowley (in 1684) remarked that the doves of the Galapagos were so tame that they would alight upon the arms and hats of visitors. Such extreme tameness had been lost by the time of Darwin's visit, although the birds were clearly still very naive, as the following anecdote (recounted by Darwin) shows:

> in Charles Island, which had then been colonised about 6
> years, I saw a boy sitting by a well with a switch in his hand,
> with which he killed the doves and finches as they came to
> drink. He had already procured a little heap of them for his
> dinner; and he said that he had constantly been in the habit
> of waiting by this well for the same purpose.[33]

Interestingly, such tameness is not universal among island-inhabiting animals, for as Darwin notes, Du Bois, who visited the tiny Indian Ocean island of Bourbon (now Reunion), found that although most of the birds were so tame as to allow themselves to be killed with a stick, the flamingos and geese were not. It is quite likely that these were migratory species, or that there were sufficient individuals arriving from areas inhabited by humans that the island populations had retained a wariness of human predators.

It may seem a little far-fetched to attempt to generalise from the habits of birds and turtles on tiny islands to the past habits of the Australasian fauna. Yet other examples of tame 'wild' animals suggest that this phenomenon is not restricted to places such as the Galapagos Islands with their limited faunas.

François Péron, French zoologist and one of the great unsung heroes of Australian biology, visited Australia in 1801–1803 and made important observations at a number of locations. He had the great good fortune to visit Kangaroo Island just prior to its settlement, and King Island (in Bass Strait) shortly after British sealers took up residence there. Both Kangaroo and King Islands are fascinating places, for they are some of the only places on Earth where humans have become extinct.

Both islands were settled by Aboriginal people during the last ice age when the sea-level was lower than present and when the islands became joined to the mainland. Between 11 000 and 8000 years ago the rising sea cut off both islands, along with their humans inhabitants, from the mainland. The subsequent history of the people of these islands will be explored more fully later, but here, the important fact is that humans had been absent from both islands for several thousand years by the time of Péron's visit. This amount of time appears to have been sufficient for the fauna to have lost its fear of humans.

On 10 December 1802, Péron stepped ashore on King Island, which lies in the western end of Bass Strait. It was a cold and inhospitable place, with Péron's stay being made more fraught when his ship was blown away from the island by a gale. This left him entirely dependent for survival upon six convicts (two of whom were Irishmen deported for their political opinions) and an Hawaiian woman, who were left there to gather the oil of elephant seals (*Miorunga leonina*). These wretches were living in squalid conditions in a few shanties made of bark. Remarkably, although they had virtually no supplies apart from strong liquor, they were all fit and well. The reason for their rude good health appears to have been the abundance and diversity of fresh meat available to them.

Péron records that the flesh of the now-extinct King Island emu (*Dromaius ater*), even when consumed without bread, biscuit or such like, was truly exquisite. They and the 'kangaroos' (as he called the red-necked wallabies, *Macropus rufogriseus*) were little trouble to catch. But the easiest meat of all to obtain was that of the now-extinct wombat (a diminutive form of the common wombat *Vombatus ursinus*), for Péron records that it had been domesticated by the sealers! He writes that:

> the wombat, reduced to a domestic state by the English sealers, searches, during the day, for the food that he needs, and at night returns to the hut that serves him as a shelter. A gentle and stupid animal, he is prized for the delicacy of his flesh, which seemed to us to be preferable to that of all the other animals in this region.[96]

To anyone familiar with wild wombats on mainland Australia (especially those living outside national parks) this is strange behaviour indeed, for wombats are extremely secretive and shy animals.

While the tameness of the land animals provided the sealers with plentiful food, the behaviour of the elephant seals made them a particularly easy resource to harvest. The elephant seals spent much of the year feeding in the subantarctic ocean, and only came ashore on uninhabited King Island to breed. Thus they had no experience whatever of people. The enormous males, weighing up to four tonnes, could be formidable opponents when they recognised a threat, for they fought ferociously amongst themselves. Tragically, they took so little notice of the sealers that they looked on indifferently while their fellows were slaughtered, even allowing themselves to be driven to a convenient killing spot. Within a few years the entire breeding colony of these breathtaking creatures was exterminated. They have never returned.

The behaviour of the animals on Kangaroo Island, which was also uninhabited when Europeans arrived, is just as strange. Matthew Flinders visited in March 1802, just a few weeks before Péron. He reported that upon sighting the island:

> *every glass in the ship was pointed there, to see what could be discovered. Several black lumps, like rocks, were pretended to have been seen in motion by some of the young gentlemen, which caused the force of their imaginations to be much admired...*[31]

The following day the laugh was on the other side, for the lumps were discovered to be kangaroos, which were incredibly tame and docile. The famished crew killed them easily with small shot and even sticks. Flinders mused that:

> *Never perhaps had the dominion possessed here by the kanguroo been invaded before this time. The seal shared with it upon the shore, but they seem to dwell amicably together... The seal indeed, seemed to be much the most discerning animal of the two; for its actions bespoke a knowledge of our not being kanguroos, whereas the kangaroo not infrequently considered us to be seals.*[31]

This difference in behaviour between seals and kangaroos is precisely what one would expect, as the seals (the Australian sea-lion *Neophoca cinerea*) forage in coastal waters and doubtless often swam to the mainland, where they would have experienced human hunters.

When François Péron stopped briefly on Kangaroo Island a few weeks later, he related that:

> with only one...dog...[called Spot, which the French had acquired in Sydney], we caught in a few days, such a large number of big kangaroos that it seemed probable to us that a few such dogs, abandoned on the island, would suffice to wipe out the race of these innocent animals.[136]

The behaviour of some animals in national parks shows that a degree of tameness can develop in a relatively short period of time if humans do not hunt animals. Although this, along with the animals of King and Kangaroo Islands, are cases of animals that have 'forgotten' to be wild, they show how vulnerable to humans naive animals can be. The sorry record of faunal extinction on King Island (the emu, wombat, elephant seal and spotted-tailed quoll are all long gone) also reveals just how rapidly such animals can be exterminated.

There are some who would argue that these examples of 'tame' wild animals relate only to island populations where there are no significant predators. But this is not entirely true, for the small birds of the Galapagos were doubtless preyed upon by the native goshawk; while the fox-sized spotted-tailed quoll (*Dasyurus maculatus*) would have hunted the smaller species of King Island. Perhaps what is more important is that humans are so different from other predators. They have evolved from the largely herbivorous apes and look, smell and behave entirely differently from the carnivores of the dog, cat and bear families. I think that it is quite likely that until experience teaches them otherwise, animals with no direct knowledge of humans fail to recognise them as predators at all.

Not all animal species become secretive in order to avoid humans. Some species have altered their behaviour in other ways in order to cope with human hunters. African *Cercopithecus* monkeys, for example, occasionally mob and attack humans in much the same way that smaller birds will mob an eagle. The behaviour of rats shows yet another alternative. The native murids of Australia vary in their disposition. A great many of the old endemics (which have been evolving in Australia for over four million years) will only make weak attempts to bite if handled clumsily. Others, particularly the more recently arrived *Rattus* species, bite more readily, but all flee or freeze if confronted by a human. Introduced rats show a markedly different behaviour. The propensity of a cornered brown rat (*Rattus norvegicus*) to launch itself at its human tormentor is well known. Doubtless this alarming behaviour has been strongly selected for, as a cornered rat responding in virtually any other way would normally have only seconds to live.

In the case of the fauna of Australia, the process of recognition and adaptation to humans as predators would have been doubly difficult. This is because although mainland Australia had some very formidable predators, all of the larger ones were reptiles. Humans are even more different from reptiles than they are from members of the dog and cat families. We do not know how long it takes for a species to learn to avoid people, nor do we know whether other predators can significantly 'precondition' various species to avoid humans. Darwin's examples of naive behaviour persisting for over a century among birds suggests that the process of adaptation to a human presence may be a slow one. Humans are such efficient predators that if behavioural change is not adopted after a couple of encounters, then extinction is the inevitable result.

If Darwin and Péron had observed in their tame island animals a glimpse of the world as it once was, how can we envisage Australia on the day the first humans waded ashore? Were the gentle diprotodons as 'domesticated' and obliging as to sleep beside their killers as the wombats of King Island did? Did the early Aborigines ride the great horned Meiolania turtles of Australia as gaily as Darwin rode atop his Galapagos tortoise?

CHAPTER 18

THERE AIN'T NO MORE MOA
IN OLD AOTEAROA

Of all of the extinctions that have occurred in the 'new' lands, none was so striking or is so well-documented as that of New Zealand's moas. As outlined in earlier chapters, 12 species of moa, weighing between 20 and 250 kilograms, inhabited New Zealand until about 600 years ago. They were New Zealand's ecological equivalents of antelope, rhinoceros and kangaroos and occupied a wide variety of habitats, from forest to alpine tundra. Before the arrival of the Maori sometime between 1000 and 1200 AD, they were abundant. Is their demise evidence of a human-caused blitzkrieg extinction event? And does it offer us a model for what happened to the megafauna of Australia and the Americas so long before? As I will show below, I think that the answer to both of these questions is a definite 'yes'.

A remarkably large body of evidence exists concerning the fate of the moa, for over the length and breadth of New Zealand, but particularly in prime moa habitat in the south-east of the South Island, are found Maori cooking sites which are literally packed with moa remains. Hundreds of sites are known. Some consist of only a pile of gizzard stones and a knife, indicating the spot where a moa was killed and gutted. Others consist of a rockshelter where a moa haunch was cooked, while yet others were the final resting place of tens of thousands of moa, and cover tens of hectares.

One of the most extraordinary sites was discovered among sand-dunes at Kaupokonui in the Taranaki District of the North Island. There, the remains of at least three species of moa, along with 55 other species of bird (many now extinct) have been found in and around ovens. Piles of uncooked and articulated heads, necks (some broken in such a way as to suggest that they had been wrung), ribs, vertebrae and pelves mark butchering sites.

Analysis of the site suggests that the wastage of meat was enormous, which indicates that protein was available in surplus at the time. Gizzard stones are rare, suggesting that the great birds were gutted where they were killed, their innards being discarded before the body was carried to the butchering site. The piles of uncooked heads, necks and other parts had clearly been left to rot, while only the leg bones are often found in oven pits, indicating that the haunches were the preferred meat.

Another great butchering site has been found near Wairau Bar in the north of the South Island. It has been estimated that nearly 9000 moa were killed and almost 2400 eggs destroyed, at this site alone. At yet another site, Waitaki Mouth in the Otago District, it is estimated that between 30 000 and 90 000 moa were killed. Several other large sites exist.

Several things are clear in an examination of these sites. The first is that they were occupied by very large numbers of people. Indeed, following the extinction of the moa, such dense aggregations of people were never to inhabit these areas again until after the arrival of Europeans. The second is that meat was in superabundance and that much was wasted. Entire moa legs have been found baked in ovens that were never opened. Piles of discarded remains included parts of moa bodies that contained large amounts of meat, while whole bodies were not infrequently left to rot. Typically, about a third of the meat available in moa carcasses was never used. The archaeologist Cassells gained the impression from his excavations of the Kaupokonui site that 'the waste is astounding'.[1]

Archaeologists have long argued over the significance of this evidence. Does it indicate, they ask, that humans exterminated the moa and most of New Zealand's other extinct, flightless birds through overhunting? Many researchers would suggest that it does. Atholl Anderson has calculated that, based upon the population density of large birds elsewhere, that New Zealand may have been able to support about 70 000 moa at any one time. If the moa from the excavated sites were killed over a relatively short period, the number of remains indicates a high degree of overhunting, for at least 30 000 and possibly 90 000 moa met their demise at the Waitaki site alone.

The evidence of substantial waste of moa meat at butchering sites suggests that moa were not difficult to hunt, for if great effort went into obtaining them, then surely they would have been utilised as fully as possible. Remarkably, absolutely no evidence of specialised hunting tools has been recovered from the excavated villages of the moa hunters. This may be because the moa hunters had to do little more than walk up to the enormous birds and spear or club them in order to obtain a meal.

Some researchers have suggested that moa became extinct within a century of hunting in any one area. In any case, all moa appear to have become extinct within 300–400 years of the arrival of the Maori. Calculations of the rate of increase in the Maori population indicate that sufficient people would have been present in New Zealand within a few hundred years of settlement to cause the extinctions. If, for example, there were several hundred people present in New Zealand in 1200 AD and the population grew at a conservative rate of about one per cent per year, then the number of people would rise from thousands to tens of thousands within 400 years of settlement; which is just when the last moa were being eaten. Atholl Anderson, in his splendid work on the moa titled *Prodigious Birds* sums up as follows:

> *Clearly there is still much to learn about moa extinction, but that it was caused, in various ways, by the human colonisation of New Zealand, can hardly be in doubt.*[1]

Although abundant evidence of the moa hunters of 700–400 years ago can still be found in New Zealand today, this evidence is rapidly being destroyed. In places where nineteenth century observers reported the ground to be 'white with moa bones', all trace of bones has now long vanished. Coastal erosion is removing the moa hunting sites of the Taranaki coast at 20 metres per year. The site at Rakaia Mouth in the South Canterbury region of the South Island is rapidly succumbing to erosion by the sea and wind, while the greatest site of all, at Waitaki Mouth, North Otago, has, through coastal erosion, been reduced from over 120 hectares last century to less than 50 hectares today.

How much evidence of moa hunters, one wonders, will be left in another thousand years, or in another 60 000 years? Imagine how much more would be destroyed if the sea-level fell over 160 metres, then rose again and if there was little accumulation of sediment to cover bones; and if rockshelters, because of their acidic soils, did not usually preserve fossil bones. These are the conditions that have prevailed in Australia since the extinction of its giant marsupials, birds and reptiles. Not surprisingly, evidence of their fate is far harder to come by than that of the moa.

CHAPTER 19

LOST IN THE MISTS OF TIME

It is generally true that more ancient events have left less evidence in the fossil record than more recent ones. This is because the Earth is dynamic and through erosion and other geological processes, is constantly destroying the old and creating anew. For this reason there are probably many hundreds of thousands of species of now-extinct animals of which every trace has vanished. Evidence of even quite recent events can also be quickly destroyed. What clear evidence would ever be found, for example, that thousands of whales were killed off the coast of New South Wales, were it not for material preserved in museums? For over a mere 150 years the dynamic coastal environment has obliterated almost all traces.

It is quite possible that the restless Earth has already destroyed all of the evidence relating to how humans affected Australia's giant marsupials, birds and reptiles. The search for such evidence, including the search for butchering sites—or even a single bone with butchering marks on it—has so far been unsuccessful. Indeed, no certain evidence has ever been found of any kind of interaction between humans and the megafauna in Australia. Remarkably, we do not even know when the megafauna vanished, for there are no unassailable dates which document their passing.

In one way, this lack of evidence is, itself, a contribution to the debate. Many have argued that it suggests humans and megafauna did not interact, perhaps because of differing ecological requirements, or through lack of interest. But there is another interpretation—for the lack of evidence is just what one might expect if the extinction of New Zealand's moa is a good model for what happened to the megafauna of Australia. The reasoning is as follows. If the marsupial giants vanished within a century to 500 years of the arrival of people at any one locality, this gives a very narrow time window for archaeological deposits to be laid down. With few deposits to start with, the dynamic nature of the Earth may have ensured that almost all evidence has been destroyed over the tens of thousands of years that have passed since then.

The lack of dates documenting the extinction of the megafauna has also provided fuel for the debate. It suggests to some researchers that the extinctions occurred long ago. They argue that had it happened less than 30 000 years ago, then surely some dateable bones of giant marsupials would have been found by now, for the remains of humans and many still-living species have been firmly dated to the last 30 000 years using the C^{14} technique.

The most controversial findings concerning the timing of megafaunal extinction have emerged from excavations made over the last few years. Professor Richard Wright of Sydney University excavated a site at Tambar Springs on the Liverpool Plains of central eastern New South Wales. There, he found, buried in mud which accumulated around the springs, small bone fragments of giant marsupials and tools made by Aboriginal people that were deposited over a time span of many thousands of years. Indeed, bone fragments from giant marsupials occur in deposits as recent as 6000 years old. On this basis, Wright has suggested that giant marsupials and humans coexisted on the Liverpool Plains for a very long time indeed.

If these findings are validated they will revolutionise concepts of Australian prehistory. There is no doubt, for example, that humans have been present in Australia for 40 000 years or more. It is also becoming clear that the giant marsupials had vanished from many areas by 30 000 years ago. If Wright is right, then the giant marsupials must have endured in their tiny refugia on the Liverpool Plains for millennia without spreading outside of it. Yet the idea that the Liverpool Plains was a refuge for megafauna is unlikely, as it is no more fertile nor well watered than many other regions of Australia. There is no clear boundary around the Liverpool Plains. If the giant marsupials had survived there then surely they would have wandered away from it in favourable times and established populations elsewhere.

I think that Wright is wrong in his interpretation of the Tambar Springs deposits, for the scraps of giant marsupial bone may have been redeposited. If this has happened, then very old bone fragments of giant marsupials could have become mixed with more recent human artefacts and animal bones. This interpretation is strengthened by the fact that there is no evidence of human cut marks or other modification on any of the bones.

Interestingly, Wright attempted to date his bone fragments using a new and relatively untried technique involving radioactivity. The dates, of 40 000 years and more, are probably unreliable. But they may also indicate that the situation at Tambar Springs is not as clear-cut as it first appears.

As the site at Tambar Springs is unique in suggesting that the megafauna survived into relatively recent times, we must go back tens of thousands of years before finding other sites relevant to the megafaunal extinction debate. The dating of older sites is bedevilled by exactly the same problems faced by archaeologists who wish to date early human remains. This is because almost all of the evidence is so old as to be at or beyond the limits of the conventional radiocarbon dating technique.

There are two sites that appear to be exceptions and these are worth examining in detail. Most important is a swamp deposit found near the small town of Lancefield near Melbourne in Victoria. The other site, Spring Creek near Hamilton, is located in Victoria's southwest. The site at Lancefield Swamp is an extraordinary one. It was discovered in the summer of 1843 when Mr James Mayne excavated a well in the swamp to water his thirsty cattle. The deposit of fossilised bones of giant marsupials that he encountered is only the second found by Europeans in Australia. Following its early fortuitous discovery, the exact location of the site was lost. Then, in the 1970s the geologist Rob Glennie (the same Rob Glennie who was instrumental in rediscovering the first Victorian dinosaur locality) tracked it down.

Excavations in 1974–76 revealed that vast numbers of bones lay buried several metres below the surface of the swamp. The bones of a gigantic ancestor of the modern eastern grey kangaroo (*Macropus giganteus*) are by far the most abundant, being represented by remains from tens of thousands of individuals. Its bones make up more than 90 per cent of all remains found. Bones of the giant wallabies of the genus *Protemnodon* make up another seven per cent, while diprotodons, the giant bird *Genyornis*, short-faced kangaroos and a few other species make up the remaining three per cent.

Curiously, very few young individuals of any species are represented and it is partly for this reason that palaeontologists suggest that the animals died during a prolonged drought. They reason that poor conditions had forced the animals to stop breeding several years before. The surviving adult animals were then forced to congregate in the area of the swamp for food. Finally, having eaten all available plant matter, they died of starvation.

About 10 years after the Lancefield site was rediscovered, a second fossil deposit was located in the swamp, about a kilometre from the first. It was discovered fortuitously on 'Ash Wednesday', 1983. Firefighters, desperate for water to douse the bushfires, dug an enormous hole to the base of the swamp. As well as water, they discovered bones, literally by the tonne. By the time I visited the site in 1985 the mullock heap left by the backhoe was white with bones, including the remains of a much greater variety of species than was recovered from the first excavation.

The Lancefield site assumed enormous importance when a large stone tool was found among bones at the original excavation. It appeared that scientists had finally discovered evidence for the coexistence of humans and the giant marsupials. Attempts to date the deposit took on a new urgency, but unfortunately the bones themselves were unsuitable for radiocarbon dating, having lost almost all of their organic content. No other material in the bone layer could be dated.

The chance of obtaining a date seemed to improve when the researchers located a channel which ran under the bone bed. It was filled with the bones of a variety of animal species, both large and small, along with some dateable charcoal. Being directly under, and sealed off by, the bone bed, the charcoal had to be older than the overlying bone bed and stone tools. Imagine everyone's joy to get back from the laboratory not one, but two radiocarbon dates suggesting that the charcoal in the channel was about 26 000 years old! This, it seemed, settled an age-old question, for many assumed that it proved that humans and giant marsupials had coexisted less than

26 000 years ago at Lancefield. Since it was believed at the time that humans had been present in Australia for at least 35 000 years, the dates implied that giant marsupials and humans had coexisted for nearly 10 000 years in Australia. Many researchers were satisfied that a New Zealand-style rapid extinction had not happened in Australia.

Unfortunately, such convenient scenarios rarely stand the test of time. Now, the evidence from the Lancefield site is being looked at ever more critically. Work currently being undertaken by Sanja Van Huett of Monash University may have solved the puzzle, for her findings suggest that the bones in the bone bed have been washed in from elsewhere. I suspect that they were washed into the site in an already fossilised state.

Presently, my best guess to explain the formation of the Lancefield site is as follows. A very long time ago, certainly longer than 26 000 years ago, a vast number of giant marsupials perished somewhere in the vicinity of Lancefield Swamp during a drought. Their bones were buried and preserved. Then, about 26 000 years ago, a small stream cut into swamp sediments and when it filled it preserved the remains of fauna that inhabited the area at that time, as well as a few redeposited bones of the long extinct megafauna. Some time later, perhaps at the height of the last ice age, when it was cold and dry, erosion exposed a vast area of the ancient bone bed which had entombed the remains of the megafauna. Flash floods then washed many bones into the part of the swamp where the channel had formed earlier, leaving a layer of redeposited bones and clay that covered over the deposits laid down by the small stream. Aborigines, perhaps attracted to the spectacle of the numerous large fossil bones, visited the site and left at least two stone tools. In 1843 a farmer with hundreds of head of thirsty stock dug a well—and thus our modern investigations began.

The second site, located at Spring Creek some 160 kilometres west of Lancefield, has a lot of similarities with the Lancefield site. I had been led to the place by Lionel Elmore, a retired western Victorian farmer and naturalist. He knew of a sedimentary deposit that occasionally yielded fossilised shark teeth and shells. They had formed some 10–5 million years ago, when much of Victoria's Western District was under the sea, the majestic Grampians mountain range then being an island. The rocks outcropped in a little creek, just below a picturesque waterfall only a metre high, in the fertile sheep-grazing country of Victoria's basalt plains.

After searching for fossilised shark teeth for a few hours, I felt the call of nature and, for the sake of modesty, decided to wander downstream a few metres. While enjoying the sound of my own tiny waterfall I espied a strange looking object sticking out of the black mud of the bank. When I removed it from the clay, I realised that I held in my hand the limb bone of a long-extinct kangaroo. Excavations subsequently revealed that at some time in the past, but many millions of years after the area had risen from the sea, a small pool had accumulated below the waterfall. There, a sticky black clay had formed and entombed the remains of at least a dozen species of marsupials. As at Lancefield, the most abundant species was the gigantic ancestor of the living eastern grey kangaroo. Its remains make up 75 per cent of all fossil bones found. As with Lancefield, the remains of young animals are very rare in the deposits, suggesting that breeding had ceased some years before the animals died. Thus it appears that drought may also have killed animals at Spring Creek.

No clear evidence for the presence of humans has been found at Spring Creek, but beautifully preserved plant remains (including fruit with shrivelled skin still adhering), beetles and freshwater mussels (preserved where they entered their last aestivation beneath a bank of clay) are all exquisitely preserved. Moreover, the remains of some of the extinct giant marsupials are in excellent condition. Unlike Lancefield, where the skeletons have been completely jumbled up, Spring Creek has yielded a number of articulated remains, including the hand of the gigantic eastern grey kangaroo and a partially articulated leg and complete skull of a giant wallaby of the genus *Protemnodon*. Clearly, these bones lay where the animals died, for if they had been moved from elsewhere the associations between parts of a single skeleton would have been lost. Furthermore, very little of the bone is weathered, so its exposure time to the atmosphere was short.

Examination of the plant remains by a botanist revealed a fascinating story. The most common seeds in the deposit came from two species of *Pimelia*, commonly known as rice flowers. These species are absent from western Victoria today and one of them presently grows only in alpine areas, the nearest being in eastern Victoria. This strongly suggests that the Spring Creek site dated to at least the last ice age, when conditions were cooler. I was therefore delighted when, upon submitting some plant remains to a laboratory for dating, I received back a date suggesting that the site was about 19 800 years old. I was immensely proud to have excavated one of the most recent, if not the very most recent, occurrences of giant marsupials and published my findings in a paper titled *The Spring Creek locality: a late surviving megafaunal site*. As time went on, however, I became more worried about the fact that I had obtained but a single radiocarbon date and that it was from plant matter and not the fossil bones.

Finally, in 1992 Professor Peter White and I returned to the site to obtain more bone for dating. Just a few weeks before writing this I received the first results back from the laboratory. The dates from the bones are all much older than the first one received on the plant material—but even this may not be reliable, for two dates obtained from parts of the same bone vary greatly.

Given the existing evidence, it is difficult to know how old the Spring Creek site really is. But it is clear that it can no longer be heralded as unequivocal evidence for a late-surviving giant marsupial fauna. Thus, I must join Wright in probably being wrong about my interpretation of a deposit containing the bones of extinct marsupial giants.

Despite the doubt cast upon the age of these two important sites, both of which have been used to argue for a long overlap in Australia between humans and giant marsupials, sites are still being excavated which suggest a late-surviving megafauna. One of the potentially most informative sites is that of Cuddie Springs near Brewarrina in western New South Wales. There, the abundant remains of giant marsupials have been found along with human artefacts in a series of sedimentary layers dating to 30 000 and 19 000 years ago. Unfortunately, the full details of the site have yet to be published and so its significance is not yet clear.

In counterpoint to these sites, which suggest a late survival for the Australian megafauna, is evidence which has come from a series of well-documented archaeological excavations in Tasmania, the south-west of Western Australia and western New South Wales, which suggest that the megafauna was already extinct in these areas 35 000 years ago.

These sites include both cave and lakeside dune deposits and are well dated, each with tens of radiocarbon dates from throughout the deposits confirming their age. All have produced large amounts of bone, the cave sites from the south-west of Tasmania being particularly rich. Excavation of one cave site in the Florentine Valley produced over 200 000 pieces of bone weighing 30 kilograms—from just one cubic metre of sediment! These sites are all remarkable for one thing: despite their well-established age and the abundance of faunal remains (with the exception of a few redeposited pieces in a Western Australian cave) not one scrap of bone attributable to the marsupial megafauna has been found in them.

Some archaeologists have long argued that this may be because the megafauna were too heavy to carry back to a cave or to a campsite beside a lake. But many megafaunal species, including the giant echidnas and short-faced kangaroos, were in fact quite small. Indeed, several were smaller and lighter than the grey kangaroos whose remains are so abundant at some sites. Furthermore, there is abundant evidence of Maori carrying enormous moa carcasses considerable distances to centralised butchering sites. No-one has ever explained what might have prevented Aborigines from doing so. These data make the explanation seem weak, and the only satisfactory argument offered so far for the absence of megafaunal remains from these sites is that the megafauna was already extinct by the time the sites accumulated. These sites are, I feel, strong evidence that the megafauna had become extinct, at least in southern Australia, before 35 000 years ago.

What can be said in summary then, about the timing of the extinction of Australia's megafauna? I must admit that we have little idea of precisely when it happened, but there is growing evidence that it occurred before 35 000 years ago, at least in southern Australia. Furthermore, at present we have no clear evidence about the nature of interaction between humans and megafauna, for we have no kill sites and very few sites where there is possible evidence for humans and megafauna coexisting. This might indicate, as some researchers suggest, that humans and megafauna avoided each other. But it is just as likely to have resulted from a human-caused extinction event like that of the moa in New Zealand, which was so rapid that it has left virtually no trace, more than 35 000 years after the event.

A final line of evidence concerning the cause of extinction derives from analysis of the vanished animals themselves. The megafauna was a particularly diverse group of species which shared only a few features in common. They were drawn from all environments; some species inhabited the periglacial tundra of the highest mountains of New Guinea, while others were found in its mossy montane forests and steamy lowland jungles. A great variety of species inhabited Australia's forests, grasslands and deserts. One thing that all species had in common, however, is that they were relatively large. The smallest species were either very slow-moving mammals or flightless birds.

The few large marsupials that survived to the present include the very fastest kangaroos and wallabies and the recently extinct thylacine. The only other large survivors are wombats, which can burrow for protection, and the koala, which hides in trees. All of this suggests that a predator, rather than a change in climate, was responsible for the extinctions. Climate change does not select against slow-moving species, nor for the survival of burrowing or tree-dwelling species. But more recent extinction events show that human hunters can rapidly cause the extinction of the largest, slowest and most obvious species in any habitat.

I feel that the weight of evidence is now clearly in favour of a very rapid, human-caused extinction for the Australian megafauna. Furthermore, the pattern of extinction seen in New Zealand and on other Pacific islands in the past few millennia are good models for the events that occurred in Australia more than 35 000 years before.

Extinction was only one of the features of the last 60 000 years of Australasian history. Equally interesting and informative is the story of the 'time dwarfs', as I have christened the surviving large marsupial species of Australia.

CHAPTER 20

TIME DWARFS

Old fishermen are famous for their tales of whopper fish, most of which got away. I have always viewed such stories with a large dose of scepticism, but recent research by fisheries experts suggests that the old codgers may not be such terrible fibbers after all. For scientists have found that in many intensely exploited fisheries, the average size of individual fish is decreasing. They think that this is because the tiddlers are thrown away or get through the net, while the larger fish are caught. This selects for the survival of early maturing dwarfs because these individuals breed and spread their 'dwarf' genes, while their larger cousins are caught before reaching sexual maturity. Unfortunately, as the average size of fish decreases, yesteryear's tiddlers become today's whoppers. So the cycle is perpetuated, the average size of fish decreasing over time and our octogenarian anglers being viewed with more and more disbelief!

This trend is of course quite alarming to fisheries biologists, as the productivity of the fisheries, as well as fish ecology, can be severely affected. They may be interested to know that fish are not the only organisms to have shrunk in average body size through time. Many well-known Australian marsupials are 'time dwarfs' too.

My own interest in these matters dates back more than a decade when, as an MSc student, I realised that the eastern grey kangaroos (*Macropus giganteus*) that exist today are much smaller than their ancestors of 40 000 years ago. I wondered whether intense hunting by humans could have affected the size of kangaroos as well as fish?

The factors that constrain the size of living organisms are many and complex. At the most fundamental level are the structural constraints. The efficiency of the breathing organs of insects, for instance, places an upper limit on the size that they can achieve. Thus, the heaviest insect, the South American beetle *Megasoma actaeon*, weighs only 200 grams. Likewise, the physiology of leaf digestion places a lower size limit (about half a kilogram) on leaf-eating mammals such as ringtail possums (*Pseudocheirus* species).

Many other more subtle factors influence size. Where two ecologically similar species coexist, the less dominant one will often be smaller than in areas where it alone exists. Latitude can also have an effect, as generally mammals from higher latitudes are heavier than those from lower latitudes, but often have shorter appendages (to conserve body heat). Thus, the koalas (*Phascolarctos cinereus*) and eastern grey kangaroos of Queensland tend to be smaller than those of Victoria.

Nutrition is also important in determining body size. It is well-established that there was an increase in the stature of Japanese men and women when a more nutritious diet was introduced to Japan after the Second World War. Conversely, poor-quality plant food can favour the evolution of very large animals. This is because herbivorous animals need enormous, complex stomachs to act as fermentation vats if they are to gain sufficient nutrition from poor-quality feed. Thus, as food quality decreases, herbivores of larger size are selected for. Low-nutrient food may therefore select for an increase in the average size of an evolving lineage of animals. A good example are Australia's extinct diprotodons. Over the past 15 million years, as Australia dried out and fodder quality became increasingly poor, these animals increased in size.

Eastern grey kangaroos have also increased in size over most of the last two million years. But then, beginning about 60 000–35 000 years ago, they started to rapidly decrease in size. At about the time this size decrease began, the diprotodons and many other large marsupials became extinct. Fossil deposits covering the past 35 000 years from Devil's Lair, south-western Western Australia, reveal that the size decrease in grey kangaroos continued after the time of the diprotodons' extinction and that it has continued to the present.

The oldest western grey kangaroo (*Macropus fuliginosus*) fossils from Devil's Lair are between 27 000 and 19 000 years old, and are 13 per cent larger than a modern Victorian sample of eastern grey kangaroos that I used as a standard in my study. In contrast, fossils from Devil's Lair dating to between 19 000 and 12 000 years old are only 10 per cent larger. This indicates something of the rate of size decrease over time.

The very largest eastern grey kangaroos existed in Victoria towards the end of the Pleistocene period. Their remains have been found alongside the bones of the extinct megafauna at Lancefield and Spring Creek. Some individuals must have been truly enormous, for their teeth alone are 30 per cent larger than living Victorian grey kangaroos. This suggests they were about twice as heavy, being more stocky than living animals. The largest probably weighed nearly 200 kilograms and, when stretched upright, stood about three metres tall.

So spectacularly different are these fossil remains from the bones of modern animals that they were long considered to represent a different species. Only after careful research into changes in kangaroo size through time has the true situation been revealed.

The grey kangaroos were not the only species to shrink in size over the last 40 000 years. The American palaeontologist Larry Marshall and co-author Joe Corruccini have shown that the teeth of red kangaroos (*Macropus rufus*) have declined 30–35 per cent in length, wallaroos (*Macropus robustus*) by 26–27 per cent, agile wallabies (*Macropus agilis*) by 15–16 per cent, swamp wallabies (*Wallabia bicolor*) by 15 per cent, yellow-footed rock-wallabies (*Petrogale xanthopus*) by 9–10 per cent, Tasmanian devils (*Sarcophilus harrisii*) by 16–17 per cent and spotted-tailed quolls (*Dasyurus maculatus*) by 5–6 per cent. Furthermore, the Director of the Queensland Museum, Dr Alan Bartholomai, has shown that the teeth of koalas (*Phascolarctos cinereus*) have shrunk by about 10 per cent. In fact, so comprehensive is the list of 'time dwarfs' among the larger surviving Australian mammals that it is difficult to find species that have not shrunk substantially. As far as I have discovered these include only the three living species of wombats and, to a lesser extent, humans.

An important point to note about the 'time dwarfs' is that, by and large, the larger their ancestors, the greater has been their shrinkage in size. Also, the smaller species of marsupials (five kilograms and less in weight) show no reduction in size, while the larger ones that did not dwarf (including the diprotodons) are now extinct.

Most scientists have argued that this shrinkage in size, which also affected larger mammals on other continents, had something to do with the ice age and a concomitant drop in the quality and quantity of fodder. This may explain some of the decrease in size, but other factors also appear to be at work. The teeth of Aborigines, for example, have shrunk by a mere seven to nine per cent over the past 30 000 years. This is a very small amount, given their body size, when compared with the shrinkage of similar-sized marsupials. The nine per cent size reduction experienced by humans may have been caused by climate change or some other factor. But the additional dwarfing (up to 30 per cent) experienced by similar-sized marsupial species, appears to be due to something else.

We now know that the 'time dwarfs' continued to decrease in size long after the last ice age passed, about 15 000 years ago. The climate theory also ignores the fact that there have been 17 ice ages over the past two million years. Why should large marsupials start to dwarf only during the last one? Furthermore, it is hard to imagine a change in the quality of food that would affect kangaroos, koalas and Tasmanian devils, would marginally affect humans, but not affect wombats at all.

I feel that the overexploitation hypothesis might be better at explaining the marsupial 'time dwarfs', although several factors must be established before it can be accepted. Firstly, is there any evidence that the dwarfed species were under intense hunting pressure from Aborigines? Secondly, is there any evidence to suggest that Aborigines might select the largest individuals preferentially? Thirdly, are there any 'independent experiments' such as islands where humans do not exist and animals have increased or at least not shrunk in size? And finally, does the fact that humans have shrunk little and wombats not at all, have have any bearing on the question?

Analysis of prehistoric Aboriginal refuse dumps is of little help in determining how heavily various species were hunted. If there are many bones of a particular species in a deposit, this could be interpreted as resulting from heavy hunting; but then, if the species were abundant, the hunting may not have had a significant impact on the population overall. Alternatively, if few bones are present, it is difficult to determine if the species was naturally rare, hunted into rarity, or simply taken infrequently.

Perhaps the best information available on hunting pressure is that from the early European contact period. If it can be shown that various species increased in numbers after Aborigines ceased to hunt them, then it is likely that hunting was having a considerable impact on their population. The best information is available for the koala and the larger kangaroos.

Dr Roger Martin of Monash University has put forward a persuasive case that hunting by Aborigines was a significant factor in keeping the numbers of koalas and tree-kangaroos (*Dendrolagus* species) low and in limiting their distributions. He points out that reports of koalas in the writings of very early settlers are rare. But by the mid-nineteenth century, koalas were being reported frequently.

The best documented example of this trend is the information presented by Harry Parris,[106] whose forebears settled on the Goulburn River in Victoria in the 1870s. Relying on his family's recollections and the published accounts of other early settlers in the district, he reconstructed changes in koala abundance in the Goulburn River district since white settlement. He found that koalas were not mentioned in any written accounts at all prior to 1850, but that occasional sightings occurred from the early 1850s onwards. Sightings had become abundant by the late 1860s and there were thousands of koalas in some areas between 1870 and 1890. He observed that this increase coincided with the annihilation of the resident Aboriginal population and suggested that it was their hunting that kept koala numbers low. Martin adds that a decline in dingo numbers may also have had some impact.

There is even a little evidence that Aboriginal hunting suppressed the numbers of the smaller, more difficult-to-hunt possums. Some early evidence is quaintly put by the Reverend Peter MacPherson, who travelled regularly from Geelong to Ballarat in the 1860s and '70s. He noticed that between 1862 and 1874, a fine stand of eucalypts growing at Bruce's Creek north of Meredith was gradually dying off. He sought an explanation, considering drought, bushfires, poor soil and swampy ground, but rejecting each in turn as not squaring with the evidence. He finally consulted the Aborigines, who gave him the answer he sought. 'Too much big one possum' they reportedly said. Apparently, in the days before white settlement, the local tribe had been in the habit of catching many possums each day for

Eastern grey kangaroo

Wallaroo

Tasmanian devil

Swamp wallaby

Koala

Agile wallaby

Some of the time dwarfs. The inked-in drawing is of the larger ancestor, the outline an average-sized living animal. (*Courtesy Peter Murray*) their food, but now they hunted no more. MacPherson saw that in times past:

> the tooth of the blackfellow operated upon the possum,
> and the tooth of the possum operated upon the leaves of the
> eucalypt.[84]

With the disruption of Aboriginal lifestyles, the possums had proliferated, endangering the trees which were their food source.

The situation regarding the larger kangaroos appears to have been similar to that described for the koala. Early explorers such as Thomas Mitchell considered the sighting of a kangaroo as a noteworthy event in areas where today hundreds can be seen in a single morning. John Gould wrote in the 1840s that the Government must act to protect the red kangaroo (*Macropus rufus*), for:

> if this be not done, a few years will see them expunged from
> the fauna of Australia. [55]

Gerard Krefft noted that in 1862 the red kangaroo was very poorly known to the general public, even its continued existence being doubted, as:

> [one] enlightened critic was pleased to designate this species
> as ante-diluvian. [78]

Clearly, red kangaroos were much rarer early last century than they are today.

The increase in numbers of the larger kangaroos cannot be attributed solely to a relaxation of human predation, for it is possible that the extermination of the dingo and the provision of water and improved pasture in some areas has had a large impact. But the importance of the cessation of Aboriginal hunting should not, on this basis, be discounted as being unimportant.

Ecologist Robert May of Oxford University has identified a phenomenon, now called the 'predator pit', that may help explain how Aboriginal hunting pressure could have suppressed the larger marsupials.[91] The model suggests that once the population of a prey species falls below a certain level, a flexible predator can stop it from recovering. It does this by switching much of its attention to another prey species, thus keeping its own numbers high. The small amount of predation it exerts on the species in the 'predator pit' is still, however, enough to prevent it building up to the point where its numbers can increase rapidly again. The small size of a population caught in a 'predator pit' may also help genetic changes (such as that for early maturing dwarfs) to spread rapidly. Humans are certainly flexible predators, so the theory fits the facts well.

The 'predator pit' phenomenon may also explain why hunting by Europeans has been unsuccessful in reducing kangaroo numbers: perhaps by the time that European hunting had begun in earnest, Aboriginal predation had ceased long enough for kangaroos to have climbed out of the 'predator pit', their numbers reaching a point where they were not easily suppressed again.

We must now address the question of whether Aboriginal hunting techniques favoured capture of the largest individuals in a population. Some interesting anecdotal data exist to support the notion that it did. Krefft, writing of the red kangaroo in 1862, notes that:

> when disturbed, the old males cover the retreat of the fleet females who are off first, so that specimens of the latter sex are rare, the dogs generally stopping the progress of the rear-guard of the red 'old men'.[78]

Old male red kangaroos are the largest individuals of all. Daisy Bates notes that in Western Australia, Aboriginal men would track an individual kangaroo for days, exhausting it enough so that they could kill it. Surely, when expending such enormous amounts of energy, it would be sensible to track the largest individual available. Whether larger koalas were hunted preferentially is less well-known, although it makes intuitive sense that smaller koalas would be harder to see and thus less vulnerable to predation than large ones.

A fortuitous piece of evidence helped decide the matter for me. While completing my MSc on kangaroos I was lucky enough to visit Kangaroo Island, South Australia and to examine some fossil grey kangaroo bones found there. To my great surprise I found that these samples differed from those of all other regions of Australia in that the modern animals were, on average, *larger* than the 20 000-year-old fossils! At first I thought that this might be due to the fact that they were from an island, for I was aware that island populations of mammals are often unusual in size. For example, in the past the islands of the Mediterranean have been home to sheep-sized elephants and hare-sized hedgehogs. But as it turns out, it is a general rule that larger animals shrink on islands and only species much smaller than grey kangaroos show an increase in size. Initially I was bewildered. But then I learned that until 11 000 years ago Kangaroo Island had been part of the mainland and until 2500 years ago, it had been inhabited by Aborigines. Could it be that the eight per cent size increase in the Kangaroo Island kangaroo (*Macropus fuliginosus fuliginosus*) occurred only after human hunting had ceased? Unfortunately, the fossil record of kangaroos on the island is not yet complete enough to test this hypothesis. But it will be a critical test of the 'time dwarfs' theory when and if material finally does become available.

Finally, can we make sense of shrink-proof wombats? Wombats are very large marsupials, reaching about 40 kilograms in weight. Thus they should have shrunk dramatically. It is possible that their burrowing habits protected them from severe predation. Another alternative is that they *have* shrunk, but that this has been difficult to detect because their ever-growing teeth (from which size estimates are made) are difficult to measure accurately. But if this is so, the average size difference must be small, as it is not so easily detectable as it is for all other species.

Incidentally, if we humans did create the 'time dwarfs', then the issue has great significance for modern wildlife management programs. Large males are currently favoured targets in the commercial culling programs of the three largest kangaroo species. Is this leading to a further decrease in size? At what point are these various species disadvantaged by a further size decrease? Should we be culling the smallest adult males in an attempt to increase average size? Whatever the answers to these questions, fisheries experts and wildlife management authorities would do well to look closely at the fossil record before implementing culling decisions.

CHAPTER 21

SONS OF PROMETHEUS

We must now address the role of fire in Australian environments. Fire is one of the most important forces at work in Australian environments today, yet this has not always been the case. For the role of fire has changed in Australia, largely, I think, as a result of megafaunal extinction and the dwarfing of the surviving large marsupial species.

It is true to say that the rise of fire has transformed Australia. Yet its effects have been modified through Aboriginal control of the firestick. When control was wrested from the Aborigines and placed in the hands of Europeans, disaster resulted.

Because the historic role of fire in ecosystems is so much better understood than its prehistoric role, it is best to begin with an examination of fire as it was used by Aborigines when Europeans first arrived in Australia.

The use of fire by Aboriginal people was so widespread and constant that virtually every early explorer in Australia makes mention of it. It was Aboriginal fire that prompted James Cook to call Australia 'This continent of smoke'. Tasman, as early as 1642, saw smoke billow into the sky for days at a time, as did other early explorers. But it was that most poetic of explorers, Ernest Giles who, during his travels in Central Australia, gave us the most vivid image of the inseparability of fire and Aborigines:

The natives were about, burning, burning, ever burning; one would think they were of the fabled salamander race, and lived on fire instead of water.[52]

In a sense, Giles was right, for fire was the staff of life for the Aborigines. Through its judicious use, they wrested their daily meat and bread from an otherwise ungiving land, they fought their enemies and expelled their pests.

Giles was writing of the arid inland. Some researchers have doubted that fire was as important to the lives of coastal Aborigines as it was to those living in the centre. But it is clear that fire was important to Aborigines living everywhere. The journals of Joseph Banks, written as he sailed along the east coast of Australia in 1770, are full of accounts of fire after fire seen as the *Endeavour* sailed northwards. Many were campfires, but others were immense and more perplexing to the English.

As a result of the extensive Aboriginal use of fire, the plant communities seen and recorded by Banks and other explorers were very different from those that exist at the same location today. Curiously, although substantially reported upon in the early literature, this fact is not well-recognised today, even among biologists and those responsible for land management. Perhaps this is because the works of Banks and others are no longer widely read. There is also a tendency to believe that early landscape painters did not render scenes accurately, but instead tended to paint them as English landscapes.

The lack of recognition of this change has critical implications for the management of national parks and wilderness areas today. This will be discussed further at the end of this book.

I first became aware of just how dramatically vegetation communities along the east coast of Australia had changed when I read Banks' journals and visited the places he described. Among the most striking changes that I could detect had occurred near Bulli. On 27 April 1770 the *Endeavour* passed close by the shore near modern-day Bulli, just north of Wollongong. Although Banks could not land, he came within a few hundred metres of the shore and left a good description of the vegetation. He wrote:

> *The countrey today again made in slopes to the sea... The trees were not very large and stood separate from each other without the least underwood; among them we could discern many cabbage trees [the cabbage palm Livistona australis] but nothing else which we could call by any name. In the course of the night many fires were seen.*[104]

I had this description in mind when I stood, on a hot February day, atop the lookout at Bulli Pass. Below me stood a magnificent patch of temperate rainforest which stretched for several hundred metres out from the escarpment edge, towered over by magnificent cabbage palms. Beyond it stood a dense and tall forest of eucalypts, their crowns extending unbroken to the expanding urban sprawl and strand vegetation at the ocean's edge.

It seemed inconceivable to me that the open woodland which Banks described was the same place that I now saw. But then I took a path that dropped off the escarpment, and I began to examine the forest in detail. Most of the rainforest trees were small, the only very tall species being the magnificent cabbage palms. High up on their trunks, they bore the unmistakable signs of fire. Yet it was clear that fire had not touched the rainforest for many years. Finally, I came across a solitary, ancient eucalypt, standing among the smaller rainforest plants. All that was left of the original trunk was a charred stump a metre in diameter. But from its base two suckers had sprouted. Now they had reached a diameter exceeding that of the original stump itself. Clearly, this was a last survivor of Banks' open woodland 'without the least underwood'. As remarkable as it seems, the altered fire regime of the last 200 years had seen rainforest and dense eucalypt forest establish on what in Banks' time was clearly an open woodland.

There are many other examples of how vegetation has changed over the past 200 years. When Banks reached Botany Bay he described the land as 'cliffy and barren without wood'. Where Kurnell stands today, he recounts watching from the *Endeavour* as Aborigines retreated up the hill and hid among the rocks. The dense vegetation growing on the spot today would obscure a party of hundreds before they had retreated a few metres from the shore.

The journals of members of the First Fleet are also full of references to events that only make sense if vegetation has altered grossly since 1788. John White records walking 20 miles (32 kilometres) westwards from Sydney without becoming lost, while Watkin Tench covered a greater distance and returned, without commenting upon any difficulty in navigation. Furthermore, White and Phillip often walked to Botany Bay without becoming lost. It was, apparently, an easy walk. White says of the land around the Harbour:

> every part of the country, though the most inaccessible and
> rocky, appeared as if, at certain times of the year, it had been all
> on fire.[140]

He says of Frenchs Forest that 'the trees were very high and large and a considerable distance apart, with little under or brush wood'.[140] Kartzoff, a writer with the NSW Forestry Commission, notes that by the 1960s the 'very high and large trees'[73] had been replaced with stands of much smaller trees with dense undergrowth. He also states:

> The vital fact remains that the evidence of all early observers
> indicates that the countryside [around Sydney] was covered by an
> open forest of very large trees with little undergrowth. This type of
> vegetation is unlikely to lead to, or result from, the type of major
> bush fires that occur nowadays.[73]

In the mid-nineteenth century, Surveyor General for New South Wales Sir Thomas Mitchell recorded the change of this vegetation type to the one that occupies the area today. He said:

> Kangaroos are no longer to be seen there [Sydney]; the grass is
> choked by underwood; neither are there natives to burn the
> grass…[and that]… the omission of the annual periodical
> burning by natives, of the grass and young saplings, has already
> produced in the open forests nearest to Sydney, thick forests of
> young trees, where, formerly, a man might gallop without
> impediment, and see whole miles before him.[100]

Just how widespread changes of this magnitude were throughout Australia is still hotly debated by biologists and archaeologists. Nonetheless, there is evidence from places as distant as Tasmania, north Queensland and Arnhem Land that modern vegetation communities differ dramatically from those of 200 years ago. Some of the best evidence for Queensland comes from Tam O'Shanter Point, north Queensland, which William Carron, botanist to the ill-fated Kennedy expedition, described in 1848:

The open ground between the beach and the swamp varied in width from half a mile to three or four miles. It was principally covered with long grass, with a belt of bushy land along the edge of the beach.[81]

Today, the vegetation is dense rainforest with emergent eucalypts. Likewise, in Tasmania, many areas which 200 years ago supported an open vegetation now support dense temperate rainforest or sclerophyll forest. I will leave the last word in this matter to a weary young Charles Darwin, who in 1836 described the simplified, fire-loving plant communities that then dominated Australia:

The extreme uniformity of the vegetation is the most remarkable feature in the landscape of the greater part of New South Wales... In the whole country I scarcely saw a place without the marks of a fire; whether these had been more or less recent—whether the stumps were more or less black, was the greatest change which varied the uniformity, so wearisome to the traveller's eye.[33]

The reasons as to why the Aborigines lit so many fires was, at one level made immediately apparent to many early visitors, yet at another level it was to remain unknown until recently. Joseph Banks had only one extended stay on the east coast of Australia. That was the enforced stopover at the Endeavour River, near present-day Cooktown, where the hull of the *Endeavour*, which had been holed by collision with a coral outcrop, was repaired. There, Banks saw Aboriginal fire put to an ingenious, and for Banks, unexpected use. Frustrated at the lingering presence of Cook's party, the Aborigines surrounded their encampment with fire and drove the Europeans into their boats before the flames and smoke. Cook's crew, terrified and discomfited, returned fire with gunfire.

The members of the First Fleet suffered similar assaults and John White, Surgeon General to the infant colony, records that when camped at a spot where the grass was:

*long, dry and sour, [he] set it on fire all around, for fear that
the natives should surprise us in the night by doing the same, a
custom in which they always seem happy to indulge themselves.*[139]

Despite these early experiences, the extent to which Aborigines
used fire as a weapon only became clear to the Europeans as settlement advanced. Thomas Mitchell preferred campsites which had
been recently burned, for he noted that:

*in the case of hostility, on the part of the natives [an attempt to
burn out the enemy] is usually the first thing they do.*[100]

Aborigines also used a 'scorched earth' policy to deprive expeditions of supplies and used fire to cover their own retreat after an
attack. Once the Europeans became established, the use of fire in
Aboriginal resistance became more subtle. Often, fires would 'get
away' at the wrong time of year and deprive pastoralists of feed for
their stock. For this reason, most pastoralists forbade the lighting of
fires by Aborigines on their properties.

In the early days of the Sydney settlement, even agricultural areas
were not safe from Aboriginal fire. In 1800 New South Wales' second
Governor, John Hunter, wrote that:

*fire in the hands of a body of irritated and hostile natives may
with little trouble to them ruin our prospects of an abundant
harvest [and] they are not ignorant of having that power
in their hands.*[67]

Despite the dramatic role that fire played in the conflict between
Aborigines and Europeans (and doubtless between various Aboriginal
groups themselves), only a tiny minority of fires would have been lit
for this reason. The vast majority of fires would have been lit by
Aborigines to obtain their daily subsistence, for fire was of para-
mount economic importance in releasing nutrients into Australia's
woodland and grassland ecosystems.

The full implications of the use of fire by Aborigines were first
realised in 1969, when Professor Rhys Jones published a brief yet

seminal article entitled *Fire-stick Farming.*[71] In it he listed the uses to which fire was put by Aborigines; including amusement, signalling, to clear ground to facilitate travel or kill vermin, hunting, regeneration of plant food for both humans and kangaroos, and expanding human habitat by limiting the extent of southern rainforest (which was largely unuseable by Aborigines).

Jones coined the phrase 'firestick farming' to describe the overall impact of the Aboriginal use of fire on the Australian landscape. His paper challenged the view that there was such a thing as a wholly natural ecosystem in Australia and, perhaps for the first time, mounted a serious challenge to the concept of *terra nullius*, for he saw Aborigines as farming the land, albeit through the use of fire. He also raised arguments concerning the role of fire in national parks, and of how fire there might be managed. These ideas are now familiar to many Australians, but in 1969 they were revolutionary.

Although Jones was the first person to bring the information relating to Aboriginal burning together, others had earlier seen parts of the picture. Thomas Mitchell saw the critical links more clearly than any of his contemporaries when he wrote in 1848:

> *Fire, grass and kangaroos, and human inhabitants, seem all dependent on each other for existence in Australia; for any one of these being wanting, the others could no longer continue. Fire is necessary to burn the grass, and form those open forests, in which we find the large forest kangaroo; the native applies that fire to the grass at certain seasons, in order that a young green crop may subsequently spring up, and so attract and enable him to kill or take the kangaroo with nets. In summer, the burning of the long grass also discloses vermin, bird's nests, etc. on which the females and children, who chiefly burn the grass, feed. But for this simple process, the Australian woods had probably contained as thick a jungle as those of New Zealand or America, instead of the open forests in which the white men now find grass for their cattle, to the exclusion of the kangaroo.*[101]

Elsewhere in Australia this interdependence of people, fauna, flora and fire was even more marked. A critically important aspect of firestick farming for people in the drier parts of Australia was doubtless the maintenance of diverse and abundant communities of medium-sized mammals such as bandicoots and wallabies. For the Aboriginal people of these regions, small mammals were the backbone of their economy. Yet, as will be explained in more detail below, a very specific fire management regime was needed in order to maintain the diverse vegetational mosaic that the medium-sized mammals favoured. If this fire regime were ever interrupted, vast wildfires could rage, obliterating both the vegetational mosaic and its mammals.

It is now time to consider how fire has transformed Australia's vegetation and soils. Our understanding of this aspect of Australian prehistory has recently undergone an enormous revolution. In order to comprehend its significance, it is necessary to recap briefly on how things were understood in the past.

Most Australians know that their land was once much wetter than it is at present. Many have doubtless heard that rainforests once grew in Central Australia and that the magnificent palms and cycads which can be seen in gorges near Alice Springs are relics of this distant age. By examining fossil deposits in central and inland northern Australia, palaeontologists have concluded that this vast drying out began some 15–10 million years ago and had affected most of the continent by two million years ago. They thus imagined that the first Aborigines stepped ashore on a dry continent, where rainforest was restricted, much as at present, to a few tiny refuges on the east coast.

Now new studies, which are still largely unpublished, show that this scenario is wrong, for we now know that until the last 100 000 years or so, rainforests blanketed vast areas of eastern Australia. They may also have been extensive across the north of the continent. These forests grew and flourished under conditions that were not much wetter than today's. Researchers have thus had to face the possibility that not only did the arrival of the first Aborigines dramatically alter Australia's fauna, but that it transformed its flora also.

By far the most important information concerning these events has come from four long sediment cores. Two were drilled in sediment that accumulated in the waters of the continental shelf off the Queensland coast. A third was taken from a small volcanic lake in an area now surrounded by rainforest on the Atherton Tablelands. The fourth was taken from the bed of Lake George, a large depression near Canberra which is erratically filled with water. All four cores show a rather similar trend. Although the Lake George core comes from the driest area it exemplifies some key changes best and so will be considered here first.

The sediment core taken from Lake George includes a record of changing plant communities and lake conditions over the last 700 000 years.[122] This long record has accumulated because the Lake George depression is slowly subsiding and as it does so, sediment accumulates. Preserved in that sediment is pollen and the remains of plants which reveal the nature of plant communities growing near the lake at the time, as well as samples of pollen blown in from more distant plant communities. Only the very top part of the core has been dated using the radiocarbon method, so the ages of the rest of the core have to be determined by indirect means.

The core reveals a fascinating picture of vegetation change, all of which is accounted for by a slowly altering climate; until one reaches the uppermost section. Then, at a point that some would argue is 120 000 years old and others 60 000 years old, an abrupt change takes place. Suddenly, the number of microscopic charcoal fragments present in the sediment increases dramatically. At the same time, almost all fire-sensitive species, including southern beeches (*Nothofagus* species), southern pines (*Podocarpus* species), tree ferns (*Cyathea* species), sheoaks (*Casuarina* species) and primitive plants such as *Lycopodium* abruptly vanish and are replaced with fire-promoting plants such as eucalypts. This abrupt change is unique in the 700 000-year-old history of the deposit and does not correspond to a change in climate.[122]

A dramatic increase in charcoal frequency is also seen in the sediment core recovered from Lynch's Crater, a lake situated near the western edge of the Atherton Tablelands in north-east Queensland, which today is surrounded by rainforest. In this case the increase in charcoal is thought to date to about 38 000 years ago, but again the precise dating of the change is uncertain.

Throughout most of the Lynch's Crater core, three basic plant communities alternate in dominating the site as climate has changed. In the driest times sclerophyll forest establishes in the area. In slightly wetter times rainforests rich in southern pines (particularly *Araucaria*) dominate, while in the wettest periods (such as today) rainforests typical of the Atherton Tablelands today take over.[75]

A dramatic change in plant communities takes place at about the time that the charcoal particle frequency increases in the core. The *Araucaria*-dominated 'dry' rainforests all but vanish from the record, even though climatic conditions are suitable for them. Indeed, one genus of southern pine (*Lagarostrobus*), actually becomes extinct in Australia, its last record being in the Lynch's Crater core before charcoal particles increase. Dr Peter Kershaw of Monash University, who has studied this core, suggests that an increase in fire tipped the balance in favour of sclerophyll forest species and against the fire-sensitive southern pines.[75]

The final evidence comes from two recently obtained sediment cores taken from the continental shelf east of Townsville and Cairns. Because sediment is accumulating continuously on the ocean floor, these cores cover an enormous 10 million year period of Australian history.[88] Although they are not yet fully analysed, they show that over this time rainforest (both dry and wet types) has dominated the east coast, that sclerophyll forest was very limited in extent and that eucalypts were rare.

One of the major preliminary findings is that the cores reflect an extraordinary stability in vegetation communities, for throughout the entire 10 million year period there are few changes in east coast plant communities as reflected by the pollen preserved in these cores. This may be because the cores accumulated pollen from a large section of coast. This would mask small-scale change, but give an accurate idea of vegetation change on the grand scale.

There is one exception to the emerging picture of vegetational stability, for in the very last fraction of time, which researchers tentatively think may represent the last 100 000 years or so, a dramatic shift occurs. The pollen of the southern pines dwindle rapidly (as they do at Lynch's Crater) and the pollen of sclerophyll forest plants become abundant. Curiously, the pollen of mangroves become an important component of the assemblage for the first time.[87] As with the other cores, none of these changes can be related to a changing climate.

Although charcoal particles have not yet been reported from the cores, these dramatic and unprecedented changes intimate that—possibly around 100 000 years ago, but maybe as little as 38 000 years ago—fire had become important in Australian environments. At this time the fire-sensitive southern pines and the 'dry' rainforests which they towered over are abruptly lost, while the fire-promoting sclerophyll forest species take their place. The surge in mangrove numbers is fascinating, for it suggests that massive amounts of sediment was building up at river mouths, providing the muddy flats that mangroves need to thrive. This is consistent with fire baring the soil, much of which was lost through erosion into the catchments. Such events have happened in recent times, such as when ecologists feared vast sheet erosion in the wake of the Sydney bushfires of 1994. But this is the first clear evidence of such events on the large scale from the prehistoric period.

These long sediment cores provide clear evidence that until recently, rainforests—particularly 'dry' rainforest communities—were a dominant element of eastern Australian plant communities. Today, as a result of fire, they have all but vanished, while 'wet' rainforest communities have been banished to a few tiny fire refuges.

Evidence that rainforest communities were also once widespread on the western slopes of the Great Dividing Range comes from the plant fossils preserved at Lake George. But there is other striking evidence—this time from modern plant communities—that rainforest plants once thrived in quite dry areas west of the divide.

Brigalow is a spindly, nondescript kind of acacia which forms dense scrubs on fertile, clayey soils in inland northern New South Wales and Queensland, where rainfall is between 500 and 750 millimetres per year. Brigalow scrubs are fascinating because of their fire-resistant properties.[112] Almost everything, from the way the dead leaves fall to the ground, to the form of the trees themselves, seems to be designed to exclude fire. This fire resistance makes them almost unique among the vegetation types growing in the drier parts of Australia.

Strange plant communities known colloquially as 'bottletree scrubs' and 'bonewood scrubs' can be found growing in the middle of patches of brigalow scrub and in other, specially fire-protected areas. Hugh Lavery described them as having:

*a general atmosphere of mild gloom and darkness as in a
rainforest, except that everything is much drier.*[81]

These plant communities are quite diverse and include many of
the plant genera which typify rainforests in Australia today. They also
support mammal species—such as mosaic-tailed rats (*Melomys
cervinipes*) and long-nosed bandicoots (*Perameles nasuta*)—most often
associated with rainforest at similar latitudes. In truth, these plant
assemblages are 'dry' rainforest communities surviving today in semi-
arid regions simply because fire is totally excluded by the extraordinary brigalow.

Throughout the north and east of the continent, other rainforest
relics can be found. Some are small patches growing on areas protected by topography from fire, while others are just a few trees,
perched among rocks or in other situations where they cannot be
burned. Even older relics, such as the palms and cycads of Palm
Valley near Alice Springs show how such rainforest plants can survive
long-term in tiny refuges. With the exception of these clearly older
relics, I think it likely that many of the isolated rainforest plants and
communities in northern and eastern Australia formed part of one
continuous 'dry' rainforest belt as little as 100 000 years ago.

To return to the sediment cores and their interpretation. They
clearly show that the role of fire in Australian ecosystems changed
dramatically at some time in the recent (in geological terms) past.
Precisely when this event happened is uncertain, with various dates
suggesting that it occurred between 100 000 and 38 000 years ago.
As explained earlier, there is uncertainty over when humans arrived in
Australia, but the best guess is that they arrived between 60 000 and
40 000 years ago.

It is not unreasonable to suggest that there may be a link between
the appearance of people and this change in vegetation. But what,
precisely, was the nature of that link? The Aboriginal system of firestick farming had clearly evolved over tens of thousands of years and
had resulted in a new equilibrium being established in Australian
ecosystems. Could early firestick farming have caused the rapid

change in plant communities seen in the fossil plant record?

The most popular of various theories put forward to explain how fire was able to alter its role so dramatically in Australian ecosystems suggests that when the Aborigines arrived in Australia they began lighting fires, and by increasing the frequency of fire they encouraged the growth of fire-loving plants.

There are some very serious problems with this simple hypothesis. The major one concerns the ability of Aborigines to increase the importance of fire simply by lighting more blazes. Research has shown that there are more than enough natural ignitions (such as lightning strikes) over most of Australia to consume the standing crop of fuel before it can be reduced by decomposition or herbivores.[112] Under such conditions, the lighting of more fires simply increases the frequency and lowers the intensity of fires but does not result in the burning of more material. Indeed, the lighting of more fires can, paradoxically, downplay the effects of fire. This is because the standing store of fuel is burned before there is enough to support really hot and devastating blazes.

How, then, can the remarkable change in fire frequency in prehistoric Australia be explained? Because the natural fire frequency is so high, the only way that the consumption of plant matter by fire could be increased is through increasing the standing fuel load. Something must have happened, therefore, that left more combustible plant matter lying around. While an increase in rainfall could accomplish this, it would also militate against fire. Furthermore, rainfall has increased and decreased many times in Australia's past without producing such an effect. Further, as noted above, the change seen in the sediment cores does not correlate with a suitable climate change.

There is only one other conceivable way in which the standing fuel load could have been increased. This is through the accumulation of vegetation that would normally have been recycled through the guts of large herbivores. I suspect that it was this change, not increased fire lighting by Aborigines, which holds the answer to the puzzle.

On the African savanna today it is well documented that herbivores consume vast amounts of vegetation and thus alter vegetation communities. Imagine what would happen in a place like the Serengeti Plains if all of the large herbivores were removed. Imagine also that the numbers of the surviving medium-sized species were kept so low that they had no significant impact on the vegetation. What would happen? Without doubt, the uneaten vegetation would build up dramatically. Given the right conditions, vast wildfires would rage.

I think that there is good evidence that something very much like this happened in Australia some 38 000 or more years ago. We have seen evidence of great extinctions which carried off every large herbivore in Australia and which suppressed the populations of the few surviving medium-sized species. In effect, in Australia the medium- and large-sized herbivores became ecologically insignificant as consumers of vegetation. This event must have left vast amounts of uneaten vegetation on the landscape. I feel that it is in these piles of vegetation, not in the lighting of fires by Aborigines, that the genesis of Australia's recent fire history is to be found.

If this scenario is correct, a few additional facts must be explained. Where, for example, did Australia's fire-loving plants come from? And why did megafaunal extinction not produce such dramatic effects elsewhere in the world? The answers to these questions can be traced back in part, I think, to those formative influences on the Australian environment: ENSO and poor soils.

The riddle of the origins of Australia's fire-loving plants is relatively easy to answer, for there is one Australian environment where fire seems to have long played a formative role. This is the heathlands. Pollen and charcoal sequences preserved in sediments which accumulated in Darwin Crater, in the west of Tasmania, demonstrate that fire has been frequent in the heath that grows near the site for several hundred thousand years.[26] This is long before humans arrived in Australia. Therefore, such fires must have resulted from natural causes.

But why, of all communities, should heathlands support fire? Australian heaths grow on some of the continent's poorest soils. The plants that grow there must conserve their nutrients very carefully and cannot afford to have their leaves eaten by herbivores. Thus they laden their tissues with chemical defences. In addition, because of the poor soils, the leaves of heathland plants are not very nutritious to start with. This combination of factors means that heathlands generally cannot support large herbivores.

Without herbivores to consume dead plant matter, fire can play an enlarged role as a consumer of plants and so, in these most nutrient-poor of environments, a slow coevolution between fire and heathland plants has been taking place for millions of years. As a result, heathland plants show a remarkable array of adaptations to coping with fire. Some are post-fire specialists. A great many others, including *Banksias* and *Hakeas*, have woody seed pods that release seed only after they are burned. Others have underground rhizomes or lignotubers, where they store energy and nutrients which are used to resprout after a fire has passed through.

Because of its poor soils, Australia has always had large areas of heathland; more indeed relative to land area than the other continents. Fossils show that it has been in existence for millions of years. With the extinction of the megaherbivores and the subsequent spread of fire, it seems likely that many species of fire-promoting plants that evolved in these communities escaped from their heathland cradles. They, along with the plants preadapted to fire from other habitats, were set to conquer a continent.

Prior to this, much of Australia supported more diverse plant communities than exist in many areas today. Even away from rain-forests, plant assemblages included many fire-sensitive species such as southern pines and tree ferns. They could exist in such areas because fire was then rare. By removing the fire-sensitive species, fire forced a simplification of plant communities.

Along with simplification came a greater segregation between fire-sensitive and fire-tolerant plant communities. Today, the boundary between these plant communities is often extraordinarily sharp, particularly in northern Australia, with the fire-sensitive plants being found only in a few refuges. These sharp boundaries are forged by fire, for the factors that exclude fire often have very sharp margins. They might be, for example, a change in slope or soil-type, or a water course.

The implications of these fire-induced changes for life in Australia were profound, for they would reinforce the effects of ENSO and old soils, acting to further impoverish Australian environments. In order to understand how this happened, we must go back a moment to the Australia of the diprotodon, or to the Serengeti Plains of Africa, for it is there that we can see how great is the interdependence of plant and animal in undamaged ecosystems.

In the Australia of 100 000 years ago there were over 50 species of medium to large specialised herbivorous marsupials. There were also several gigantic herbivorous birds and turtles. Each species would doubtless have had its own favourite food type. Working in concert, their browsing and grazing probably maintained a complex vegetational mosaic which supported all of them and allowed a diversity of plants to coexist.

What is more, whenever an animal ate a plant, the nutrients were returned quickly to the vegetation, for within a day or so they would reappear, composted and laced with nitrates, in the form of dung. If the large animal communities that exist elsewhere are any guide, this dung would have been the lifeblood of guilds of now-extinct Australian dung beetles. Fighting avidly for their share, the dung beetles would have buried and consumed the droppings, hastening the recycling of nutrients to the plants.

In such an ecosystem nutrients can be recycled spectacularly quickly. Thus, even though the soil may be relatively poor, the rapid turnover of nutrients compensates. Because rapid turnover of nutrients is critical to the success of the system, it is not in the plant's interest to lace its leaves with toxins which would inhibit herbivores, for it is far better to keep the nutrients moving. Even more critically, very few nutrients are lost in this process. It is a tight, fast and self-contained nutrient-cycling system.

When compared with coevolved guilds of large herbivores, fire is a far inferior way of recycling nutrients. It promotes plants that originated in the nutrient-starved heaths. There, plants must lace their leaves with chemicals in order to defend from browsing herbivores the few nutrients which they have accumulated. These toxins may also inhibit the breakdown of plant matter by decomposition, so nutrients are recycled much more slowly than in other environments, being released from dead plant matter only by fire.

Because of this, extremely poor soils promote a nutrient-hoarding strategy, which in turn encourages fire. Even worse, when fire does finally consume the plant matter, making its nutrients available to living plants, the nutrients leak out of the system. It has been estimated, for example, that for every hectare of grassland burned in the Katherine region of the Northern Territory, four-and-one-half kilograms of nitrogen is lost as nitrous oxide due to combustion.[69] On Fraser Island it has been calculated that between 30 per cent and 51 per cent of sulphur is lost through volatilisation from sclerophyll forest as a result of fire.[70] Other nutrients are lost because they are converted into inorganic compounds in the ash and, if heavy rain follows fire, any remaining nutrients are easily washed into watercourses and carried off. Worse, the nutrients are not alone in being vulnerable to loss through water transport. For after a fire has bared the soil wind can strip it away in massive sheet erosion.

The situation can be summarised thus. Large herbivores return nutrients to the soil quickly and with a bonus. Fire returns nutrients to the soil only after a long period—and then at a considerable loss. As a result, fire and poor soils act to promote each other. Together, they can produce an ecosystem which is spiralling ever downwards as nutrients become fewer while fires become more important.

The result of this cycle is an accelerated selection for scleromorph plants, which can survive in nutrient-poor soils. A self-reinforcing cycle of soil impoverishment, soil drying and soil exposure is then initiated. Much water is lost through runoff in such situations before it can be returned to the skies through transpiration. This lowers effective rainfall. Degradation can go so far that even if fire can be stopped, the soil is so impoverished that it can no longer support the kinds of plants needed to feed large herbivores. Thus, the change can be made almost irreversible.

Furthermore, the presence of fire-loving plants can make the re-establishment of fire-sensitive species impossible, for the fire promoters can cause blazes that will kill the fire-sensitive species before they become established. As a result of these changes I have no doubt that fire has made Australia—originally the most resource-poor land—an even poorer one. Fire saw the fat of the land slowly flushed onto the floodplains and into the estuaries, where today it supports swamp and mangrove. It is no accident that such areas supported the greatest density of Aboriginal occupation at the time of

European settlement. The bones of the land, along the ranges and away from the sea and rivers, in contrast, supported tiny populations.

As dramatic as these results of fire are, they are not the end of the story, for recently geologists have found evidence that even Australia's climate may have been altered as a result of fire. Dr Gifford Miller of the University of Colorado has been studying the conditions that prevail when Lake Eyre is filled with water.[98] The water that fills Lake Eyre falls as rain far to the north in the catchments of the Cooper Creek and Diamantina River. He found that if the northern Australian monsoon is strong enough, then rainfall extends far enough south in northern Australia to fill the catchments. Months later, the water which fell as rain a thousand kilometres to the north finally reaches the lake, which slowly begins to fill.

For most of the last few million years the occurrence of 'lake full' conditions have corresponded well with periods of strong monsoon activity worldwide, but 11 000 years ago a marked anomaly occurred. The only time over the past 50 000 years that was favourable to fill Lake Eyre occurred 11 000 years ago. Then, for a brief period, conditions were excellent, yet the lake remained dry. How, Miller asked, could this have happened?

The answer seems to lie in the nature of the vegetation that has covered northern Australia over the past few million years. A canopy of broad-leaved 'dry' rainforest species, such as survives in tiny fire refuges across the north of Australia today, could, if they were more widespread, enhance rainfall by up to 60 per cent and push rainfall much further south. This is because the plants and the soils they protect retard the runoff of water. Through the leaves in their dense canopy they release vast amounts of the trapped water as moisture into the atmosphere. During the wet season, the winds blow in from the coast. As a result, the moisture transpired by the plants is formed into clouds and blown southwards to fall again as rain.

This appears to have been the situation that occurred in northern Australia before the last ice age. Then, with ice age aridity, a dry-adapted flora established across the north. At the end of earlier ice ages the dry-adapted plant communities seem to have given way to 'dry' rainforest plant communities, but at the end of the last ice age eucalypt woodlands established instead. As suggested earlier, I think that this change resulted elsewhere in Australia from a change in fire regime. This, in turn, resulted from megafaunal extinction caused by human hunting.

As a result, most of northern Australia is covered with eucalypt woodlands today. After rain, the water drains rapidly away, for the plants and thin soil cannot hold it. The release of moisture to the atmosphere through the narrow eucalypt leaves is insufficient to form significant clouds. As a consequence, the rainfall gradient between coast and inland is incredibly steep in northern Australia. Without the help of broad-leaved plants, the rainfall does not normally penetrate far enough south in sufficient quantity to fill the drainages that flow into the lake.

Miller's work has received support from a recent study of variability in the runoff of water in Australia.[40] The study found that Australia has the greatest runoff variability of any continent. Much of this is due to ENSO, but the variability was found to be even higher than ENSO alone could account for. The researchers found that the additional variability which they detected was due to the scleromorph plants. The water which they allow to escape as runoff, rather than use in transpiration, was the origin of the additional runoff variability, continent-wide.

The idea that before the dramatic change in Australia's fire regime, Australia's northern coast supported various 'dry' rainforest communities, just as the east coast did, must remain speculative. This is because, so far, no sedimentary cores containing pollen and charcoal, like those found along the east coast, have been discovered in northern Australia.

Miller's findings are quite extraordinary for, combined with the ideas developed here, they suggest that the extinction of the megaherbivores may have altered the climate of an entire continent. Such enormous climatic change has never before been postulated to have resulted from an extinction event.

The reasons why Australia seems to have been uniquely affected by the rise of fire among the world's continents are probably to be found in three factors. The first is that the Australian environment was preadapted for such change. Because of its poor soils, it already supported extensive and diverse fire-loving plant communities that could take advantage of the changes brought about megafaunal extinction.

The second factor concerns the nature of the extinction event itself. The extinction of Australia's large mammal fauna was much more extensive than that which occurred on any other continent. If we compare the number of genera of large mammals lost on the various continents, we find that Australia lost 94 per cent, North America 73 per cent, Europe 29 per cent and Africa south of the Sahara five per cent. It is possible that sufficient large herbivores survived on the other continents to continue to play an important role in recycling plant matter in most environments.

The final factor that may have uniquely affected Australia is time. The Australian environment has experienced at least 40 000 years of adaptation to the lack of a megafauna. In North America and Europe the major extinction events happened a mere 14 000–11 000 years ago, while on other landmasses they occurred even more recently. Australian environments have thus had far more time than environments elsewhere to adjust to fire-dominated regimes.

There is no doubt that the environmental degradation resulting from fire, affected Aborigines as well as all the other living things. It seems likely that following the extinction of the megafauna there was little that Aborigines could have done to entirely avoid the consequences. Firestick farming was doubtless the best way to mitigate the effects of fire, for by lighting many small, low-intensity fires the Aborigines prevented the establishment of the vast fires that stripped soil and nutrients most dramatically. They also prevented the loss of small remnant patches of rainforest in places like the Northern Territory, which provided critical resources at certain times. Most importantly, perhaps, firestick farming supported the diverse communities of medium-sized mammals that Aborigines were so dependent upon for food. The story of the evolution of these animal communities and their eventual extermination is a fascinating one. It will be examined in the following chapter.

CHAPTER 22

WHO KILLED KIRLILPI?

The Warlpiri Aborigines of the Central Australian desert knew the desert bandicoot (*Perameles eremiana*) as *kirlilpi*. *Kirlilpi* is just one of 23 native mammal species to have become extinct in Australia since the arrival of the Europeans. This amounts to almost 10 per cent of the Australian mammal fauna in existence in 1788. Tragically, many of these extinctions have occurred in the last 30 or 40 years, at a time when growing environmental awareness might have helped to save them. These historic extinctions of Australian mammals are the worst that the world has experienced in the last 500 years, accounting for just under one-third of all mammal extinctions that have occurred worldwide over that time.

A decade ago just about everyone was certain that they understood the cause of these extinctions. The standard line went that the 'primitive' marsupials had given way to the 'superior' placental predators and herbivores, such as the fox and rabbit, that Europeans had brought with them from their homeland. But evidence has slowly been accumulating which suggests that this assessment was wrong.

Perhaps the most compelling piece of evidence that tells against the old theory concerns the peculiar specificity of the extinction event. It carried away virtually all of the mammals that weighed between 50 grams and five kilograms which inhabited the drier regions of Australia. But it did not affect similar-sized rainforest-dwelling species and few of those inhabiting the wet sclerophyll forests. Neither did it affect the larger nor smaller mammals of the drier habitats, nor birds, reptiles or frogs. Indeed, some areas where the middle-sized mammals have suffered most still support the most diverse lizard assemblages to be found anywhere on Earth.

If foxes, cats and domestic stock alone were responsible for these extinctions, it is hard to understand why the impact fell solely upon the middle-sized mammals in the drier regions, for foxes and cats are widespread and both take a wide variety of prey, including birds and reptiles. In other parts of the world feral animals have exterminated a wide range of vertebrates, from wrens to crested pigeons and snakes and lizards.

Interestingly, it was not just the 'primitive' marsupials that were affected by extinction, for over half of the now-extinct species were native rodents of the family Muridae. This family contains the most successful of all mammals—the rats and mice that have adapted to life in human settlements and have spread to almost every landmass on Earth.

As these anomalies have become more widely known, researchers have begun looking farther afield for explanations for these peculiarly specific extinctions. The most critical work in this regard has been done by Dr Ken Johnson and his colleagues at the Arid Zone Research Institute near Alice Springs, and Dave Gibson of Conservation and Land Management, Western Australia.[51] They have been able to demonstrate that the extinction of middle-sized mammals occurred in areas where domestic stock never reached and where foxes are virtually unknown. They have also shown that many extinctions happened as late as the 1960s and that they followed the departure of Aboriginal people from their tribal lands.

They suggest that Aboriginal firestick farming was an important factor in maintaining suitable conditions for the middle-sized mammal species. This was because Aborigines continually burned small patches of habitat. This created a mosaic of old, dense, unburned vegetation which provided shelter, with newly burned patches which were rich in newly sprouting leaves and shoots. This mosaic attracted a wide variety of herbivorous species. Given such conditions, the middle-sized mammals are better able to withstand the effects of drought, predation and competition for resources from introduced species.

Also vitally important was the fact that, because the country had been patch burned, widespread wildfires could not take hold. When firestick farming ceased in Central Australia, huge amounts of vegetation built up over the entire country following good rains. This supported vast wildfires. In 1974–75 they burned 120 million hectares in the arid zone. While such fires never burn everything in their path, inevitably leaving a few small islands unburned, they effectively destroy the mosaic of variously aged plants necessary for the survival of the middle-sized mammals.

This new hypothesis explained many things that the earlier ones could not. In particular it explained why the middle-sized mammals of the arid zone were so vulnerable to change. The larger species, such as the red kangaroo, along with the birds, are able to travel long distances. When a fire burns their habitat they simply move on. The very tiny mammals can find refuge in the few unburned patches left by the fire. The reptiles cannot move, but they can aestivate (remain inactive) for months under the ground or in termite mounds until some vegetative cover returns.

These options are not open to the middle-sized mammals. They are too small to migrate and must eat and find shelter daily. Furthermore, they need considerable habitat areas in order to meet their daily requirements of food and shelter.

Other researchers have found that most of the damage to populations of middle-sized mammals is not done during a fire, but after it. Their studies have shown that the majority of middle-sized mammals in drier areas survive the initial blaze. The real problem occurs later, when they must forage exposed among the ashes for the few remaining food items. Predators know that animals are vulnerable at this time and they flock into newly burned areas. Such areas—and the unburned bush surrounding them—can briefly support enormous numbers of foxes, cats and birds of prey. Not only do they destroy the native middle-sized mammals that inhabited the burned area, but through force of numbers they can destroy most individuals in the surrounding unburned habitat as well, thus preventing recolonisation of the burned area.

Interestingly, it is not only introduced predators that do this, for one study of brush-tailed bettongs following a fire in the south-west of Western Australia showed that about half of the population fell victim to the western quoll (*Dasyurus geoffroyi*), a native marsupial predator, while the others were consumed by foxes. The large numbers of foxes and cats that exist in many habitats in Australia, however, greatly add to the problem.

The elegant argument put forward by Johnson and others[51] also explains why the species of the arid zone were affected so much more severely than those of the wetter regions. In the flat, arid zone, there are few topographic features to prevent a fire burning a vast area to ashy uniformity. The more hilly east coast and south-west make it more difficult for fire to burn everything in a given area, for ridges, gullies and waterways often provide refuges for considerable stands of vegetation.

While this hypothesis has great power in explaining the causes of the extinctions of the middle-sized mammals of Australia's drier regions, it begs a very important question concerning the origins of firestick farming. The fossil record shows that virtually all of Australia's middle-sized mammals have been in existence for a million or more years, which is far longer than humans have been present in Australia. If firestick farming was so crucial to their existence, how did they survive before Aborigines began their burning practices?

The answer lies, I think, in the ecology of Australia's long-vanished megafauna. Before humans arrived, the 60 or so species of large vertebrates then inhabiting Australia were the main consumers of vegetation. They would also have consumed sufficient plant matter to prevent large fires most of the time.

After their extinction, fire became the main consumer of Australian vegetation. Without human interference the fire pattern in Australia would probably have been one of vast, periodic wildfires that ravaged huge areas of the continent. Indeed, this is precisely the kind of fire regime that has emerged over much of the continent since Aboriginal firestick farming ceased. Such a fire regime would, 40 000 years ago, have led to the demise of the middle-sized mammals. This would have been a terrible blow to the Aboriginal economy, particularly in the arid and semi-arid regions, where middle-sized mammals were the lifeblood of the various tribal groups.

It seems entirely possible that firestick farming initially evolved as a response to the threat that the natural fire regime posed to the middle-sized mammals. If so, it was a highly effective response, for it maintained an assemblage of more than 20 middle-sized mammal species over much of the continent for over 35 000 years. But it was a fragile system, entirely dependent upon the understanding and actions of a people who had coevolved with it. The arrival of a new people, with no understanding of the ecology of their new homeland, was to prove the undoing of 35 000 years of conservation effort.

CHAPTER 23

WHEN THOU HAST ENOUGH
REMEMBER THE TIME OF HUNGER

The preceding chapters may suggest to some that the history of the 'new' lands is an unending chronicle of the destruction of ecosystems by people and that people are all-powerful in the struggle against nature. But this is far from being true, for from the very moment that humans set foot in the new lands, the unique and terrible forces that had shaped the lands themselves began to shape the people and their culture. This process of coadaptation and coevolution eventually absorbed the newly arrived humans into the complex web of a functioning and integrated ecosystem.

People have been a feature of the ecosystems of Meganesia (Australia, New Guinea and Tasmania) for at least 40 000, and possibly 60 000 years. This almost unimaginably long history has led to a highly coevolved and complex relationship between Australian Aborigines and New Guineans and their ecosystems. In contrast, the first inhabitants of Tasmantis arrived in New Caledonia only 3500 years ago. They only arrived in New Zealand 1000–800 years ago.

This difference in timing gives us a splendid opportunity to examine the way that humans become reconciled to their environments. The short history of the Tasmantans allows us to see the future eaters at the low point in their history, at the time of ecosystem collapse and resource shortage following catastrophic resource overexploitation, while the longer history of the Meganesians tells an intriguing story of long-term human accommodation to a unique land. Because of the pioneering work done by New Zealand's archaeologists and because of a unique written record by prominent Maori and early European explorers, New Zealand offers by far the most complete history of any of the lands of Tasmantis. And it is here we will begin.

The first Maori—the moa hunters—were lucky people. They appear to have brought few diseases with them, for the weeks or months they spent aboard their canoes travelling from the north-east would have acted as an effective quarantine. When they arrived in New Zealand they found a land packed with birds, seals and fish that had never seen a human before. There was no need to garden extensively and hunting was surely the most leisurely of activities. Unlike many Polynesians, the Maori did not have bows and arrows. Indeed, the moa hunting sites are conspicuous for the absence of specialised hunting weapons of any kind. There is also no evidence of weapons of war or fortifications. It seems that there was such a superabundance of food and other resources just waiting to be harvested that there was little for people to fight over.[1]

The first Maori may well have carried pigs with them to New Zealand, for Polynesians carried pigs throughout the rest of the Pacific as far as Hawaii. They also probably arrived carrying chickens, which are known as *moa* throughout Polynesia. Curiously, both pig and chicken were absent by the time that Cook landed in 1769 and the bones of neither species is found in the archaeological record. It seems likely that, having seen a 'real' moa, the Maori did not bother keeping or caring for chooks or pigs and that their domestic animals were all eaten shortly after arrival.

As sensible as that action may have seemed at the time, the loss of domestic animals was to have a major impact on future generations, for the only surviving domesticated species possessed by the Maori was the dog. Had they retained pigs or chickens, the time after the moa ovens grew cold may have been much less unpleasant for the descendants of the moa hunters.

With the extinction of the moa, the Maori were forced to rely upon other, more difficult-to-obtain resources. The smaller birds bore the brunt of the hunting effort on land and many either became extinct or suffered dramatic range retractions. The huia, for example—that remarkable bird in which the beaks of the male and female are so different in shape—was nearing extinction by the time of European contact, surviving only in a tiny portion of its former range. Likewise the takahe, that gigantic relative of the swamp hen, had vanished from everywhere except for a few lonely valleys in

Fiordland, while the meaty kakapo, formerly the most common of birds, had been exiled to the remote mountains and remained tolerably abundant only in Fiordland.

New Zealand's marine mammals had also suffered a dramatic decline due to overhunting.[3] Prior to the arrival of the Maori at least three species of seals had bred throughout the length of the New Zealand archipelago and several others were regular visitors. The three tonne elephant seals were early victims of overexploitation and their breeding colonies vanished quickly. They were followed rapidly into local extinction by the half-tonne sea-lion (*Phocarctos hookeri*). But the smallest species, the 200 kilogram New Zealand fur seal (*Arctocephalus forsteri*), hung on—just. By 1600 AD all breeding colonies of fur seals on the North Island and northern South Island had been exterminated. By 1800 AD they had gone from everywhere north of Dunedin on the east coast and Fiordland on the west, only those in the most remote south-western corner of New Zealand surviving.[3] There is no doubt that had things continued as they were, they would soon have been eliminated from this last icy refuge by Maori hunters.

Even fish populations were affected by overexploitation. Snapper (*Chrysophrys auratus*), for example, was fished early on in the South Island, but then vanishes from the archaeological record, suggesting a local extinction. The average size of snapper declines throughout its distribution in New Zealand with time, suggesting that the fishery was being overexploited by the Maori. Interestingly, the average size of fish hooks also decreases, although whether in response to a decrease in fish size, or a shortage of moa bone (the most important raw material for making hooks, which was becoming scarcer as the time since the moa extinction elapsed) is not clear.

It might seem scarcely believable that the Maori could overexploit such a difficult-to-obtain resource as snapper, but archaeologists have estimated that the northern Northland Maori would have taken about 1200 tonnes of snapper per year. The modern commercial fishery, with its 100 vessels, lands only 1000 tonnes per year and this probably exceeds the maximum sustainable catch, as stocks have been declining since the 1980s.[3] Even such humble resources as shellfish show signs of overexploitation, for as more desirable foods vanished people turned increasingly to less desirable food sources. Large limpets (*Cellana denticulata*) all but vanish from the Coromandel Peninsula and many other mollusc species decline in size throughout New Zealand.[3]

While humans had been devastating the larger land-based species, the Pacific rat (*Rattus exulans*) or kiore as it is known in New Zealand, was wreaking havoc among the smaller fauna that people might eventually have turned to in order to sustain themselves. The largest frogs, the tuatara and some of the smaller birds, were doubtless their victims, for many such species survive today only on rat-free islets.

By the sixteenth century the human population of New Zealand had been growing exponentially for at least 300 years, as a result of making an easy living from seals, moa and other flightless birds. When these had been eaten out, they were left with frighteningly few resources. Their agriculture was based essentially on tropical species which grew best in the warmer parts of the North Island. The loss of chickens and pigs left them without suitable domesticates upon which to base an economy centred upon animal husbandry, while the kiore had destroyed the smaller wild game. Never again could people congregate in such numbers—and so unprotected—as in the great villages of the moa-hunting days. Throughout much of the country people were forced to live in small communities, for that was all that the land could support.

By the fifteenth century the Maori living in the north had begun to build great *pa*, or forts. One function of the *pa* was to protect stores of sweet potato that would grow in the milder climate there, but which had to be protected over winter. Another, more fundamental function was to protect people themselves from predation by other hungry humans. Because of the lack of resources, the sweet potato, a humble staple elsewhere, became an exalted food in the north of New Zealand. And this despite the fact that most tubers never exceeded the diameter of a human finger![76] For want of alternatives, people were forced to survive upon the root of the bracken fern—reckoned inedible or a famine food elsewhere—but prized in New Zealand. By the early historic period it had become the staple crop of the Maori, even though in the south it was rarely of edible quality. There, the sugary base of the New Zealand cordyline

(*Cordyline australis*) was relied upon as a major food.[3] Everywhere, the flesh of dogs and rats was highly valued, while the declining fish and shellfish resources provided most of the protein.

Some remarkable adaptations seem to have developed in response to these dire shortages. One, which may or may not have been a deliberate strategy, resulted in the transformation of New Zealand ecosystems. At some time after colonisation, the Maori began to burn the forest which at that time almost completely covered the New Zealand hinterland. Fire was to play a rather different role in New Zealand to that which it played in Australia. In both places fire had been present for millions of years, but in Australia there are sufficient natural causes of fire ignition (lightning etc.) to consume the standing crop of fuel. This does not appear to have been the case in New Zealand. There, human-lit fires are at least 10 times more common than natural ones. Therefore, in New Zealand, control of fire lay in the hands of people. It did not escape to dominate a land as it did in Australia.

The Maori practice of setting forest fires lead to catastrophic soil erosion and biological impoverishment. As disastrous as this was for native vegetation, animals and soil, it was probably beneficial to the Maori. This is because after the extinction of moa and other large birds and the destruction of the smaller fauna by the rat, the mountain forests and scrubs—once a place of enormous resource richness—contained little that people could eat.

Fern root of a quality that can be eaten by people, only grows on the better soils. Soil erosion encouraged by fire stripped the hills of their nutrients and transported them into the valleys. There, on the lowland flats, the precious root of the bracken fern could grow to edible size. Furthermore, fire encourages the growth of the cordyline and it had assumed a critical importance as a food source in the south. Soil erosion would also have released nutrients into the estuaries, where they became available to larval fish and other marine organisms, which were the main source of protein for the Maori after the extinctions on land.

But such desperate and wasteful strategies cannot continue for long—even in a rich land like New Zealand—for soil and forest are not inexhaustable. By the seventeenth century the Maori, their options limited by the decisions of the past, were facing economic and cultural crisis. Driven by hunger, the peaceful lifestyle of their Polynesian ancestors was under a threat as dire as was their food sources. The land of milk and honey once possessed by their ancestors was, as the Maori would say 'Ka ngaro, i te ngaro, a te Moa'—lost as the moa is lost.

By the time Europeans first sighted New Zealand, the resource crisis was in full swing. By the early colonial period it had assumed disastrous proportions. We know this because we are extraordinarily fortunate in having some written works which describe how things were in New Zealand before European contact. And by the early colonial period, we have an abundance of accounts, by both Maori and Pakeha, which document life as it then was.

The first European account of New Zealand dates to 1642, nearly 200 years before the colonial period. It gives a unique and fascinating insight into Maori society at the time. On 18 December 1642, the Dutch navigator Abel Janzoon Tasman sighted New Zealand, a mere three weeks after having been the first European to set foot in Tasmania. He recorded his only prolonged meeting with the Maori as follows:

On the 19th dº early in the morning a boat of these people with thirteen heads in her came within a stone's throw of our ship. They called out several times, which we did not understand... As far as we could see these people were of average height but rough of voice and build, their colour between brown and yellow. They had their hair tied back together right on top of their heads, in the way and fashion the Japanese have it, at the back of their head, but their hair was rather longer and thicker. On the tuft they had a large, thick white feather... The skipper of the Zeehaen sent his quartermaster back to his ship with the cockboat, in which six rowers, in order to instruct the junior officers not to let too many on, should the people want to come on board, but to be cautious and well on their guard. As the cockboat of the Zeehaen was rowing towards her, those in the canoe nearest us called out and waved their paddles to those lying behind the Zeehaen but we could not make out what they meant. Just as the

> *cockboat of the Zeehaen put off again, those who were lying in front of us, between the two ships, began to paddle towards it so furiously that when they were about half way, slightly more on our side of the ship, they struck the Zeehaen's cockboat alongside with their stern, so that it lurched tremendously. Thereupon the foremost one in the villain's boat, with a long, blunt pike, thrust the quartermaster, Cornelis Joppen, in the neck several times, so violently that he could not but fall overboard. Upon this the others attacked with short, thick, wooden clubs and their paddles, overwhelming the cockboat. In which fray three of the Zeehaen's men were left dead and a fourth owing to the heavy blows, mortally wounded. The quartermaster and two sailors swam towards our ship and we sent our shallop to meet them, into which they got alive. After this monstrous happening, and detestable affair, the murderers left the cockboat drift, having taken one of the dead in their canoe and drowned another.*[130]

This is hardly the kind of welcome that the moa hunters would have extended to such interesting visitors. That Maori living in the seventeenth century would initiate an unprovoked attack upon the first strangers to touch New Zealand's shores for 400 years suggests that the archipelago had already been transformed from a paradise to a war zone. The retrieval of the victim's bodies (the only person taken to shore alive was seen to be decapitated there) further suggests that by 1642 cannibalism was already occurring. Certainly, by the late eighteenth century, the bodies of those killed in war were a prized source of food.

Incidentally, the implacable hostility evinced by the Maori prevented Tasman from landing and making more detailed observations of New Zealand and her people. Had he been able to observe the land, it is just possible that he could have provided us with a first-hand account of the very last of the moa and much more of great interest besides. As it was, he named the place Murderer's Bay and sailed away to the north in search of a landing place where he could safely obtain fresh water. He never found one.

Cook's accounts of his first contacts with various Maori groups on the North Island, written over a century later in 1768, read like one continuous and violent confrontation. Everywhere he saw the effects of war, including destroyed *pas* and he noted that people were always on their guard, weapons at the ready. Of one *pa* he said:

> the situation is such that the best engineer in Europe could not
> have choose'd a better for a small number of men to defend them-
> selves against a greater, it is strong by nature and made more so by
> Art.[27]

He witnessed cannibalism on this and subsequent visits and became embroiled in some violent acts. He, along with many other early European visitors, were greeted with the refrain '*Haere mai ki uta kia patua*'—quite literally 'Come ashore and be clubbed'. It was only when the superiority of the European armoury was conclusively demonstrated that the aggression ceased. Once friendly relations were established, however, one of the first things requested by the Maori was help in destroying their neighbours. They were willing to trade almost anything for such help and Cook commented:

> I might have extirpated the whole race, for the people of each
> Hamlet or village by turns applyed to me to destroy the other, a
> very striking proof of the divided state in which they live.[27]

We do not have to rely solely upon accounts written by Europeans to gain an impression of the state of Maori society between the seventeenth and mid-nineteenth centuries, for several important accounts of the times were written by Maoris themselves. The most extraordinary of all is the biography of the great chief Te Rauparaha (?1770–1849) written in Maori by his son Tamihana, who completed the work in 1869.[22] Tamihana was one of the first Maori to learn to write and he was witness to many of the events he records. His work is the *Beowulf* of Maori literature and deserves to be more widely known.

Tamihana records fascinating insights into Maori warfare before European weapons and foodstuffs greatly changed Maori life. His writings make it clear that the flesh of the victims of war were a valued resource. Indeed, it appears likely that without the protein-rich fuel that they represented, the many long and strenuous raids undertaken up and down the coast by chiefs such as Te Rauparaha would not have been possible. Human bodies were certainly the principal food source during these long voyages.

Tamihana's writings make it clear that warfare was endemic in New Zealand at the time and that any strategy would be entertained provided it resulted in victory. Clearly, Te Rauparaha was a man for his time, a dynamic and brilliant chief, leading his people from victory to bloody victory against their neighbours.

Here are just a few incidents recounted by Tamihana. The first probably took place about 1820, the night after Te Rauparaha won a great victory, but expected a further attack on the morrow:

When one of Tangahoe's children cried in the night Te Rauparaha commanded that it be slaughtered. He said to Tangahoe 'Strangle that child, for I am that child' and Tangahoe strangled it... The bodies of the slaughtered were covered in a heap until later, when the party would return and dismember them. Te Rauparaha and his sons at last arrived at the pa of Ngati Tama. There Te Rauparaha told how he had won the battle, saying that Tukataro, Karewha and ten other men were killed and that they should go back to eat the bodies. Then the Ngati Tama set off, with Te Rauparaha also going to show where the dead were and to get the body of Rangihaeata's child, who had fallen into the water from Topeora's canoe. Having found the body he carried it, so that the father might see it as he was grieving for his child. The Ngati Tama cut up the bodies of the slain and carried them to their pas to be cooked, as was the Maori custom.[22]

Slightly earlier, Te Rauparaha's relatives had been attacked in the following way:

a war party of 800 Nga Puhi led by Pomare and Te Waero, was at Hauraki and had invested the pa at Te Totara, the main stronghold of the Ngati Maru. There were 600 men within the pa, as well as 1800 women and children. Although firmly blockaded the pa did not fall, so the Nga Phui resorted to cunning and

*said to the chiefs of Ngati Maru: 'We should make an end to our
hostility; let us make peace'. The Ngati Maru chiefs agreed. When
the Nga Phui entered the stronghold they slew the Ngati Maru
and captured the pa.*[22]

In 1821 Te Rauparaha decided to build his *pa* on Kapiti Island, a
natural and impregnable fortress. From there he carried out numerous successful raids. In one:

*About 500 people were killed in the battle, four pas were cap-
tured, and a thousand women slain [while in another] three pas
were captured, and 200 men and 800 women and children were
killed, while others were brought to Kapiti as slaves.*[22]

Not surprisingly, given the state of continual warfare that Maori
society was embroiled in by the early nineteenth century, one of the
first uses that Europeans were put to was as a weapon. Not all Europeans were as moral as Cook in these matters and Tamihana records a
successful attempt by Te Rauparaha to use Europeans against his enemies. In 1830 the ship *Elizabeth* arrived at Kapiti. She was the first
European vessel to call there. Tamihana recalls:

*He [Te Rauparaha] asked the captain whether he would take him
to Whangaroa to get revenge for the murder of his companions by
Tamaiharanui. The captain disapproved, but Te Rauparaha persisted, saying: 'The reward for taking me and my men would be a
full load of prepared flax'. Because of his persistence the captain
finally agreed. Upon arrival at their destination Tamaiharanui
was tricked into boarding the vessel. There Te Rauparaha tied his
hands and took him and his family to another cabin... When it
was dark the 70 warriors got in the canoes and went ashore. They
entered the villages at dawn, and the slaughter began. Two hundred men were killed and perhaps 300 or 400 women and
children; many were also brought aboard as slaves. Heaps of men
were left dead ashore. The ship returned to Kapiti and Te
Rauparaha had a very happy heart. Out in mid-ocean
Tamaiharanui and his wife decided that their daughter should be*

strangled to prevent her becoming a slave. When night came they secretly strangled her and threw her body into the water through the porthole in their cabin... The day the ship came in [to Kapiti] became known also for the thumping of men's feet in the war dance. To us who were at home it seemed like an earthquake.[22]

Tamihana concludes the biography of his great father with the following observation:

I have heard the elder chiefs of the tribes of Kawhia, Maunga-tautari, Rotorua, Tauranga, Hauraki, Te Ati Awa and even the Nga Puhi and Waikato say that they never knew an old man who could equal Te Rauparaha as a warrior. Although there be a thousand chiefs none would be equal to Te Rauparaha, for none could acquire as much land as he did.[22]

The extent to which the Maori had become militarised by the first half of the nineteenth century is vouchsafed by almost innumerable examples. F.E. Maning, a self-described Pakeha Maori, recorded an account of the war against Heke by the British and Tamati Waka (Thomas Walker) Nene in 1840. He writes:

I will say here that though the native language [of the Maori] is, as might be supposed, extremely deficient in terms of art or science in general, yet it is quite copious in terms relating to the art of war. There is a Maori word for almost every infantry movement and formation. I have also been very much surprised to find that a native can, in terms well understood, and without any hesitation, give a description of a fortification of a very complicated and scientific kind, having set technical terms for every part of the whole—curtain, bastion, trench, hollow way, traverse, outworks, citadel, etc., etc.—all well-known Maori words, which every boy knows the full meaning of.[86]

He also records the extraordinary resistance put up by the defenders of some *pas* during this war. Indeed, during the Maori defence of the *pa* Puapekapeka, the British learned their first lessons in trench

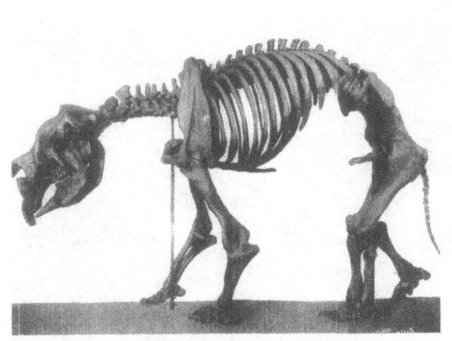

The skeleton of a diprotodon. At up to two tonnes in weight the diprotodons were the largest of all marsupials. Until 60 000 years ago they inhabited Australia's arid and semi-arid regions. (Photo: E.M. Fulda, courtesy American Museum of Natural History)

A reconstruction of diprotodon. While the general proportions are correct, the exact shape of the fleshy parts of the nose remain speculative. (Courtesy American Museum of Natural History)

Lake Callabonna, northern South Australia. Now in the arid heart of the continent, conditions were only slightly wetter when diprotodons browsed its margins over 60 000 years ago. (Photo: N.S. Pledge)

A diprotodon trackway on Lake Callabonna, northern South Australia. Such delicate and evocative reminders of Australia's megafauna are rarely preserved. (Photo: John Mitchell)

A reconstruction of a large moa (*Dinornis*), photographed in 1903 at Dunedin along with Maori medical students. (Photo: Augustus Hamilton, courtesy National Museum, Wellington)

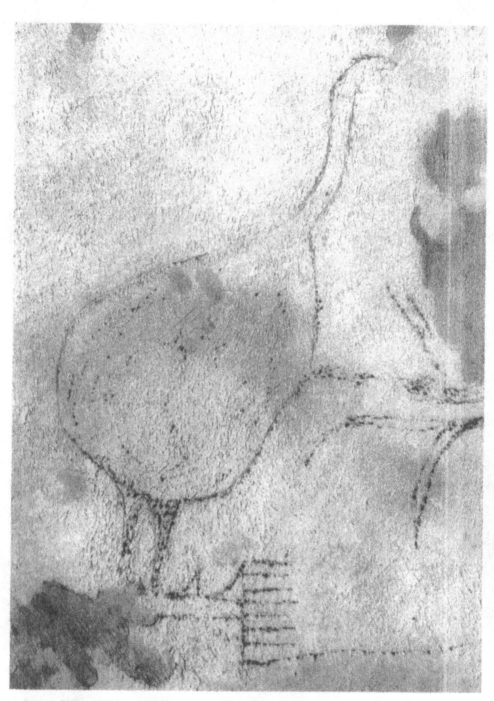

Between 500 and 1000 years ago, a Maori made this crude drawing of a moa on the wall of a rock shelter on New Zealand's South Island. (Photo: Theo Schoon, courtesy Michael Trotter)

Early European settlers reported seeing tracts of land 'white with moa bones' at some places. This photograph of a moa butchering site at Waitaki River, taken in 1936, gives some idea of what the early settlers saw. (Courtesy Michael Trotter and Beverley McCulloch)

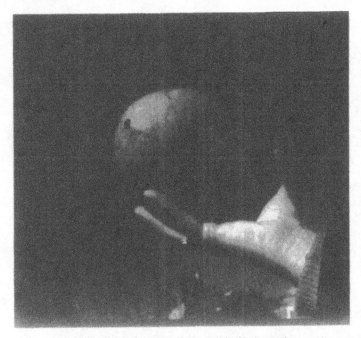

A moa egg found at Wairau Bar, New Zealand. It had been used as a water container. Very few complete moa eggs survive. (Courtesy Michael Trotter and Beverley McCulloch)

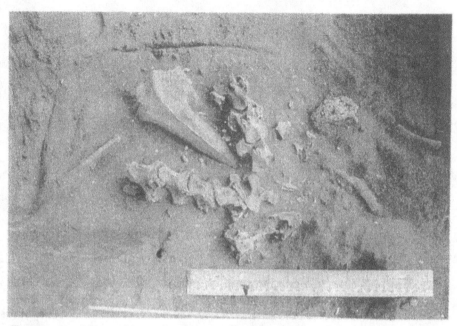

The leg and neck bones of a moa found in a moa-hunting site. The articulated nature of the neck bones suggest that the upper parts of the body of this moa were wasted. (Courtesy Richard Cassels)

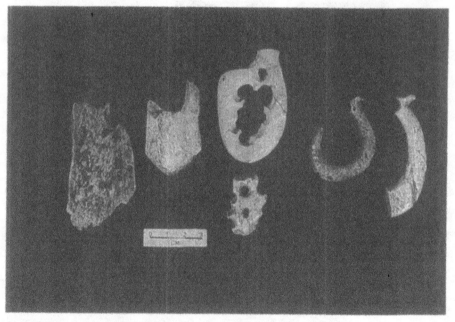

Long after the moa vanished their bones were of great importance to the Maori, for moa bone was prized for the manufacture of fish hooks. These fragments are at various stages of manufacture. (Courtesy Richard Cassels)

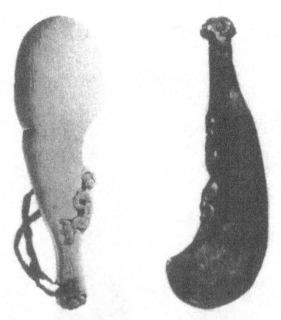

A patu (club) carved from bone (left) and wood (right). Such clubs were both weapons and symbols of status. Club-wielding Maoris taunted Cook with the refrain *'Haere mai ki uta kia patua'* —Come ashore and be clubbed. (The British Museum, London)

A fortified settlement or *pa*. It was a fort built in an inaccessible location such as this, that prompted James Cook to say 'the situation is such that the best engineer in Europe could not have choos'd a better ... it is strong by nature and made more so by art'. Drawn by Cook from a sketch by Parkinson. (By permission of The British Library)

Tamati Waka (Thomas Walker) Nene (1890) chief of the Ngati-Hoa and ally of the British in the Flagstaff War of 1840. His enthusiasm for the conflict was so great that he took to the field before the British had organised their troops. (Oil on canvas by Gottfried Lindauer, 1839–1926, Partridge Collection, Auckland City Art Gallery)

A badge of status or mourning, the highly valued tail feathers of the Huia adorn the head of this unidentified Maori woman. The photograph was taken in the 1880s. (Alexander Turnbull Library, Wellington)

Europeans' first vision of the Maori. In Abel Tasman's pen and ink drawing of 1642, Maori warriors are shown intercepting the Zeehaen's cockboat and killing Cornelius Joppen and three other members of Tasman's crew. A shocked Tasman called the place Murderer's Bay. (Mitchell Library, Sydney)

A Macassan prau (1839). These vessels, resembling miniature galleons, have been visiting the northern Australian coast for centuries. (Lithograph by Le Breton in Dumont d'Urville, Voyage au pole sud et dans l'oceanie ..., *Atlas Pittoresque*, National Library of Australia)

Probasso, captain of the Macassan trepang fleet encountered by Matthew Flinders off Arnhem Land in 1803. Probasso had first visited Australia in the early 1780s and had made seven or eight voyages to Australia by the time he met Flinders. On one visit he had been speared in the leg by Aborigines. (Pencil drawing by William Westall, 1781–1850, National Library of Australia)

A Macassan trepang processing station, Raffles Bay, northern Australia, in 1839. These substantial settlements were visited annually by Macassan traders and fishermen from the island of Sulawesi. (Lithograph by Le Breton in Dumont d'Urville, Voyage au pole sud et dans l'oceanie ..., *Atlas Pittoresque*, National Library of Australia)

View of Malay Bay from Probasso's Island. (Engraving by William Westall, 1781–1850, Rex Nan Kivell Collection, National Library of Australia)

Part of the Tahitian naval fleet, which in 1774 consisted of 160 great war canoes and 160 smaller vessels. Cook was entranced by the skills of the sailors and the magnificence of their martial displays. (National Maritime Museum, Greenwich)

Easter Island as James Cook saw it in the late eighteenth century. Within a few years the great statues were all thrown down and the human population in dramatic decline. The bones in the foreground were prophetic of the island's bleak future. (National Maritime Museum, Greenwich)

Inhabitants of New Ireland, Bismarck Archipelago, as Tasman saw them in 1642. Amidships is one of the great propeller-like blades used by the New Irelanders for shark fishing. (Mitchell Library, Sydney)

The *Endeavour* careened at the mouth of the Endeavour River, north Queensland, in 1770. It was here that Aborigines used fire to drive away the Europeans. (By permission of The British Library)

One of the earliest representations of New Guineans by a European. This group was painted by an anonymous Spaniard during the early seventeenth century. (Mitchell Library, Sydney)

A New Guinean hunter, Telefomin area. Descendants of the first future eaters, New Guineans had, by the nineteenth century, developed ecologically attuned cultures. (Photo: T.F. Flannery)

NOUVELLE - HOLLANDE : ÎLE KING.

L'ELÉPHANT-MARIN ou *PHOQUE À TROMPE.* (Phoca Proboscidea, N.)

Vue de la Baie des Eléphants.

The magnificent elephant seals (*Miorunga leonina*) of Sea Elephant Bay, King Island, 1803. Shortly afterwards, the colony was exterminated and has never re-established. The generic name Miorunga comes from the Tasmanian Aboriginal word *miouroung*, and is one of the few Tasmanian words to survive in use to the present. (Engraving by Victor Pillement, 1767–1814, Voyage de d'ecouvertes aux terres australes, 1807, Rex Nan Kivell Collection, *Atlas François Péron*, National Library of Australia)

'Cold morning', an Aborigine of southern Victoria, with his family camped near Portland, Victoria, in 1845. The presence of dogs and the diversity of artefacts contrasts with the limited possessions of the Tasmanians. (Watercolour by G.F. Angas, South Australian Museum)

A Tasmanian family as François Péron and his companions saw them in the summer of 1802. The young woman standing second from left is probably Ouray Ouray. (Engraving by Francois Denis Nee, 1732– 1817, Voyage de d'ecouvertes aux terres australes. *Atlas François Péron*, 1807, Rex Nan Kivell Collection, National Library of Australia)

A man and woman of Van Diemen's Land. These are the first known portraits of Tasmanians. Contact with Europeans was disastrous. (John Webber, by permission of The British Library)

A group of Aborigines photographed at Coranderrk Aboriginal Station, near Healesville, Victoria, around 1877. Note the possum-skin cloaks. Such cloaks were sewn together using bone needles and were presumably worn by Tasmanians until some 3500 years ago, when bone needles ceased to be made there. (Albumen print by Fred Kruger, 1831–1888, presented by Mrs Beryl M. Curl, National Gallery of Victoria)

Victorian Aborigines with possum-skin cloaks and a bark canoe, around 1879. (Albumen print by Fred Kruger, 1831-1888, presented by Mrs Beryl M. Curl, National Gallery of Victoria)

Some of Victoria's few surviving Aborigines, photographed participating in the Jerail ceremony at Bairnsdale, eastern Victoria around 1884. Note the bark bands, similar in appearance to the ammunition bands of soldiers, worn across the chest. (Photo: A.W. Howitt, courtesy D.J. Mulvaney)

A group of Queensland Native Police, Dawson River, late 1850s–1860s. The Native Police were among the most feared of all troops by Aborigines living on the frontier. (John Oxely Library, Brisbane)

Beechmont, 1904–05. European destruction of Australia's forests was often hastened by ill-conceived plans for settlement. Here a rainforest full of valuable timber is being destroyed to create poor-quality grazing land. (John Oxley Library, Brisbane)

Timber-getting in the Atherton District. It is said that cedar cutters felled three million super feet of red cedar into the Barron River where it was smashed to pulp by the Barron Falls. (John Oxley Library, Brisbane)

ENSO has long held sway in Australia. Here ENSO-induced floods spill over into Brisbane in 1893. (John Oxley Library, Brisbane)

An Anglican church in 1872 near Ipswich, Queensland, made from bark. Settlers and gold miners stripped entire forests of trees for their bark, resulting in massive and unnecessary forest destruction. (The Hume Collection, Fryer Library, University of Queensland)

warfare and underground bunkers from the Maori. They were to turn these tactics to their advantage in the First World War. Some Maori were crucial allies of the British. Indeed, Tamati Waka Nene began prosecuting the war of 1840 on behalf of the British before European troops even took to the field! The military ineptitude of the British troops was such that Maning recorded that:

> now they [the Maori] think less of fighting Europeans, and are less afraid of them, than their own countrymen.[86]

Perhaps we should leave the last words on warfare to the Maori themselves. The following extract is from a *tangi* composed at the beginning of the eighteenth century by Makere of the Ati-Awa:

> I will go to the ridge of Okawa
> I will pluck out his liver
> That will show these men
> What I mean when I speak of revenge.
> Where is the man that could kill you?
> Where is the hand that defiled you?
> No! The gods
> willed you to die,
> Tore out your heart and lungs,
> Splintered bones and spattered brains
> Like vomit, ribs picked clean
> And blood oozing through the stones
> Of the feast.
> Let your foul cousins taste
> The sweetness of the their ancestress
> In thy breast. Mairie-i-rangi
> Will lie like a stone in their belly.[76]

The process of environmental ruin and concomitant social up-heaval suffered by the Maori is not unique to New Zealand, for virtually every Pacific island whose archaeological record has been examined has a similar, if less extreme story to tell. Only one case is more dramatic. That concerns distant Easter Island, which is, along

with New Zealand, the only temperate land colonised by the Polynesians. Why the colonisers of these two temperate lands were
particularly susceptible to such extreme ecosystem collapse is not
clear. It may be, however, that Polynesian agriculture was based upon
essentially tropical crops, which could not thrive in temperate climes
and thus could not buffer the decline that inevitably occurred in wild
resources. It may also be that temperate ecosystems are less robust
than tropical ones. Whatever the case, the story of Easter Island is
worth telling, for it illustrates starkly and in great detail just what it
means to be a future eater.

Easter Island is the most remote speck of inhabited land on Earth.
Over 3500 kilometres west of the coast of Chile, nearly 4000 kilo-
metres south of the Galapagos Islands and 4000 kilometres east of the
Marquesas Islands; its nearest neighbour at 2250 kilometres to the
west is lonely Pitcairn Island, where 50 or so descendants of the
Bounty mutineers still eke out a lonely existence. Despite its isolation,
Polynesian voyagers had discovered Easter Island by about 900 AD and
made it home. Of all of the domestic animals owned by the Polynesians, only the chicken survived to become a major part of the
Easter Island economy. Lying at 27° south (about the latitude of
Brisbane) it is just outside the tropics and is rather small, covering
only 166 square kilometres. It is not particularly suited to agriculture,
the sweet potato being the major crop.

The first outsider to see Easter Island arrived some 800 years after
it was settled by Polynesians. On 5 April 1722, at about 5.00 p.m.,
Jakob Roggeveen, captain of the Dutch ship *Afrikaansche Galei*, sighted land, naming it Paasch (Easter) Eylandt, for it was then Easter. The
following morning, the sea being too rough to permit close inspec-
tion, he stood his ship offshore. Then, to his surprise, he saw a tiny
canoe crossing the five kilometres of open sea towards his vessel. It
was, in his own words, bearing a:

*Paaschlander, a well-built man in his fifties, with a goatee beard.
He was quite naked, without having the least covering in front of
what modesty forbids being named more clearly. This poor person
appeared to be very glad to see us.*[5]

This was hardly surprising, considering how infrequent visitors to the island had been!

On the following day the Dutch came ashore and encountered one of the enduring marvels of the island. Roggeveen records that the Paaschlanders:

> set fire before some particularly high erected stone images... these stone images at first caused us to be struck with astonishment, because we could not comprehend how it was possible that these people, who are devoid of heavy thick timber for making any machines, as well as strong ropes, nevertheless had been able to erect such images, which are fully 30 feet high.[5]

During his brief visit Roggeveen noted two important features of the island that were to intrigue people for centuries. His first critical observation concerned the lack of trees on Easter Island, for he noted that it was covered in coarse grass and nourished virtually no woody plants at all. The second was that the island was quite literally littered with enormous stone carvings.

So great an effect have these carvings had upon the European imagination that they have been accepted by thousands as evidence for visits of extraterrestrial life forms to planet Earth, as was first postulated by the fantasy writer Erik von Daniken in his book *Chariots of the Gods*. They have prompted others, such as Thor Heyerdahl, to sail the Pacific in primitive vessels in attempts to prove that the island had been settled by the expert stonemasons of various Central and South American cultures. For many years such ideas flourished, for no-one could offer a plausible opinion as to how the great statues had been transported and erected.

In truth, the entire situation of the Rapanui, as Easter Islanders are known, is paradoxical, for by the time of European contact they had almost no natural resources, yet had produced one of the most striking cultures in the entire Pacific. They were the only Polynesians to develop a kind of picture writing, known as *Rongorongo*, which to this day remains undeciphered. They were great artists and builders and their famous statues, or *moai*, were of course unparalleled in the entire Pacific. Despite its uniqueness, the mystery of Easter Island has slowly been solved through patient archaeological and historical research.

Only two European expeditions, those of Roggeveen in 1722 and González in 1770, saw the statues of Easter Island in their original glory—standing in rows—complete with their pumice-stone hats and shell-inlaid eyes. By the time James Cook arrived in 1774 they had been largely thrown down and smashed, perhaps because the ancestors seemed to have deserted the Rapanui, or perhaps they were victims of the petty warfare that raged over the few resources that the island could still provide.

From Cook's time on—and perhaps before—starvation was endemic on the island, as is graphically illustrated in the almost skeletal *moai kavakava* carvings that were produced there during the nineteenth century. Warfare, mass destruction and probably cannibalism were all too frequent during this period. Defeated people lived for decades in caves, emerging under cover of darkness to forage for the tiny winkles and other shells that had become so important to everyone as a food source, or to steal chickens from the castle-like, fortified henhouses of their neighbours.

A remarkable pollen sequence preserved in the crater of an extinct volcano has revealed that Easter Island was not always treeless and barren. The record has been described by Paul Bahn and John Flenley in their book *Easter Island, Earth Island*, as 'truly dramatic, and one of the most striking records of forest destruction anywhere in the world'.[5] It reveals that before the arrival of people Easter Island was, taking into account its temperate location, much like any other Pacific island. The toromiro tree (*Sophora toromiro*), which is used throughout Polynesia to make rope, was dominant in the virgin forests. Even more important to its early settlers, the island was thickly covered with a large palm, closely related to the prized Chilean wine palm (*Jubea chilensis*). The Chilean wine palm is the largest palm in the world. It has a smooth trunk which is up to a metre in diameter and 20 metres tall. These make perfect rollers for transporting large stone statues. They also make ideal structural members for building, while the fronds can be used for thatching. If left standing, they can be milked for a sugary sap which is delicious and nourishing if eaten raw. Crystallised, it can be used in cooking, or fermented, made into alcohol. The palm was clearly an incredibly valuable tree to the first inhabitants of the island. Its loss through overharvesting must have been a major blow to them.

Easter Island's forests had been completely destroyed by 1400 AD. With them went the native birds and other edible species they housed. The soil eroded rapidly and lost fertility, but unlike in New Zealand, there was no coastal plain to catch the runoff, nor bracken fern to thrive upon the transported nutrients. There were also no estuaries which could have benefited from the transported nutrients. As a result of deforestation, people were no longer able to build with timber. Instead, they lived in caves or rude boulder huts. Worse, they could not even build substantial canoes from which to fish or travel. The few highly prized yet pathetically small canoes present in Roggeveen's time (such as the one that carried the single Paaschlander five kilometres from shore) were highly unseaworthy, being made from small pieces of driftwood fastened together.

At the height of Rapanui culture, Easter Island supported about 6000–8000 people. This suggests that the human population had undergone exponential growth much as early Maori populations had. Clearly, on a small and isolated island such as Easter, something had to give once population had built to this level. By the middle of the seventeenth century, the ability of the island to support its population had been vastly reduced, and society was suffering. At the time of Roggeveen's visit in 1722 only about 2000 people called Easter Island home. By 1877 there were just 111 human beings left, their numbers having been reduced by both a declining environment and slave traders from Chile.

Cut off from the world without even the means of escape that a canoe might afford, a dwindling resource base had, since the fourteenth century, been slowly strangling the Rapanui. By the time of Roggeveen's visit things had clearly progressed much further than they had in New Zealand by the time of Tasman's visit. It seems likely that by 1642 the Maori population, while suffering severe resource depletion, was yet to experience the major population collapse which the Easter Islanders had clearly suffered by 1722. With too many people, but still sufficient resources to deploy armies, the Maori were entering a fiercely aggressive stage. By the eighteenth century, this stage had by and large passed for the Rapanui, but the continuing environmental collapse had kept conflict boiling.

In contrast to the Maori, the Rapanui lacked the resources to oppose a landing by Europeans. The difference between the Maori and Rapanui in the late eighteenth century may have been like the difference between meeting a merely hungry man and one wasted by starvation.

Although both Maori and Rapanui represent an extreme in the Polynesian experience, there is no doubt that the unique Polynesian lifestyle and the Polynesian preference for small islands meant that this pattern was repeated over and over wherever they went. This is because the early Polynesians, rather like people of the modern world, were addicted to frontiers. Their great passion was to discover new, virginal lands, overexploit them briefly, then move on.

Surprisingly, there is evidence that even on Pacific islands a balance can be struck between people and their ecosystems which allows a more peaceful lifestyle to develop. New Caledonia had been settled for 3500 years when James Cook sighted it in September 1774. He found a 'friendly, honest and peaceful people',[27] who showed a mixture of Polynesian and Melanesian ancestries. I have no doubt that they, like their Maori and Rapanui relatives, underwent social trauma brought about by resource depletion. For New Caledonia also has its large extinct animal species, which are the hallmarks of overexploitation and accompanying environmental collapse. But after a history four times as long as that of the Maori, they had finally found stability. It may be that the poor soils and ENSO-influenced climate of New Caledonia had encouraged people to cooperate in order to survive and thus hastened the process. As Shakespeare wrote, sweet indeed are the uses of adversity.

The habit of future eating has recently become an almost universal one for humans. It is not unique to humans, however, for it is seen in many creatures that are transplanted out of their original biological context. Some of the very best examples are from our own region, the introduction of red deer (*Cervus elaphus*) into New Zealand providing a classic example.[23] By 1923 about 1000 red deer had been released at 50 sites throughout New Zealand. They spread rapidly and lived in high densities on the New Zealand pastures that lacked defences against deer. For 10 to 30 years following their release, the population remained below carrying capacity at the release sites. During this phase the deer were healthy and well-fed, growing to a larger size than their ancestors in Europe had. They also reproduced and grew more rapidly. For about three to 13 years after this, the population stayed in equilibrium with its environment and while the animals were not as large as before, they remained healthy. But the population continued to grow and within a few seasons the carrying capacity of the pasture was exceeded. The number of deer dropped, as did their average size, level of health and rate of reproduction. The pasture was grossly altered and would never revert to its original form. It was only with the onset of heavy culling that health of the populations was restored; but at lower density than originally, for the capacity of the pasture to support them had been damaged.

The recent eruption of cane toads (*Bufo marinus*) in eastern Australia has followed a similar course. First introduced in the 1930s to control cane beetle, cane toads spread rapidly along the Queensland coast. By the 1970s and 1980s they had reached plague proportions in settled areas. People were horrified to find enormous, bloated toads eating pet food out of their dog's bowl, while nearby beloved Rover lay dead, having been poisoned by the toad he had bitten. Alarm bells rang everywhere and people became convinced that the toads represented a dire threat. Then, the late 1980s saw starving and dying toads appear in many areas. In some places their average size and numbers have decreased and it now appears that the worst of the plague is over in these areas.

The difference between cane toads, red deer and humans is that humans can easily transfer from one resource to another when a particular resource nears exhaustion. This makes them much more able to thoroughly destroy a resource, for they can continue to exploit its last remnants while they gain most of their needs from other resources.

It is probably no accident that future eating is such a feature of the Australasian environment. Here, evolution selects for animals that use little energy and work together to maintain and circulate the few nutrients that are available. They therefore usually breed slowly and have few defences against introduced species. Thus, any introduced organism can easily upset the ecosystem.

CHAPTER 24

ALONE ON THE SOUTHERN ISLES,
WEIRDS BROKE THEM

I would dearly love to be able to travel back in time to the Australia of 50 000 years ago, for there are so many questions which I fear we may never be able to answer, the passage of time having erased all clues. Take, for example, the nature of the early Meganesians themselves. Despite much speculation, we still have no clear picture of what the first inhabitants of Meganesia were like. Even quite fundamental questions, such as how many different racial groups invaded Meganesia at that early period, are unanswered.

Various researchers have suggested that Australia was settled from South-East Asia by several distinct kinds of humans. For several decades it appeared that the archaeological record was providing evidence which supported this hypothesis, for 10 000-year-old skeletons found at Kow Swamp in northern Victoria and elsewhere were from very robust people, while some skeletons (including those over 32 000 years old) from western New South Wales were much more gracile. Researchers interpreted this as meaning that two distinct groups of people lived side by side, without interbreeding, for tens of thousands of years in Australia. This was a discovery of enormous apparent importance, for not only did it provide evidence for multiple invasions of Australia, but the fact that two distinct human groups occupied the same land yet did not interbreed, indicates to a biologist that they were different species!

Although the debate about whether more than one group invaded Meganesia is far from over, the evidence supporting the case for multiple invasions is now looking decidedly thin. Recent reinterpretation of the bones suggest a less startling, but nonetheless fascinating story. It has been suggested that the robust skeletons are those of men and the gracile ones of women. This makes sense because before 10 000 years ago, people all around the world were seven to nine per cent larger than they are today. The size difference between the sexes was also greater than at present. The early Australians seem to represent an extreme in sexual size difference that was in part related to their large size.

Unfortunately, biochemical studies, which have been so helpful in determining the origins of the other people of the Pacific, are of no help in determining how many groups colonised Australia from South-East Asia, for apparently it all happened so long ago that the genetic trails have been erased by subsequent genetic change.

But the archaeological record holds one intriguing piece of evidence that suggests a single common culture for the first Meganesians. This is the discovery, in sites as distant from one another as northern New Guinea and Kangaroo Island off southern Australia, of huge stone axes with a distinctive 'waist' pecked into them, presumably in order to take bindings. These axes are of enormous antiquity; in the case of the examples from the Huon Peninsula, Papua New Guinea, at least 40 000 years old.[140] The use to which these impressive implements were put is still speculative, but if they were anything like modern axes they were probably used to cut notches in trees, shape wooden implements and clear saplings from around favoured food plants. They appear to have been a uniquely Meganesian invention and hint that a uniform, rapidly expanding population may have covered the length and breadth of Meganesia soon after an initial invasion before 40 000 years ago.

A few aspects of the lifestyle of the first people to inhabit Meganesia are evident. It seems reasonable to suggest that they had a well-developed language—as advanced as any spoken in the world today—and that they lived in small groups. We know also that they lacked domesticated animals and that they probably practised only rudimentary forms of plant curation. They presumably built only simple shelters, while their watercraft were doubtless basic, perhaps consisting of rafts, bundles of reeds, or bark and wood canoes. Beyond this, there is very little we can say about the technology or culture of these, the first people to venture from the great Afro–Eurasian homeland of humanity.

Likewise little is known concerning the spread and adaptations of these people immediately following their arrival in Australia and New Guinea. But this has not prevented archaeologists from speculating about how humans dispersed over the land. One of the most detailed hypotheses concerning this has come from Dr Sandra Bowdler of the Department of Archaeology, University of Western Australia. She has suggested that as the first Australians were an essentially maritime people, they would have spread very slowly, first taking up life along the coasts and inland rivers. From a biological perspective this is a singularly unlikely hypothesis, for it suggests that people would have ignored such attractive resources as the naive large kangaroos, tortoises and other fauna that existed over the entire continent.

It seems far more likely that people would have simply followed the available resources—including the megafauna—and when they were exhausted, moved on. Indeed, if the experience of New Zealand is any guide, the present dependence of some Aboriginal people upon marine and estuarine resources such as molluscs probably followed the exhaustion of other, more easily obtained foods such as large animals. It may also have been hastened by the beginning of large-scale erosion, which flushed nutrients into estuarine systems, enhancing their productivity.

I would also suggest that, given human curiosity, the early Meganesians probably spread quickly over the entire region. This was certainly the case in New Zealand, where studies indicate that the Maori had explored much of New Zealand within a few generations of landfall. In Australia smaller, less mobile species than humans—including rabbits and foxes—have taken less than a century to cross the continent, while the European invasion itself had (with the exception of the New Guinea highlands) penetrated the whole of Meganesia within a century or so of first settlement. All of these factors point to the likelihood of a rapid and complete invasion by the ancestors of the Aborigines.

We can imagine then, the first Australians spreading rapidly into all of the continent's environments that afforded them a living. Their spread would have been made easier by the great herbivore trails that doubtless crossed the land, linking watering and feeding sites. These trails may have been particularly important in penetrating the dense jungles of New Guinea. There, the all but impenetrable high-elevation forests offer few resources to humans. But once people living in the lowlands crossed that barrier, they would have entered the subalpine grasslands, which are rich in game and a perfect hunting ground.

Following this rapid invasion, the people of Meganesia must have begun to differentiate, to specialise in making a living in the particular environment that they found themselves in. Put another way, the land began to shape the people. Much of the early part of this process is unknown, but the last 20 000 years or so reveals some extraordinary trends in cultural evolution.

The Aboriginal people who occupied Tasmania, Kangaroo Island and the islands of Bass Strait, represent one extreme in the process of the land shaping a people. Beginning about 12 000 years ago they found themselves cut off by rising sea-levels. Because they lacked seaworthy vessels they were completely isolated. Their story is a most extraordinary one, for the relatively small size of their island homes drove their evolution in a particular, inexorable and fateful direction.

The outcome of this local evolution was depressing; for all populations, with the exception of the Tasmanian Aborigines, were to become extinct thousands of years before the arrival of Europeans. Humans survived on Flinders Island until about 4700 years ago, while Kangaroo Island supported people for a little longer, until about 2250 years ago.[37,80] The factors that caused these extinctions are the same factors that made Tasmanian Aboriginal culture so distinctive. They have severe implications for all Australasians today.

The reasons for the prehistoric extinctions of the people of Flinders, King and Kangaroo Islands becomes apparent when one considers the relationship between island size and human population. Flinders Island in Bass Strait, for example, is some 1800 square kilometres in area. It is, by Australian standards, rich in resources, for game is abundant and the cold sea is relatively bountiful. Nonetheless, it was, if we judge from the density of Aboriginal people in Tasmania, capable of supporting a human population of about 400 individuals. At 3890 square kilometres, Kangaroo Island is even larger. But its resources were more limited than those available to the Flinders Islanders and its carrying capacity is likely to have been below 500 people.

The human population of both islands was thus probably near or below the critical size (about 500 individuals) that geneticists predict is necessary for long-term survival of any mammal population. Isolated populations smaller than this suffer continual loss of genetic diversity and also become susceptible to genetic disorders through inbreeding. Their small size also makes them susceptible to chance events, such as an imbalance in the birth rate (for example, a preponderance of boys), disease and disaster (for example, mass food poisoning). These factors dictate that such small, isolated populations are doomed to extinction.[37]

The sole surviving Aboriginal population inhabiting a temperate Australian island was that living in Tasmania. Tasmania is large enough to support some 5000 Aborigines living traditional lifestyles.[37] This is some 10 times more than the absolute minimum size necessary for long-term survival. But is a population of 5000 large enough to maintain a complex material culture? Recent archaeological discoveries suggest that it was not.

When the first reports of the Tasmanian Aborigines reached Europe they created intense interest. Europeans thought that their simple tool kit and lifestyle meant that they were a very primitive people. For a very long time after, it was widely believed that these apparently truly primitive people had survived in their remote corner of the world because they had not had to compete with more advanced races.

The French savants of the Baudin Expedition, who observed the Tasmanians in 1802, were amazed that even though the Tasmanians lived in an often bitterly cold climate, they lacked clothing. Extraordinarily, they also lacked the ability to make fire. Mannalargenna, one of the last of the Tasmanian Aborigines to live a traditional life, told of what would happen if a group's fire was extinguished. He said that people had no alternative but to eat raw meat while they walked in search of another tribe. Significantly, one of the universal laws among the Tasmanians was that fire must be given whenever requested, even if the asker was a traditional enemy who would be fought after the gift had been given.[39]

The French were also struck by the fact that the Tasmanians did not eat fish, even though they were abundant in Tasmania's coastal waters. François Péron records that when members of the Baudin Expedition offered some fish which they had caught, the Tasmanians expressed amazement and horror. This was not an isolated instance, for earlier, in 1777, members of Cook's third expedition recorded that Tasmanians reacted with horror or ran away when fish were offered to them.[37]

There are some other quite extraordinary features of Tasmanian culture. The Tasmanians, for example, had no hafted implements (such as axes), no implements made of bone, no boomerangs or spear throwers, no dingos and no microlithic stone tools. Indeed, their entire tool kit seems to have consisted of about two dozen kinds of objects. The largest was an unusual kind of watercraft made of bunches of reeds that was used to reach islets. The reeds trapped air bubbles which gave them buoyancy. They were useful for travelling short distances only, for they became waterlogged and sank after a few kilometres.[37]

It is easy to see how the limited material culture of the Tasmanians could seduce the savants of the eighteenth and nineteenth centuries into classifying the Tasmanians as the world's most primitive people. Not surprisingly, the anatomists of the day had an almost insatiable demand for corpses. Through dissection, they hoped to find additional evidence supporting the idea that the Tasmanians were primitive; maybe even a kind of living missing link. Needless to say, these anatomical studies yielded no such evidence. Because of the cultural gulf between Tasmanian and European, the explicit instructions of the deceased regarding funeral arrangements were often ignored in order to provide material for such studies.

Until very recently, many people found no reason to doubt the conclusions of nineteenth century science concerning the Tasmanians. But detailed archaeological research, much undertaken only in the last few decades, has now shown conclusively that there was nothing primitive about the Tasmanians at all. They were, instead, a highly specialised offshoot of the Australian Aborigines, whose culture evolved under the extraordinary constraints that 10 000 years of solitude would place on any small band of humans.

The most striking evidence concerning the evolution of the culture of the Tasmanians has come from the study of campsites occupied over the last 7000 years. Deposits that date to 7000 years ago or more are full of bone tools, including awls, reamers and needles. There seems to be little doubt that these implements were used for sewing, probably to make skin cloaks similar to those used by the Aborigines of southern Australia right up until the nineteenth century.

The variety of bone tools found in Tasmanian middens dwindles with time, until eventually, about 3500 years ago, the last of them disappear from the archaeological record. This suggests that stitched clothing was lost from the material culture of the Tasmanians at about this time.[37]

Interestingly, the older archaeological sites show that fish—although despised as a food in historic times—once formed an important part of the Tasmanians' diet. Evidence from some sites suggests that fish made up about 10 per cent of their diet in the past. In all, the remains of at least 31 species of fish have been found in ancient refuse dumps, including rocky reef and open water species. Then suddenly, about 3500 years ago, the remains of fish cease to appear in refuse dumps.

Unfortunately, archaeological evidence for the use of wooden artefacts is largely lacking, but there is good reason to believe that Tasmanians once possessed boomerangs, spear-throwers and probably hafted tools, for all were present in Australia before rising sea-levels isolated the Tasmanians. They too, must be counted among the artefacts lost during 12 000 years of isolation.

The archaeological record also reveals that the Tasmanians deserted various parts of their homeland in the face of climate change. They flourished in the tundra-like south-west of Tasmania right through the height of the last ice age, but the abrupt cessation of occupation of cave after cave in the south-west about 12 000 years ago indicates that they deserted the region when the climate warmed.[16]

This may seem paradoxical, but the warmer conditions allowed the dense tangle of temperate rainforest to creep out of the valleys and replace the buttongrass plains. While they look hostile to modern visitors, buttongrass plains support a great abundance of animals, particularly wallabies, that could have been hunted by the Tasmanians. The rainforests, in contrast, had lost most of their animals of huntable size during earlier ice ages, when they shrank to such a tiny area that viable animal populations could not be sustained. To the Tasmanians, the rainforests were virtual deserts, so dense as to be impenetrable, and lacking worthwhile game even when a hunter fought his way in. Strangely, fire was not used to limit the rainforest's advance as it was elsewhere in Australia. Perhaps in the wet southwest the rainforest was just too dank to burn, or the human population too scattered to have an impact.

Although the loss of various artefacts, food items and habitat may suggest a people in decline, it is too simplistic to imagine Tasmanian society as being on a one-way journey towards extinction. For some indigenous technologies did develop during their isolation. The archaeological record, for example, shows that people started to visit nearby islets some 4000 years ago. This suggests that the unique Tasmanian reed canoe had been developed by this time.

Various explanations for the simplification of the material culture of the Tasmanians have been put forward by archaeologists. But as Jared Diamond has pointed out,[37] most seem to be attempts to explain away the obvious. Some suggest, for example, that Tasmanians ceased to eat fish because they contained too little fat and that a high fat diet was necessary in chilly Tasmania. Others have suggested that spear throwers were not an advantage in the Tasmanian environment. These explanations lack credibility, for Tasmanians survived perfectly well for thousands of years on a diet that included fish, while the Aborigines of equally chilly southern Victoria continued to thrive on fish right up until the nineteenth century. Similarly, no-one has suggested how it could be that spear throwers were of no use in Tasmania, yet they were invaluable in Victoria. Tasmania, like Victoria, has a great variety of environments and the more open ones are splendidly suited to hunting with spear and thrower.

The most plausible explanation seems to lie in the unique isolation and small population size of the Tasmanians. The theory goes something like this. A small group of people is less likely to come up with technological innovations than a larger group. If the group is completely isolated, then new ideas cannot reach it. Because of this, innovation in material culture is slowed. Because the population is small, activities and knowledge may be lost simply through the early death of skilled people before they can pass their skills to the next generation.[37]

Losses such as that of clothing and the ability to make fire may have resulted from rare, early deaths occurring over a long period of time. The 5000 Tasmanians lived scattered in small groups. It may be that only one or two people in any one group had all the skills necessary to make bone needles and prepare skins. Over 12 000 years there is a high chance that the few such specialists in any one area would, at some stage, die before they could pass their skills on. Repeated chance events like this might have led to the loss of many skills that require specialised knowledge. These may have included the ability to make bone needles and thus clothing, fire-making equipment, hafted tools, boomerangs and spear throwers.

If the population is small enough, there may be strong evolutionary pressure to dispense with high-risk activities. This is because risks that are acceptable for larger populations can threaten the very survival of smaller ones. The loss of fish from the Tasmanian diet may be an example of a high-risk activity that is strongly selected against and thus lost, in small populations.

Eating fish can be a risky business, because occasionally a dinoflagellate bloom known as a 'red tide' can lead to mass poisoning. The simultaneous death of hundreds of people in a large human population is a great personal tragedy, but it poses no threat to the survival of that society because the statistical chance of losing all members of one age group or sex is tiny. Such a poisoning in a small population, however, can be a disaster for the entire group. This is because, through chance, it may kill a significant proportion of the women of child-bearing age, or all of the older and more knowledgeable individuals. In order to avoid such catastrophic events, extreme conservatism may be selected for in small societies. This is because in evolutionary terms it may be better to forego the benefit gained from eating such 'dangerous' food as fish, rather than risk an extremely rare but catastrophic poisoning event. This very necessary adaptation of course carries a severe penalty, for resources normally available to people in other circumstances are denied to small groups in already vulnerable situations.

Incidentally, the final loss of bone needles from the archaeological record occurs at about the same time that fish disappear, suggesting that these events may be related. If a large number of Tasmanians did perish due to fish poisoning some 3500 years ago, it may be that the last people who possessed the knowledge to manufacture and use bone points died in the catastrophe before their skills could be passed on.

The many losses from their material culture appear to have cost the Tasmanians dearly. Without spear throwers they may not have been able to catch the fastest game, while without the ability to make fire they were occasionally cast upon the mercy of hostile neighbours. The losses speak eloquently of the terrible tyranny that interminable isolation and small population size impose on a people. Yet these costs were only part of the story, for the demise of the Tasmanians in the face of the European invasion may have been the final price they paid for living at the end of the Earth.

The arguments used by some anthropologists and archaeologists to try to explain away these trends suggests that there is a great reluctance on the part of most people to accept that humans can be so deleteriously affected by their environment. But history is replete with examples of people who have suffered similar, if less sustained, cultural simplification. Perhaps the most telling of all is the 'dark ages' which stifled technological progress for centuries in Europe. An Anglo-Saxon Englishman has left us a touching account of what it was like to live at such a time, in a tiny fragment of poetry known as *The Ruin*:

> Well-wrought this wall: Weirds broke it.
> The stronghold burst...
> Snapped rooftrees, towers fallen,
> the work of the giants, the stonesmiths,
> mouldereth.

A stroll through the ruins of a Roman town—perhaps *Aquae Sulis*, as Bath in southern England was known to its Roman builders—inspired a barbarian from northern Europe to write these wonderful words. The town had been deserted just a few centuries before the poem was written, yet so reduced was the culture of the people living nearby that they could only interpret the grand and imposing ruins as the work of giants. Clearly, the people of southern England, some doubtless descendants of the builders themselves, had lost all knowledge of how to build so splendidly. My European ancestors lived in this state of cultural diminution for centuries.

We must also remember that 12 000 years is an almost unimaginably long time. It is as long as the entire human history of Ireland, or the entire history of agriculture. It is five times as long as the entire Christian era. Imagine taking a town of 5000 people—say Atherton in north-east Queensland—and isolating it completely for 12 000 years. The outcome would be too dismal to contemplate. The simple survival of the Tasmanians through such an extraordinary exile is testimony to their ingenuity and durability.

Perhaps the history of the Tasmanians looks strange to us because the trend towards cultural simplification in the face of a limited resource base is relatively unusual. The general trend in human evolution has been towards broadening the ecological niche, with its concomitant increasing cultural complexity and resource utilisation. Despite its relative rarity, the history of the Aborigines of Australia's southern islands has a critical lesson to teach modern Australians, for Australia is—and always will be—a small nation in a populous world. I suspect that, even today, isolation would prove fatal to Australia.

CHAPTER 25

SO VARIED IN DETAIL—SO
SIMILAR IN OUTLINE

I have taken the title for this chapter from the epigraph of a seminal work on Aboriginal lifestyles and environments called *Aboriginal Man and Environment in Australia*. It is a quote from the explorer Edward John Eyre, writing in 1845, who said that Aboriginal culture was:

> *so varied in detail, though so similar in general outline and character, that it will require the lapse of years, and the labours of many individuals, to detect and exhibit the links which form the chain of connection in the habits and history of tribes so remotely separated; and it will be long before any one can attempt to give to the world a complete and well-drawn outline of the whole.*

Time has proven Eyre correct, for despite many partially successful attempts, there is still no complete overview of Aboriginal lifestyles and cultures. These cultures are the result of over 40 000 years of coadaptation with Australian ecosystems. The experience and knowledge encompassed therein is perhaps the single greatest resource that Australians living today possess, for without it we have no precedence; no guide as to how humans can survive long-term in our strange land.

The relatively few technological changes in the archaeological record of the Aborigines may suggest a kind of cultural stasis. Several researchers have argued for change in the form of an 'intensification' of Aboriginal activity over the past 5000 years, resulting in an increase in population, but this is yet to be conclusively demonstrated. Overall, it appears that Aboriginal cultures have changed slowly over the last 12 000 years when compared with those of people living on other continents. Part of the perception of slow change, at least, is due to the very incomplete nature of the archaeological record and the fact that many of the cultural adaptations of the Aborigines do not leave evidence in the archaeological record. Where the record is more complete, or where details of lifestyles are known, there is much greater evidence of cultural change and adaptation, at least at the local level.

There are other ways of viewing this apparent stasis. While it is true that the cultural change experienced by the Aborigines over the last 12 000 years was relatively slow, this period represents perhaps only the last one-sixth of their tenure in Australia. It is entirely possible that cultural change had been extremely rapid in the millennia following initial colonisation and that things changed so little in the last 12 000 years because so much cultural adaptation happened over the previous 50 000. Changes seen over the last 12 000 years may, in this context, be seen merely as exceedingly fine-tuning of the relationship between the Aborigines and their environment.

Today, the lifestyles of all Aborigines are grossly altered by the European economy. In many regions of Australia very little is known of how people lived in the past, and what little is known is difficult to interpret. For those who recorded Aboriginal lifestyles did not see through the eyes of an historical ecologist. For these reasons, the archaeological record assumes great importance in interpreting the vital matters of lifestyle and human impact. It is, however, an extremely limited record, providing glimpses of now-vanished tribes even more elusive than the thin, wailing songs captured on wax cylinders in Central Australia by the great Professor Baldwin Spencer a century ago. More ghostly indeed than the fading sepia images of proud, naked men and women, that inform us of how life once was in Australia.

A first and most crucial deficit in the Australian archaeological record is the virtual complete absence of a beginning. Living Australians know, from their own life experiences, how good life can be for the first generations of future eaters, to whom fall the lion's share of resources. The archaeological record and early history of New Zealand speaks eloquently of how dire the situation becomes when that first, full harvest is exhausted. Yet on both of these matters, the archaeological record of the Aborigines is silent. This is presumably because these initial phases of human adaptation are brief—a few hundred years at most in the infinity of time. Unless those few hundred years lie in the very recent past, time will destroy all traces of them. In the extraordinarily long history of the Aborigines, those phases may have passed over 40 000 years ago. Thus there is only a slim chance indeed that we will ever know much about this part of Aboriginal history.

This lack of a beginning is a great pity, for adaptation proceeds fastest when ecological pressure is greatest; and pressure is greatest during the resource crash that occurs in the first millennium following settlement. Many of the most important practices and beliefs that were to be crucial in determining the course of Aboriginal cultural evolution were probably shaped during that missing millennium.

Despite its shortcomings, the great strength of the Australian archaeological record is the enormously long, stable and slowly evolving relationship it documents between a people and their land. In some rare instances, such as that of the rock art galleries of Arnhem Land, it is a detailed record. Much, however, is quite literally skeletal, consisting of stone tools and bones. It tells us very little about the cultural and social life that the people of the time must have enjoyed and little of the forces that brought about social and cultural change. Indeed, often the record is barely sufficient to detect the small, incremental changes in diet and technology that reveal ecological co-adaptation. Despite its many deficiencies, it is important to examine the archaeological record before making observations about the fundamental underpinnings of the lifestyle of the modern Aborigines.

There is little evidence for marked, continent-wide technological change in the archaeological record of the Aborigines before 6000 years ago. Perhaps the most striking change occurring before then is the loss of the great, waisted axes that have been found in archaeological deposits as far afield as Kangaroo Island and New Guinea's Huon Peninsula. People may have ceased manufacturing these axes as long as 40 000 years ago. As their function is not clear, it is difficult to interpret their economic importance, or whether they were replaced by superior tools.

At least three major changes can be detected among the stone tools that have been made in Australia over the past 6000 years. None, however, present unequivocal evidence for a major shift in the economy of a people. Instead, some may mark shifts in their spiritual and social life. The most striking of these changes concerns the development of the edge-ground axe, a clearly recognisable stone artefact with which many Australian farmers are familiar, having ploughed them from the soil of their fields.

About 5000 years ago, the manufacture of these splendid stone axes with their fine, ground edge, became popular throughout Australia.[140] Although at first sight they appear to mark a revolution in the relationship between humans and trees, this has proved not to be the case. For such axes are not particularly suitable for cutting down trees. Although commonly referred to as axes, these tools are in fact hatchets as they are wielded with one hand. A New Guinean friend who used an edge-ground axe in his youth to clear gardens, described their action (in Melanesian Pidgin) as *'chew 'im down diwai* [tree]' and laughed at the enormous effort put into felling trees before steel axes were introduced by Australian kiaps. The much tougher hardwoods of Australia would have been even less amenable to felling with such implements. Thus it seems likely that these fine axes were used primarily for shaping wooden tools and enlarging pre-existing tree hollows to reach possums or honey, rather than felling trees. Indeed, these are precisely the uses that were recorded for these axes by early observers of Aboriginal culture.

While the technological significance of the edge-ground axe may have been limited, its potential social significance was enormous. This is because superior stone suitable for the manufacture of edge-ground axes can be obtained at only a few quarries around Australia. As items of trade, edge-ground stone axes were enormously valued, for they are beautiful and much time goes into their manufacture. The quest to obtain axes often saw people undertaking journeys of hundreds of kilometres. Such journeys resulted in social ties that may otherwise never have been consolidated. As will be explained below, it is precisely such changes that may have been the most important ones in helping the Aborigines to flourish in Australia's environment.

A second technological development in stone-working technology which can be traced in the archaeological record is the adoption, in northern Australia about 6000 years ago, of bifacial stone points.[140] Their use rapidly spread into southern Australia, but by 3000 years ago they were being abandoned in the south. It was only in the north, where they originated, that their use persisted into historic times. Although it is difficult to know precisely why this kind of stone point began to be used, one possibility is that it enhanced the aerodynamic properties of spears, particularly those launched from woomeras. This theory has never been tested, and if correct, the reasons as to why bifacial points should have been abandoned in the south has never been satisfactorily explained.

A variety of other tools, such as backed blades and tiny, geometric stone tools known as microliths, spread rapidly across Australia some 4500 years ago.[140] Their function is still unclear, but many researchers postulate that microliths were used as sharpened barbs on spears. Debate still rages concerning the origins of the tiny and unusual microliths. Similar tools were in use in south Asia at about the same time. Indeed, similar tools were in use in other parts of the world for 20 000 years before this. Thus it seems possible, but not certain, that the technology was passed from Asia to Australia.

If we shift from the examination of stone to bone in the archaeological record, we find evidence for several other widespread changes. The dingo was the first domesticated animal to reach Australia. It was adopted enthusiastically by the Aborigines and rapidly spread throughout the continent. Only Tasmania, being cut off by Bass Strait, was inaccessible to it. It is difficult to assess the impact of the dingo on the economy of the Aborigines. There is some evidence that it may have had a large initial impact, for when Tasmanian Aborigines acquired dogs from European settlers, they quickly turned them to use in hunting marsupials. The Tasmanian marsupials had evolved no strategy for dealing with dog predation and thus dogs may have found it particularly easy to catch them. In this situation, dogs may have been a very valuable asset. Marsupials may have acted similarly on the mainland before they became used to dingos.

But by the time of European contact with the Aborigines of mainland Australia, the situation was very different. Early observers noted that Aborigines living in more open habitats rarely used dogs for hunting. Their presence is often a positive disadvantage, as they disturb prey before the hunter is able to stalk close enough to launch his spear. The advantage of acquiring dingos may also have been lessened once feral populations had become established. Dingos are efficient hunters of the larger marsupials and feral populations may have actually reduced the numbers of these important prey species that were available to Aborigines.

Again, although the economic value of the acquisition of the dingo remains equivocal, its social role was clear. Eric Rolls, in his book *From Forest to Sea* puts the relationship as well as anyone ever has:

> *Aborigines naturally adopted dingos as hunting aids, as companions, as warmth on freezing desert nights. There soon began the long, strange, inadvertent association of Aborigines and dogs. They treasured them, they treasure them, but they showed no rational concern for their welfare... Old men and old women welcomed them as sleeping companions—the more dingos they had the warmer they slept—so they broke their forelegs to keep them in camp. When they moved around, they might carry one or two and leave the others to die.*[118]

The other great social value of dingos was to alert people to the arrival of strangers at a camp and particularly to scare away nocturnally active dangerous spirits.

Prior to the introduction of the dingo some 3500 years ago, Australia had experienced a period of relative ecological stability. It had suffered no extinctions since the great wave that carried off the megafauna over 30 000 years earlier. Three species fell victim to the dingo on the Australian mainland, but all survived in Tasmania, where the dingo was never present. The Tasmanian tiger (*Thylacinus cynocephalus*) and Tasmanian devil (*Sarcophilus harrisii*) were probably affected through direct competition for resources and became extinct on the mainland soon after the arrival of the dingo. Their extinction may indicate that dingos had greatly reduced the prey available to both carnivorous marsupials and humans.

The only prey species that appears to have been exterminated by the dingo is the flightless Tasmanian native hen (*Gallinula mortierii*). This chicken-sized bird grazes on short grass sward, which is often maintained by grazing marsupials. Four thousand years ago it inhabited the alpine areas of south-eastern Australia. It was probably relict in these areas, for during the last ice age it was more widespread on the mainland, occurring as far north as Queensland. It may be that dingos were able to destroy it because it favoured short grasslands which are precisely the same kind of habitat that dogs perform best in. While its extinction may have been a loss to the Aborigines of the high country, it clearly had only a local impact.

A final, fascinating study which has arisen from the examination of bones is the remarkable work of Dr Steven Webb of the University of New England.[138] In the early 1980s, before so many human remains had been lost from Australian museums, Webb examined virtually every human bone in museum collections for signs of stress caused by disease and starvation. He found that there was a substantial increase in the incidence of such stresses with time. Just why this occurred remains unclear. It may be that it reflects a strengthening of ENSO, or indeed the opposite, for if ENSO weakened, human populations may increase, leading to increased competition for food and a greater chance of disease transmission.

Curiously, Webb also found that the incidence of stress varied around the continent, with the greatest levels in the central Murray River area and coastal New South Wales. Both of these areas—but along the Murray in particular—supported higher human population densities than were normal for Australia. Webb speculates that the high incidence of evidence of diseases among bones from the central Murray may have been caused by a unique lifestyle that led to osteoarthritis. Victorian groups suffered seasonal and acute stress, perhaps due to the harshness of winter there.

Clearly, Webb's study is of enormous potential importance in understanding the past. Rapidly developing technologies for examining bones, such as CAT scans, would almost certainly have seen even more extraordinary advances in this area. Unfortunately, now this will never happen, for increasingly, museums are yielding to pressure from lobby groups and are giving over Aboriginal osteological material for reburial. Those bones once belonged to people for whom it was not possible to write to their descendants about how life was for them. But through studying the ancient bones, living Aborigines can read in each the very personal story of a life led so very long ago. As bleak as it is to be a future eater, it is as nothing compared with the tragedy of obliterating one's past.

One area of Australia preserves a far more detailed and informative record of cultural change than any other. This is Arnhem Land, which shelters one of the greatest, if not *the* greatest, concentration of ancient art on Earth. There, people have been painting the walls of rockshelters for 18 000 years. Researchers are now beginning to understand the significance of various painting styles and are coming up with a chronology for the works. Some individual paintings are even able to be dated.[129]

Analysis of the Arnhem Land art galleries has revealed some extraordinary things. Some paintings dating to over 3000 years ago, for example, show people holding boomerangs. The Aborigines who inhabit Arnhem Land today do not possess boomerangs and have no traditions regarding them. This seemed like a mysterious loss until researchers noted that spear throwers were not depicted in the older art. Then it was discovered that at about 3000 years ago the vegetation of Arnhem Land became denser. Boomerangs can only be used effectively in open country and in forest the spear, especially if propelled by the ingenious woomera or spear thrower, is a far superior weapon.

There are still many mysteries regarding the Arnhem Land paintings. Some dating to about 3000 years ago show people using a short, trident-like spear. Its points are barbed and spread widely (Taçon, personal communication). No-one living today can determine its use.

Other changes recorded in the art are those concerning food resources and even the fighting of pitched battles. The study of Australia's rock art is still young, but it promises to be a treasure-trove of information when further research is completed.

It is now time to look at Aboriginal cultures as they were in 1788, to see what was recorded before they were disrupted and to determine what these early observations can add to our understanding of the ecology of the Aborigines.

The traditional European view of Aboriginal cultures has been that, by and large, life has 'stood still' for the Aborigines since they had arrived in Australia. This view was based upon the observations by anthropologists and others that Aborigines had arrived in Australia in the stone age and were still in it by the time Cook arrived. This is a very superficial and naive view. The truth, as always in such matters, is far more complex. Indeed, I suspect that many of the features of Aboriginal lifestyles that we continue to view as primitive are highly specialised responses to Australian conditions.

A fascinating study of Aboriginal settlement patterns was undertaken by the anthropologist Joseph Birdsell in the 1950s.[17] He tested the simple premise that Aboriginal population density was directly related to rainfall. He found that throughout much of the continent the basic relationship held good. But he also concluded, from reading early ethnographies, that the average size of an Aboriginal 'tribe' was 500 individuals. Remarkably, this is the same population size that geneticists postulate is needed for long-term survival of mammalian populations. Birdsell found that it was only in very unusual circumstances that larger tribes existed and that very few areas of Australia he examined had dramatically denser populations than the average. Indeed the only exceptions in terms of population density were the tribes inhabiting the Murray River drainage system, where the population density could have been as great as 40 times the average for Australia overall. This exceptional density was permitted by the nutrients and water carried to the lower reaches of the Murray River from a very large catchment. This concentration of resources created a uniquely rich environment that allowed an exceptional build up of population.

One of the things that Birdsell's study tells us, I think, is that the Aborigines of most of Australia were living a delicately balanced existence, for their population was kept very close to the limit of viability dictated by genetics. In order to maintain this number, during bad times they had to use every resource available, as well as to develop some ingenious social arrangements. An analysis of just what constraints they laboured under and what solutions they developed to overcome them, forms the bulk of the discussion below.

The most striking features of Aboriginal lifestyles to nineteenth century Europeans were doubtless the absence of agriculture, widespread nomadism, its associated simple dwellings and limited material possessions. We now know that it is rather deceptive to view these features in isolation, for they were part of an adaptational response that included religious beliefs and social customs which, as a whole, maintained a balance between Aborigines and their environment. In order to understand that adaptational response it is necessary to examine all of these factors. There is no better place to start than the lack of agriculture.

A lack of agriculture has long been cited as evidence of the 'backwardess' or 'laziness' of Aborigines. It was indeed the basis of the British legal concept of *terra nullius*. *Terra nullius* gave the British a moral right to occupy 'unused' lands and to the English, Australia appeared to be one vast, unused land. To them, the very essence of ownership constituted tilling of the soil.

I have long been uneasy with the idea that the lack of agriculture by Aborigines is a primitive trait retained from remote ancestors. This is because it seems probable that humans have been practicing rudimentary forms of agriculture for many tens of thousands of years. As I will discuss below in relation to New Guinea, the boundary between agriculture and wild harvesting seems to disappear when food-gathering practices are examined in detail. In some areas of Australia there is indeed evidence for 'plant curation'. The most important was broad-acre management using fire, but because of the multiplicity of uses that Aborigines put fire to, it is difficult to evaluate from the perspective of plant curation alone.

More exclusively agricultural in nature are the practices of small-scale curation, such as the removal of competing plant species, the diversion of small streams to provide water to certain food plants and the transplantation of useful plant species. Some of these small-scale practices are perhaps hundreds of thousands of years old. Some were, I think, part of the behavioural repertoire of the first Aborigines. What is remarkable is that in Australia, these practices, with the exception of fire, seem to have survived only in exceptional circumstances. Over most of the continent, even the relatively small investment of time and energy in agriculture that such practices entail may have been uneconomical.

There are a number of reasons why agriculture may have been down-played in the economic activities of the Aborigines. One way of understanding this is to seek examples from elsewhere in the world where people have abandoned agriculture. An extraordinary example has been recorded among people living in coastal areas of Europe some 5000–4000 years ago. There, people abruptly abandoned agriculture and began to harvest increased numbers of maritime resources, particularly seals. In this case, an increase in the biological productivity of the North Sea, with a corresponding rise in seal numbers, might have been responsible.[20] This example is interesting, for it shows that agriculture does not always pay, even where conditions are favourable. Clearly, a similar scenario could not apply in Australia, where marine resources are so limited. So other reasons must be sought.

The lack of agriculture among Aborigines was certainly not brought about by a lack of plant species suitable for cultivation; for among the indigenous flora of Australia there are yams, taro, nardoo and various grasses, relatives of most of which have been cultivated elsewhere. There are also various members of the family Solanaceae (including tomato, potato, tobacco, capsicum and other species that have been widely cultivated). In addition, various tree species such as *Macadamia*, *Terminalia* and *Araucaria* produce nut crops that are important cultivars in the Pacific and elsewhere.

It is true that Australia's poor soils militate against the development of agriculture, yet New Guinea also has only small areas of fertile soils, but agriculture there is a venerable and profitable tradition. So an argument based upon poor soils alone is not entirely convincing.

The real reasons, I suspect, lie in the effect of ENSO on productivity in Australia. ENSO brings more variability and unpredictability in weather patterns to the eastern two-thirds of Australia than are experienced almost anywhere. Even with all of the benefits of irrigation, a modern transportation system, a worldwide economic system and extraordinary storage technology, Australian agriculture is at the mercy of ENSO. Australian agricultural industries often find it hard to fill orders placed the previous year because of drought, fire or flood, all spawned by ENSO. Thus, with its two- to eight-year-long cycle,

ENSO makes life a gamble even for the most prepared of farmers.

It is easy to imagine the difficulties that Aboriginal people may have encountered had they attempted to intensify plant management into agriculture. Were they living in an area with tolerably good soils and if ENSO was kind, they may have done very well for a couple of years, increasing their numbers and investing effort in the development of a permanent camp. Then, one year, no useful rain would fall. Even if they had storable food on the scale that European farmers use to get through the winter, it would have been insufficient to save them, for the critical difference between ENSO and a European winter is predictability. European farmers know, with a margin of error of a couple of weeks, how long winter will last and they plan accordingly. We owe the great mid-winter feast—so beloved by our pagan ancestors that upon being missionised they changed its name to Christmas and continued to celebrate it—to such knowledge and planning. The fact that the Europeans could, in the face of dismal mid-winter, consume vast amounts of foodstuffs with alacrity, is remarkable testimony to the predictability of the coming spring.

In contrast, an ENSO drought might last for months—or years. It may be followed by useful rain or devastating floods. Without a tight social and economic network that spans a continent and the technology necessary to store and transport vast amounts of food, such obstacles are probably insurmountable difficulties for agriculturalists. It makes an enormous amount of sense to me to see the lack of agriculture by Australian Aborigines as a fine-tuned adaptation to a unique set of environmental problems, rather than as a sign of 'primitiveness'.

I think that the same can be said for patterns of Aboriginal settlement and resource use. Nomadism was clearly an adaptation to tracking the erratic availability of resources as they are dictated by ENSO. Nomadism has a great cost, for possessions must be kept to a minimum. The Aboriginal tool kit was thus rather limited, consisting of a number of usually light, mostly multi-purpose implements. Investment in shelter construction is likewise constrained by such a lifestyle, for there is no point in building large and complex structures when ENSO may dictate that the area be deserted for an unknown period at any time.

Before moving on, it is worth mentioning that none of the characteristics of Aboriginal culture discussed above have a genetic basis. In the past, people have argued that Aborigines were, by virtue of their supposedly primitive condition, incapable of 'advancing' by adoption of agriculture and a settled life. That this is demonstrably untrue is shown by the very early development of agriculture by their New Guinean relatives, who are genetically and technologically extremely similar to the Australian Aborigines. Indeed, at the time agriculture was developing in New Guinea, people could have walked over dry land from Tasmania to Irian Jaya.

To return to the issue of human adaptation to Australian conditions, how do we make sense of other aspects of Aboriginal lifestyles in light of ENSO? In this regard, I think that it is singularly shortsighted to try to understand the ecology of the Aborigines solely through their technology and economy. This is because the nature of the conditions that they were adapting to made the social contract of extraordinary importance and down played the significance of technology. Just why technological advances were relatively unimportant I will explain below. But now we must look at the nature of the social contract in Aboriginal society.

An important facet of the cultural life of the Aborigines is that extraordinary social obligations must sometimes be honoured. This is because in the most difficult of droughts, people abandon their land temporarily and seek refuge with neighbours. It takes remarkably strong social bonds for people to share their limited resources with guests at such times. Doubtless, warfare and bad relations between neighbours existed in Aboriginal Australia, but they appear to have been common only in those few areas where population density was high or the impact of ENSO lessened. One example comes from densely populated Arnhem Land, where approximately 25 per cent of males of reproductive age were killed in inter-tribal skirmishes. Intriguingly, the Aborigines living in the south-west of Western Australia had a well-developed system of 'payback', remarkably like that existing in New Guinea today.[57] This region is almost unique in Australia in that neither ENSO nor monsoon variability has a marked effect. It is possible that this relative climatic predictability accounts for some anomalously high population densities recorded there, as well as for the development of the payback system.

Despite these exceptions, when looked at overall, Aboriginal society clearly lacked the degree of xenophobia and constant warfare exhibited by the people inhabiting New Guinea and New Zealand. This was due, I suspect, to the development of harmonious and extensive social networks, the existence of which was imperative if people were to survive ENSO.

Other factors selected for widespread social networks. ENSO and the inherent poverty of Australian ecosystems have kept Aboriginal populations close to the lower limit for long-term genetic survival. This has made neighbours with whom one can intermarry one of the most valued resources.

Just how the social life of Aborigines was organised in light of these conditions is revealed through an examination of Aboriginal religion. Religion is one of the most omnipresent features of Aboriginal culture. Today, even in the face of rapid cultural change, the effects of traditional Aboriginal religion remain strong. They are manifested in land claims, social obligations, rules of marriage and other practices. Through sacred sites and land claims, they are having an important impact on all Australians.

In pre-contact times, religious beliefs manifested themselves in every aspect of an Aboriginal person's life. Religion was the main controller of peoples' lives and movements in a way matched only, in European culture, among the inmates of the stricter Catholic monasteries. Even now, in remote areas religious beliefs dictate whom a person will marry, when certain tracts of land are burned and when they are left idle, when a person will visit another group and even what a person will eat and when. All of these things have an economic impact, and economic and long-term religious interests coincide closely.

The reason why long-term economic concerns and religious belief coincide so perfectly in Aboriginal society, is, I suspect, because Aboriginal religious beliefs have been evolving for tens of thousands of years. In effect, they embody hundreds of generations of accumulated wisdom regarding the environment and how best to utilise it without destroying it.

As noted above, Aboriginal religion regulates the social life of a people by determining when, given the right environmental conditions, ceremonies will occur and who will take part in them and what obligations will be carried by whom as a result of them. It also determines, very exactly, who will marry whom. This is often known long before a person is born. Thus, religion tightly controls a people's social relationships and it determines where the genes of an individual will be spread. Perhaps, over 40 000 or more years of experience with Australian environments, Aborigines found that the best way to 'codify' their relationships in order to achieve maximum genetic fitness where potential partners were few, was through strict adherence to invariable religious observance.

Nomadism and low population densities means that opportunities for contact among people of different groups is very limited. For this reason, the great tribal meetings, known by Europeans as corroborees, were critically important events for Aborigines. Corroborees often involved hundreds of participants drawn from several tribal groups. There, contacts were made, social obligations renewed and opportunities for individuals to move from one tribe to another, either through marriage, or for reasons of initiation, were created.

Despite their importance, corroborees could only take place in exceptional circumstances, for enough food had to be present in one place to feed hundreds of people for weeks. In the drier parts of Australia this was a particularly severe problem, for it was only following exceptionally good rains that favoured areas contained the requisite resources. Elsewhere, people were less constrained by their environment. A few groups were fortunate enough to have a reliable resource such as the bunya pine (*Araucaria bidwilli*) of south-east Queensland, which produced nutritious nuts in abundance every three years and to which people flocked from miles around.

An interesting variation on the corroboree is the *Rom* ritual of Arnhem Land, which involves visits by the performers of the *Rom* ceremony to distant parts and results in the cementing of numerous social obligations. It is interesting to hear the words of an Arnhem Land elder concerning the impact that transport by car and plane has on the ancient visiting ceremonies such as *Rom*:

> *we're getting friends; we want to make friends. For a long time now we had no friends, too far away. But now with planes and*

cars it's all right. We are one fullblood Aborigines.[63]

Similar visiting or corroboree-like ceremonies occurred throughout Australia. They were highly successful in linking the scattered bands and in providing information about distant groups. An extreme example is provided by Daisy Bates, who records that people living at Ooldea in South Australia had detailed knowledge about the tribal relationships of individuals who lived up to 1600 kilometres away.[10]

It is all too easy to see ENSO as an unmitigated disaster for Aborigines. But in truth there were also some great benefits. The size of hunter–gatherer populations is determined by the resources available at the worst of times. This is because humans reproduce slowly and live a very long time. This reproductive strategy means that a bad season can reduce a population so greatly that it takes a long time for it to recover. During the long period that the population is recovering, the people will be living at a level well below the carrying capacity of the land. This means that many resources will be easily available. Researchers have calculated that, in normal times, hunter–gatherers utilise only 20–30 per cent of the resources available. This probably results from the occasional hard season which periodically devastates the population.

In Australia, ENSO produces such wild swings in climate that the hard times are extraordinarily hard, even though they may be experienced as rarely as once every few decades. Despite their infrequency, these events may keep populations at a lower level than they would otherwise reach, giving people a relatively easy living, with a choice of resource options, in the good times.

Because resources are normally more than sufficient for people in such situations, they have much leisure time. During their stay with the Anbarra people at Anadjerraminyia outstation, Drs Betty Meehan and Rhys Jones calculated that the Anbarra obtained about 50 magpie geese daily, which is at least one goose per adult in the camp. This food was obtained during some 10 minutes work each morning and evening by three or four men, who admittedly were using shotguns. Nonetheless, the amount of time invested in these 'wild goose chases' was a mere one or two man-hours per day.[94]

Much of the leisure time of the Anbarra was utilised in the pursuit of religious and social goals, with ritual business totalling at least 40 man-hours per day. This division of time, with much devoted to social activities, seems rather typical of Aborigines living throughout Australia.

Richard Gould, who has lived with the Aborigines of the Western Desert, notes that resources in this inhospitable corner of Australia are sufficient so that even prepared plant food is occasionally in surplus and is discarded, people saying that they are not hungry for any more of it. Yet the Western Desert Aborigines have fewer plant staples (10) than has been recorded for any other group of Aborigines, or indeed any desert-dwelling hunter–gatherers living anywhere on earth.[56] Incredibly, in times of mild drought, the number of plant species supplying staples drops to three, yet the volume of food usually remains sufficient to feed the people. In very bad times, however, things are different. The lack of food and water forces people to abandon their territories and seek refuge among neighbours.[56]

The easy life that was created most of the time by ENSO made it possible for Aborigines to invest heavily in ritual and religion. Religion, in turn, codified ecological wisdom. Advances in technology which made the gathering of food easier were, in these circumstances, secondary considerations. Because of their role in trade, stone axes may have been more important socially than economically. It seems that, as with fire, ENSO reinforces a self-perpetuating loop that drives society inexorably in one direction.

Before moving on, it is worth noting the adaptations that have occurred on a regional level among Aborigines who find themselves in a particular environment and fine-tune their ecology to local conditions. There is considerable diversity in the resources utilised around Australia, perhaps because of the ENSO-influenced low populations and resource excess that exist most of the time. Thus, in some areas, people concentrated on shellfish, in others on fish and in yet others on small mammals. The vegetable foods also varied. Yet in all, these were essentially small-scale variations, which is remarkable considering that Australia is an enormous continent, spanning environments as diverse as rainforest and desert. Before European contact it was home to some 300 000–600 000 people who spoke 250 distinct languages and who were members of hundreds of different 'tribes'. Europe, by comparison, appears to be linguistically homogeneous. Yet the diversity of lifestyles lived by the Europeans, including as it does, farmer, herder and fisherman, is far greater than that seen in Australia.

This overarching similarity of Aboriginal culture throughout the continent is most striking. It may have its roots in the broad social networks of the Aborigines, which allowed for wide diffusion of ideas and technologies. It is clear that both trade and ceremonies linked Aborigines across the continent and that songs and dances were passed from group to group, often over enormous distances. Roth, the great recorder of the ways of the Queensland Aborigines, noted that in the late nineteenth century a ceremony, the *Molonga*, which originated in the Selwyn Ranges of north-western Queensland, was passed on to groups living as far away as Adelaide and Alice Springs.[119] Likewise, stone axes quarried near Mount Isa made their way to areas as distant as the Gulf of Carpentaria and the Great Australian Bight, while pearl shell from the Kimberley was traded deep into southern Australia. There can be little doubt that people and ideas travelled with these songs and goods. People certainly knew of each other—and the general lay of the land—in areas far distant from their own territories.

These great social networks, so important in linking widely scattered people in a people-poor land, were probably the great achievement of the Aborigines. Of land and food they normally had plenty. It was people that they needed, yet neighbours often lived far away. To build 'ropes' with distant groups, some perhaps meeting on only three or four occasions over a lifetime, was the great challenge.

Finally, we must consider the relationship between Aboriginal people and their food species. Despite the early exploitation that probably led to the demise of the megafauna, by 1788 Aboriginal societies had developed a large number of rather sophisticated practices for conserving animal resources. Unfortunately, the environmental implications of many Aboriginal practices have been lost, for they vanished before the study of ecology was understood by researchers. But some striking examples have recently been recognised in regard to hunting. One of the most interesting concerns the tradition of 'story places' in the rainforests of north Queensland.

Even today, the Aboriginal people of the Cooktown area regard certain mountain summits as 'story places'. The Aborigines believe that these places are inhabited by spirits and should on no account be entered. The most important story places in the lands of the Gugu-Yalandji people of the Cooktown area are the summits of Mount Finnigan and Mount Misery. Until hunting ceased in relatively recent times, these were the only areas where Bennett's tree-kangaroo (*Dendrolagus bennettianus*) was regularly seen.[90]

Bennett's tree-kangaroo was among the most important of game animals to the Gugu-Yalandji people in traditional times. Until the 1960s they hunted it avidly with dogs. In some ways it was an easy target, for once a dog had located the tree that a tree-kangaroo was resting in, it rarely escaped the humans who climbed after it. This hunting pressure had greatly reduced the population and early zoologists often recorded just how rare tree-kangaroos were in north Queensland. Carl Lumholtz, a biologist who visited the Atherton Tablelands area in 1882–83 was determined to obtain a specimen of a tree-kangaroo. The specimens he sent back to Europe were eventually named Lumholtz tree-kangaroo (*Dendrolagus lumholtzi*) in his honour.

Lumholtz searched for months in prime habitat with a party of Aboriginal hunters, yet had this to report:

> *We searched the scrubs in the vicinity thoroughly and found many traces of boongary [tree-kangaroo] in the trees but they were all old. It could be hunted more easily here, for the reason that the lawyer palm is rare, and consequently the woods are less dense. The natives told me that their 'old men' in former times had killed many boongary in these woods.*[83]

It was only when Lumholtz entered a wild and precipitous region, where 'progress was difficult and it was almost impossible to find a suitable place to camp', that he finally saw his first tree-kangaroo.

If this were the state of affairs throughout the entire distribution of tree-kangaroos, they would long ago have become extinct due to overhunting. They survived because there were a few special places, including some prime tree-kangaroo habitat, where they remained unhunted. These were the story places of the Cooktown and Atherton Aborigines.

The development of story places was probably the very best solution available to Aborigines faced with the problem of maintaining a sustainable resource of tree-kangaroos. The usual European solution to such a problem would be to limit the number of tree-kangaroos taken, either by restricting access to them to a privileged class, or licensing and setting an annual quota. These were not really options for Aboriginal people, for they lived in scattered groups which were only irregularly in contact with one another and they lacked a centralised authority structure. In any case, the European solutions may have been inherently inferior, for in the variable Australian environment, the sustainable yield of tree-kangaroos may have varied wildly from year to year and thus have been exceedingly difficult to determine.

The story places were clearly defined and extensive areas of prime tree-kangaroo habitat which humans never entered. They acted as reservoirs for sustainable populations of tree-kangaroos, from which animals could disperse into hunted areas. This happened in a number of ways. Tired old males were kicked out of their territories in the story places by younger rivals, while the young of the year dispersed from there to find new homes. All of these individuals took up residence in the suitable yet heavily hunted habitats outside the story places and it was these animals that had, for tens of thousands of years, provided protein to the Gugu-Yalandji and other groups without driving tree-kangaroos to extinction.

Interestingly, the New Guineans with whom I have worked also had story places to protect tree-kangaroos. In 1985 I discovered a hitherto unknown species of tree-kangaroo in the remote Torricelli Mountains of Papua New Guinea. It is one of the very largest of the tree-kangaroos—a magnificent black creature with a short face and long, dense fur. Known to the Olo people as *Tenkile*, it had clearly become very rare by the 1980s, as one after the other, its story places lost their sacredness in the eyes of the local people and were hunted out. I witnessed the demise of the last story place in 1990.

It was a mountain summit called Sweipini—an eerie place of gnarled, stunted trees covered in long wisps of moss. It seems to be eternally shrouded in mist; a place where the strange calls of birds of paradise can be heard, but the birds themselves are only rarely glimpsed as they feed upon the red fruit of a palm that grows only on the mossy summits. At the centre of Sweipini, near the mountain summit itself, is a small circular lake that was once the most sacred place in the entire Olo region. The Olo believe that the lake was inhabited by gigantic eels which, if ever woken, would cause terrible weather that would destroy everyone's gardens, resulting in widespread starvation. A variety of frogs acted as the eel's sentinels. Only one old man, the most senior traditional landowner, was allowed to enter the area, for the frogs knew his face. If the frogs saw the face of a stranger, they would all begin to croak, thus waking the great eels.

Sweipini was the last refuge of the *Tenkile* in the area and was widely acknowledged as such by the local people. But in 1990 some village men decided that the huge 'devil eels' should be exorcised by the local Catholic priest, a delightful İrishman who has devoted his life to improving things for the people of the region. To comply with his parishioners' wishes and in the great tradition of the Catholic Church, he undertook the arduous walk to the remote lake. In a long ceremony he exorcised the devil eels. Everyone went home pleased, but over the next month 11 adult *Tenkile* were captured at Sweipini, including females with young. When I finally entered Sweipini in late 1990, tree-kangaroos were no longer in evidence.

Sadly, the senior land owner and guardian of the place was dismayed at this turn of events, being upset that the eels had been exorcised and the tree-kangaroos destroyed. Yet he was helpless against village pressure to resist the 'Christianising' event. With his help, I tried to convince the local people that there was real wisdom in their ancestors' decision to leave places like Sweipini to the spirits and tree-kangaroos. Once broken however, taboos are not easily repaired and it seems likely that the respect with which people once regarded Sweipini will never be fully restored.

Tragically, unless story places can be protected, or, as has happened in Australia, local people stop hunting, species like tree-kangaroos are doomed. Already, after just 50 years of hunting, the *Tenkile* has vanished from many previous strongholds. The next few decades will probably see it vanish from its last.

CHAPTER 26

A FEW FERTILE VALLEYS

Austrália and New Guinea have been settled by similar people for an equally long time, yet the human experience in New Guinea has differed fundamentally from that of the Australian Aborigines. In Australia, geology, ecology and climate have conspired against the development of agriculture and its attendant social changes. They have instead pushed inexorably for social links that united people over vast areas. In privileged parts of New Guinea, deep and fertile soils, abundant rainfall and the presence of many plant species suitable for simple cultivation were pushing people in the opposite direction. As a result, the people of New Guinea developed agriculture at a very early date. Indeed, they appear to have been among the world's first agriculturalists. They are certainly amongst the world's most successful.

The problem of defining just what constitutes agriculture is an acute one when examining the prehistory of New Guinea. Traditionally, the major crops of the region have been root crops such as taro, or suckering species such as bananas. In order to propagate these plants one simply needs to grub them up, cut off the tuber or sucker and stick the leafy top back into the ground. This simple act has probably been a part of the human behavioural repertoire for 100 000 years or more. Clearly it does not qualify a person as an agriculturalist. But what is to be said of the person who returns to the newly established plant occasionally and clears competing species (weeds) away from it? And what if they plant 10 taro tops together; does that qualify as a garden? Would it do so if they fenced the patch? Clearly the definition of agriculturalist merges insensibly into the definition of hunter–gatherer and it is impossible to say where one ends and the other begins.

Despite these problems of definition, there is good evidence that

New Guineans have been sophisticated agriculturalists for a very long time. Professor Jack Golson of the Australian National University has excavated structures that appear to have been drainage ditches on what is now the Kuk Tea Plantation near Mount Hagen in the New Guinean highlands. These drains were dug over 9000 years ago and were probably excavated in order to regulate water flow in swamps so that taro could be grown.[20] Similar ditches are still used in many parts of New Guinea today.

Today, agricultural practices in the New Guinean highlands are quite sophisticated. They include the use of ditches such as described above to drain away water and possibly cold air, terracing to prevent soil creep, the use of wooden irrigation pipes, mounding and fertilisation of plants with ashes. These agricultural practices give the highland valleys a neat and curated appearance.

Remarkably, these valleys remained entirely unknown to the outside world until the 1930s. They were cut off by dense jungle, steep mountains and hostile people who inhabited the highlands fringe. Despite the use of aeroplanes, they remained unseen because of the dense clouds that wreath the mountains that surround them. It was the gold prospector Mick Leahy who, in 1931, was the first European to penetrate this isolated world. It was a world inhabited by over three-quarters-of-a-million people who had no idea that there was anything outside their valleys. It was the last time that contact would be made with a previously unknown people on such a grand scale.

Leahy both wrote about and filmed his momentous discoveries. It is extraordinary to see the valleys, alive with people, which he captured with his old movie camera. The people appear proud—as they still do today—the main superficial difference being that ragged European clothes have largely taken over from the splendid traditional *bilas* (body ornaments). Both the outside world and Leahy were surprised by the orderly and sophisticated gardens and villages of the highlanders. Here is his impression of the gardens of the Wahgi Valley:

The green garden patches were a delight to the eye, neat square beds of sweet potatoes growing luxuriantly in that rich soil, alternating with thriving patches of beans, cucumbers and sugar cane.

Some of the gardens had picket fences, the pickets being made of straight branches two inches thick, neatly hacked off to the same height. Others were stoutly fenced with rails, each section of the fence consisting of eight or ten rails laid horizontally between stakes driven in the ground... Each group of houses has a clump or two of beautiful, feathery bamboo, a few banana trees and a grove of casuarinas, and invariably flowers and ornamental shrubs.[82]

One could hardly imagine a more different culture from that possessed by the Aborigines of Australia. Yet both sprang from people of the same stock. Indeed, these people were in contact with each other until a mere 10 000 years ago. The diversity created from such similar ancestry is testimony to the power of the environment in shaping peoples' lives.

Given the extremely long history of agriculture in New Guinea, it is not surprising that a number of plant foods appear to have originated there. Among these are certain varieties of taro, sago, some kinds of yams, banana (particularly the cooking or plantain varieties), sugar cane and various nuts. Some of these crops were adopted by people as far afield as Africa, Asia and the Pacific Islands many hundreds of years before European colonisation of the Pacific. New Guinean agriculture has thus made an important, if largely forgotten, contribution to the food crops of the world.

Many other New Guinean crops would doubtless find a ready market elsewhere were they better known. The growing tip of a small forest palm tastes like a heavenly mixture of almonds, coconut and whipped cream; while to my taste, roast *pit-pit* (the immature flowering head of a relative of sugar cane) is finer in flavour than the best asparagus. Among the fruits, a wild relative of the lychee is among the most subtle and aromatic I have ever tasted. Of nuts there is a surfeit: *Karuka, Galip* and *Ngali* all vying in texture and flavour with the best anywhere. It is just possible that as New Guinea becomes more frequently visited, these fine contributions to the world's larder will become appreciated around the world.

There is one crop which is a newcomer to New Guinea. This is the sweet potato, which appears to have revolutionised life in the highlands valleys. It is now the staple throughout almost all of the highlands, where it feeds pigs and humans alike. It probably arrived in New Guinea only 400 years ago. Like the potato in Ireland, it seems to have allowed an enormous increase in population, with concomitant change in culture and economy. The characteristics of highlands society may have become accentuated with these changes.

The development and intensification of agriculture in New Guinea may have been hastened by the difficulties of hunting. Animal food is particularly hard to come by in the dense New Guinean rainforest. Indeed, away from the fertile valleys and the coast and in the absence of agriculture, New Guinea is an appallingly difficult place in which to make a living. I have stayed in the bush with experienced New Guinean hunters when days have gone by without the capture of anything larger than a rat. The difficulties of hunting in the New Guinean rainforest are best described by an experienced hunter. Saem Majnep is a Kalam man who spent the earlier part of his life as a stone age hunter in the mountainous Schrader Range. He is now a successful author who has published several books. Of hunting for cuscus in his beloved beech (*Nothofagus*) forest, he says:

occasionally hunters have days when no animals are delivered to them (by the Kceky, the goblins). There are some days when hunters are successful, others when they find nothing and come back empty-handed.[85]

For hunters not used to climbing the towering beech trees, hunting is extremely tiring work:

The reason is that there are so many possible trees, and so many clumps of epiphytes in them, that they can climb tree after tree, until their elbow-joints feel quite weak.[85]

The acquisition of dogs by New Guineans some 2000 years ago appears to have brought real economic benefits, for a good hunting dog dramatically increases a hunter's efficiency. This is because the main difficulty faced by a hunter in the New Guinean rainforest is to locate a game animal. The forest is dense with trees and every tree is festooned with epiphytes. A man might have to climb tens of trees and check hundreds of potential hiding places before locating a possum or rat. But a dog knows by smell where the animal is. Its problem is that it cannot climb to catch it. Thus, dogs and humans make a particularly devastating combination in New Guinea. Ecologists call such a combination a 'mixed feeding flock', because such relationships were first recognised and studied among birds.

Dogs appear to have caused the extinction of mammalian prey species in New Guinea. Sometime after 3000 years ago, several species of pademelon, small wallabies of the genus *Thylogale*, became extinct in the mountains of Irian Jaya.[46] Since then, other populations of alpine wallabies have vanished. Dogs are at their most effective in open areas such as alpine grasslands and today the only wild dog populations inhabiting the New Guinean mountains are found in such habitats.

The development of agriculture supports about 1614 people per square kilometre in some highlands valleys. This is the highest rural population density found anywhere on Earth.[45] Clearly, this left New Guineans with an entirely different set of problems to those faced by most Australian societies, where a shortage of humans was the greatest difficulty to overcome.

One of the most striking features of many traditional New Guinean highlands societies is the enormous antagonism and xenophobia displayed towards members of other tribal groups. Even today, the highlands are infamous for the tradition of payback which takes quite literally the biblical injunction of an eye for an eye and a tooth for a tooth. But in New Guinea the offender himself (and they are almost invariably male) need not be the target of revenge, for any member of the group to which he belongs will suffice. Often, an old woman or a child becomes the victim, for they are the most vulnerable.

The payback system involves virtually every group in the highlands in almost continuous conflict with some other group, whether it be open warfare or simply simmering hostility lying barely concealed below the surface. So universal and frequent is violence in the highlands that injuries inflicted during payback are now a burden upon the Papua New Guinean medical system. As a result, the government recently decided that, while most medical services remain free, a stiff fee must be paid for arrow removal.

It seems likely that the payback system thrives in New Guinea because it is resources, not people, that are at a premium. Every square kilometre of the highlands valleys supports enough people to be genetically self-sustaining in the long term, while the more constant climate means that groups do not have to depend upon each other during frequent hard times. In these conditions, evolution selects for individuals who fight hardest for land, women and pigs; not those who have the broadest social ties with other clans.

There are other consequences of agricultural development in New Guinea. One of the most fascinating involves certain tribes who inhabit the highlands fringe. These are usually small communities who live in marginal environments. They live within the malarial zone, so levels of health are generally poor and virtually everyone is infected with malaria and elephantiasis. Other diseases such as tuberculosis are common. Not surprisingly, infant mortality is appallingly high in these areas and longevity limited—often being as low as 30 years. To make matters worse, soils are generally poor and slopes steep, discouraging agriculture. But big game is unusually abundant by New Guinean standards, for pigs and cassowaries abound as a result of the low human populations.

These people of the highlands fringe often live in close proximity to the dense populations of the fertile and healthier highlands valleys. In some areas, they exploited the dense human populations of the valleys as a resource. Indeed, until the 1960s some of these groups were the most feared of all New Guinea's cannibals.

I lived among one such group, called the Miyanmin, for several months in the 1980s and was intrigued by their economy. They told me that in traditional times (until the early 1970s) the year was divided into two parts: the drier part of the year was known as the pig-hunting season and the wet as the human-hunting season. The human-hunting season was particularly important for them, for it provided children as well as meat. Children were valued because high infant mortality meant that almost no infants survived to childhood in some villages. Many of my informants were brought into the village as booty after raids and are now leading members of society.

Despite the intense hatred and fear of the highlanders for the highlands fringe dwellers among whom I worked, the highlanders had no way of revenging themselves for the many raids which decimated them. For the Miyanmin lived in small, scattered communities in a vast wilderness of steep terrain and dense jungle. They were also incredibly wary, the men rising before dawn each morning to check the paths and tracks around the village for signs of footprints. As a result, the valley dwellers had to look on, helpless, while a huge toll was taken on them. Indeed, in a few decades in the mid-twentieth century, about 20 per cent of the population of one particularly vulnerable valley in western Papua New Guinea was killed and eaten by the Miyanmin fringe dwellers.

New Guinea, clearly, has major lessons for the interpretation of Australia's history. It shows that, given the right environment, people of Australoid stock are able to become highly successful agriculturalists. It also shows that, at a time when agriculture was just a distant glimmer on the south-western horizon for Europeans, New Guineans were already feeding themselves from their well-made gardens. It is also important because it shows that Australoids could develop societies that live in extraordinarily high densities. But most important of all, it shows that Aboriginal society was not primitive, nor was it the only option available to the people of Meganesia.

Part Three

THE LAST WAVE: ARRIVAL OF THE EUROPEANS

*His Majesty's commission [was] read,
appointing his Excellency Arthur Phillip, Esq.
Governor and Captain General in and over the
territory of New South Wales, and its dependen-
cies... the Governor immediately proceeded to
land on that side, in order to take possession of
his new territory, and bring about an intercourse
between its old and new masters... After nearly
an hour's conversation by signs and gestures, they
repeated several times* whurra, *which signifies,
begone, and walked away from us to the
head of the Bay.*

Watkin Tench, Botany Bay and Port Jackson, January–February 1788.

CHAPTER 27

THE BACKWATER COUNTRY

Europe—the land of my ancestors—has recently been dubbed the 'backwater country'.[20] This is quite a shock to me, for I was taught at school that Europe is the cultural centre of the world, the place where civilisation as I know it began. Furthermore, my everyday experience teaches me that Europe is ancient and the source of my transplanted Australian lifestyle. Despite this, the new view of Europe as a 'backwater' has some validity, for it stems from a new appreciation of the last 40 000 years of world history.

The story of the European peoples is an important part of the history of the 'new' lands, for unless we can understand the forces that shaped the Europeans, their technology and attitudes, it is very difficult to appreciate the nature of their impact in the 'new' lands after 1788.

The peopling of Europe presents an extraordinary contrast with the peopling of Australia. Thousands, and possibly tens of thousands of years after the ancestors of the Aborigines and New Guineans had settled in their respective homelands, the home of my ancestors was still occupied by a primitive kind of human who may even have belonged to a different species from our own. Named Neanderthals, after the Neanderthal Valley near Düsseldorf in Germany where their remains were first found, they are the true old Europeans.

Neanderthals appear to have evolved from earlier European and west Asian human-like populations. The oldest remains clearly attributable to a Neanderthal are 120 000 years old, and were found in caves in Israel. A fully modern people called Cro-Magnon, who were the ancestors of the living Europeans, existed at this time but did not manage to penetrate Europe proper until some 45 000 years ago. Abundant remains of Neanderthals, in contrast, show that they were widespread in Europe by 60 000 years ago.[58]

It seems likely that technological innovations developed around 35 000 years ago finally allowed the more advanced Cro-Magnon to oust the Neanderthals from the remaining ecological niches they occupied. As a result, the last Neanderthals died out some 35 000 years ago, after coexisting in Europe for some 10 000 years with my ancestors. Neanderthals may have managed to coexist alongside modern humans for even longer elsewhere.

Neanderthals were the first people to adapt to a truly cold climate. In the intense cold of northern Europe they often set up home in caves and rockshelters. Thus 'cave men' is a not entirely inappropriate description of them.[58] Physically, they were a quite extraordinary people. Although averaging only 1.6 metres (5'3") in height, they were extremely muscular, being far stronger than any living people. The closest that modern humans come to matching Neanderthal strength is among body builders and weight lifters. But even they are probably not as strong as the average Neanderthal.

There is absolutely no doubt that even if Neanderthals were dressed in a suit and taught to speak a modern language, they would stand out spectacularly on the street of any modern city. For they had enormous faces with a very prominent and robust nose, heavy brow ridges, and a low forehead. Despite the low forehead, their brain was on average larger than that of any living people—occupying approximately 1.8 litres as opposed to 1.4 for modern humans. Their teeth were also very large, but the lower jaw was relatively chinless. The head was supported by short and very powerful muscles that attached at the nape of the neck. Studies of female skeletons suggest that Neanderthals may have had a longer pregnancy than modern humans (perhaps up to 12 months), suggesting a different reproductive pattern.

The Neanderthal tool kit was quite basic and reveals little change with time. Neanderthals almost never made tools of antler, bone or ivory, yet these were important materials for contemporaneous Cro-Magnon people. Although in the past researchers have reported finds which suggest that Neanderthals possessed rudimentary religious beliefs, all evidence for this is now disputed, for researchers have re-examined the original information in the light of modern archaeological techniques, and found all of it extremely dubious.

This is true even of such long-accepted claims as those for cave bear cults and Neanderthal burials.[58]

The absence of clear evidence of the survival of Neanderthal genes into the modern European population is intriguing. The archaeological record suggests that Neanderthal and Cro-Magnon peoples coexisted for thousands of years in Europe. This would have allowed ample opportunity for gene flow. Yet many researchers have concluded that the archaeological record provides no evidence of hybrid populations or individuals. If they are right, it seems likely that there was some genetic barrier between these groups. This strongly suggests, in turn, that Neanderthals belonged to a different species from our own.

The Neanderthals were the very last of the primitive human types to survive. It seems possible that their specialised ecological niche of big-game hunter on the frigid northern margin of Europe and in alpine areas had allowed them to evade competition with more modern humans until 35 000 years ago. If one takes a very long view of history, this survival of relict populations really does make Europe appear to be a backwater.

The Cro-Magnon people that inherited Europe from the Neanderthals did not differ physically from ourselves. They presumably evolved their distinctive features—such as a long, narrow nose, relatively robust body, pale skin and often partially depigmented hair and eyes—in the 100 000 or so years before they displaced the last Neanderthals. It is difficult to be certain about their hair or skin colour, but cave art dating to almost 20 000 years ago which was recently found in France shows people with a fully European appearance. Our earliest well-preserved European cadaver, the Ice Man, certainly provides unequivocal evidence that the people living in Europe some 5300 years ago were identical physically to the living Europeans.

As their distinctive bodies show, very different forces have shaped the evolution of the Europeans from that which has moulded the Australasians. But the environment shapes more than people's bodies; for it shapes their culture, their beliefs and their economy. Europe is mountainous and fertile, and much of it is newly emerged from under vast sheets of ice. Admittedly it is cold, but its climate is predictably seasonal, and spring brings a burst of biological activity on a scale seen in few other environments. Furthermore, the great mammals of the forest and plains, such as horse, cattle and bison, have managed to survive in Europe. These factors have been enormously important in shaping the fate of Europeans, and thus the recent history of Australia. If we are to understand Australasian history properly, we must understand a little of the ecology of the Europeans. To do that, we must see Europe over 35 000 years of environmental change.

Some 10 000 years after the Cro-Magnons vanquished the last Neanderthals, the vast ice sheets of Europe began a final advance. This period is known as the last ice age. Living conditions throughout Europe deteriorated rapidly and the Cro-Magnons were driven from much of their land. Scandinavia, most of the British Isles, and western Siberia, were covered in ice sheets several kilometres thick. A vast, separate ice sheet also enveloped the European Alps. To the south of the ice, a forbidding polar desert, inimical to life, blanketed what little remained of Britain, the Netherlands and other parts of northern Europe. Beyond that, permafrost conditions extended as far south as southern France and the Crimea. Even in the most protected parts of Europe, forest survived only as isolated patches of scrub. This vast and dramatic climatic upheaval devastated life in northern Europe, for not only did the people leave, but all other animal and plant species had their distributions dramatically altered.

About 15 000 years ago, perhaps over just three to five years, the climate warmed and the grip of the ice began to loosen. By 8000 years ago the ice had vanished, and the geography of Europe appeared much as it does today. Life responded extraordinarily rapidly to the changing conditions and a vast variety of plant and animal species, from willows to wisent, began a great race polewards to colonise the north. So rapid was the dispersal that by 8000 years ago the distribution patterns of plants and animals were beginning to resemble those seen in historic times.

As dismal as they must have seemed at the height of the ice age, the ice sheets left an important legacy to the Europeans. For in their 10 000 years of grinding across the landscape they had pulverised countless millions of tonnes of rock, releasing nutrients and laying down deep and fertile soils. It was the enormous riches lying untapped in this new soil bed that the plants, animals and humans were rushing north to exploit.

For the plants that colonised the newly created soils, a lack of nutrients was not a great problem. Thus there was little impetus to develop a great variety of species, each with its particular ecological niche, as is seen in the heathlands growing on impoverished soils in Western Australia. In northern Europe, it was the species that could rapidly travel the thousands of kilometres from refuges in the south to colonise new soils and which could grow most rapidly after arriving, that would thrive.

The plant species that we call weeds all have certain characteristics in common that are frighteningly similar to those traits selected for after the ice age in Europe. Perhaps the most important is that they are invasive, that is, they can rapidly colonise any bare soil left after a disturbance. They can also travel over vast distances, and can dominate an environment when they arrive. As their history suggests, the plants of Europe arrived after what was in effect a great weeds sweepstakes.

The animals that survived in the new European environments were also those which were able to reach the north fastest and which were preadapted to disturbed environments. Such species are usually generalists with broad ecological niches. They can breed rapidly and disperse widely, are adaptable, and can tolerate close human settlement. Furthermore, because nutrient shortage is not a great constraint, they grow and breed rapidly, and are profligate energy users. These characteristics of its flora and fauna have made Europe a 'weedy' environment. Mobile, fertile and robust, Europe's life forms were purpose-made to inherit new lands.

Following their arrival, the plant and animal communities of northern Europe did not get a chance to form complex ecosystems with lots of endemic species and unusual coadaptations. For soon the northern Europeans followed their plant and animal companions in their quest for the riches of the new soils. By 7000–5000 years ago they had acquired agriculture and began cutting holes in the newly arrived forest in order to plant crops.[20] In an agricultural pattern known as swidden, after cropping, these patches were left for the forest to regenerate before they were cleared and planted again. This continual disturbance mimicked the repeated withdrawal of the ice, for it opened up virgin soil and again and again a series of invasive species were free to establish and compete against each other. Finally, only the most disturbance-loving, tenacious and hardy survived.

This unusual history has meant that Europe is quite species-poor when compared with Australia. But what species it does have can often be found in extraordinary abundance. It also means that the physiology of many European species is different from that of Australian ones. The Europeans are profligate energy users, for soil nutrients are abundant in Europe. In Australia, in contrast, soil infertility limits all.

These great differences between the ecology of Europe and Australia are most readily illustrated by comparing the numbers of the most energy-demanding species of all: humans and other warm-blooded carnivores. The lifestyles of its large, warm-blooded herbivores are also revealing. Europe is only slightly larger than Australia, but is home to over 660 million people, compared with Australia's 17 million. Despite its enormous human population it still has the resources needed to support 27 species of mammalian carnivores, including two species of bears, which are the largest land-based carnivores of all. As discussed earlier, the warm-blooded carnivore assemblage of Australia is pitiful by comparison, the largest of its few species (most now extinct) weighing a mere 60 kilograms.

Europe's herbivores also use energy in the most extraordinary ways. Consider the deer for example. Males characteristically grow enormous antlers—in some cases weighing many kilograms. They use them for a couple of months, shed them, and then sprout new ones next year. The energy needed to undertake this is simply beyond the energy budget of Australian mammals.

These differences help explain, for example, why European organisms have been so successful at invading the 'new' lands, but why the species of the 'new' lands have been so poor at invading Europe. It also explains why Europeans have had so little success at exterminating their introduced pest species. Not only do species like the fox (*Vulpes vulpes*) and rabbit (*Oryctolagus cuniculus*) sleep with the enemy (or at least in his modified habitats), but they have learned, over 10 000 years of intimate contact, how he thinks and acts. They are as persistent and successful as colonisers, as are their human associates.

A further critical implication of this evolutionary history is that Europeans were blind, and still largely are, to endemism and biodiversity and the importance of these features in an ecosystem. They assumed that all ecosystems worked pretty much like the European ones they coevolved with; with its few tenacious species occupying ranges of hundreds of thousands of kilometres. Australia is simply not like that. In Australia, the total range of a plant species may be as little as a few square kilometres. Yet it may have persisted there or nearby for hundreds of thousands of years. Clear that square kilometre or two and you lose a unique heritage.

In Europe, life had always been a war against hardy competitors. The Australian environment did not evolve through warfare, but rather through cooperation between species to make the most of what little there was. Such an ecosystem was as foreign to the Europeans as was the way of life of the Aborigines. These factors, as much as any, account for the extraordinary lack of sensitivity of the first European colonisers of Australia to their new environment. They clearly led to enormous environmental damage and helped form the crippled environments that Australians inhabit today.

The striking contrast in evolutionary trends between the Australasians and the Europeans reveal much about how nature has shaped the peoples of these two lands. It is now time to examine those adaptations in some detail. Because culture is so complex and myriad, it is often difficult to understand and relate it to environmental factors. Technological change, by comparison, is easy to understand, and it leaves a clear fossil record. It may nonetheless be misleading to try to understand a people through their technology. Despite this, and because of its relative simplicity, it is worth beginning a comparison of adaptation of European and Australasian peoples with an examination of their technology.

Among the fields of technological endeavour where the Australasians outshone the Europeans was maritime skill. As a result, the progress of the Europeans in expanding to outlying islands was remarkably slow when compared with that of the Australasians. Australasians had settled New Ireland and the isolated Solomon Islands by 28 000 years ago. But by 11 000 years ago the Europeans were yet to reach Ireland, and it would be another 4000 years before they would reach the majority of Mediterranean islands such as Majorca. Iceland would remain undiscovered until only 1000 years ago.

Remarkably, the Australasians were to maintain their edge in maritime skills until late in the historic period. Cook encountered the superb, double-hulled sailing canoes of the Pacific Islanders on many occasions. He notes that the canoes of the Tongans were:

double ones with a large sail that carried between 40 and 50 men each. These sailed around us, apparently with the same ease as if we had been at anchor.[27]

In April 1774 he encountered the Tahitian fleet of over 300 double-hulled canoes manned by 7760 warriors. Upon seeing 40 vessels brought to shore he enthused:

it was a pleasure to see how well they were conducted, they paddled in for the shore with all their might conducted in so judicious a manner that they closed the line a shore to an inch.[27]

The situation with land-based technologies was somewhat different. There, the Europeans seemed to have the advantage, although it was not until the development of agriculture in Europe that great differences became apparent. The Europeans underwent a rapid technological expansion after 28 000 years ago. Stone tools dating to earlier times are rather simple scrapers, knives and arrow heads. But by 18 000 years ago tools had become much more refined and elegant. They began to be made with a sophisticated technique known as 'pressure flaking'. The bow and arrow had also come into common use in Europe, and painting, as evidenced by the magnificent friezes of Lascaux and Altamura, had reached a high degree of refinement. Despite these achievements, the Europeans and Australasians of 10 000 years ago would not have differed so greatly in their land-based technology or art.[20]

Technological change seems to have gradually led to the extermination of some large European mammals, such as two species of European rhino, the cave bear, long-tusked elephant and hippopotamus. The large mammals of the tundra survived until around 12 000 years ago. Then, in a dramatic wave of extinctions, the Irish elk, woolly mammoth, and the woolly rhino all vanished. There is good evidence that some tundra-like habitats in northern Eurasia were not settled by people until 12 000 years ago, being too inhospitable for any but the most superbly kitted-out peoples to survive in. Thus, the large mammals of this habitat would have remained relatively naive of people.

Despite these extinctions, the Europeans were lucky in that their large mammal fauna had had long contact with members of the genus *Homo* and so had avoided massive extinction. Thus Europe managed to retain a substantial part of its large mammal fauna, such as aurochs, wisent, moose, horse and brown bear, into the historic period. Indeed, of all mammal genera whose members exceed 44 kilograms in average adult body weight, Europe has lost only 29 per cent of those that were present there some 200 000 years ago. This is truly remarkable when one considers how altered most European landscapes are. It stands in stark contrast with the loss of 94 percent of such animals in Australia.

Professor Jared Diamond has suggested that the survival of large mammals in Europe was responsible for the development of some important differences between the ecology of the peoples of the two lands.[36] For the muscle power and speed of animals like the horse offered Europeans some extraordinary advantages once the principles of domestication were understood.

As it happened, domestication and agriculture were to remain unknown for a considerable time in Europe, despite the fact that European environments were ideally suited to both. Agriculture arose in the Middle East some 11 000 years ago. It was only 8000 years ago, however, that agriculture finally reached Greece, and it was not until 5500 years ago that it first appears in Britain and Scandinavia.[20] This slow spread of agriculture means that had early agriculturalists, such as the Egyptians, been able to visit Britain and Australia 6000 years ago, they would doubtless have categorised both peoples as primitives—just simple hunter–gatherers.

By 6000 years ago a vitally important development had taken place in western Eurasia, for people had domesticated sheep, goats, pigs, cattle and horses. The ancestors of all of these domesticated types could have easily been lost, for all, by virtue of their large size, are vulnerable to extinction through hunting. We know this because in the Americas horses, pig-like animals, and members of the families that include goats, sheep and cattle, all became extinct when humans arrived some 11 000 years ago. Paradoxically, the tight coevolution that holds people constrained within an ecosystem where they have been long resident, in this case ultimately conferred a very great benefit. For the fact that the large mammals of Europe were successful in their arms race against humans preserved a wonderful legacy of potential domesticates for the Europeans of 6000 years ago.

Domesticated animals open up extraordinary options for people. Indeed, so radically have they altered the ecology of humans that, in places where they have long been present, they can be said to have shaped the life of the people themselves. The horse is perhaps the best example of such a profound impact. As fast as the wind, as powerful as a dozen men, it has contributed critically to many fields of human endeavour. It is in the field of warfare, however, that it reached the pinnacle of its success as a human commensal. As Jared Diamond put it 'horses became the unstoppable Sherman Tanks of ancient war'.[36]

It is likely that it was the domestication of the horse that allowed the speakers of proto-Indo–European (the ancestor of most European languages and Sanskrit among others) to spread from their centre of origin in the area to the north of the Black Sea throughout much of Europe and western Asia. Today, their heritage is spread from India to Ireland. The horse was also a critical tool in the conquest of the Americas by Europeans, and there is no doubt that Australia's rangelands could not have been efficiently exploited had it not been for the horse. Indeed, it is difficult to imagine what European history would have been like were it shaped in the absence of horses.

The acquisition of domesticated cattle, sheep, goats and pigs, likewise opened options for resource exploitation to Europeans that otherwise would have been closed to them. With control of this group of herbivores, people could gain access to the energy held in a great variety of plant tissues. This is because goats, sheep and pigs eat a great variety of foods that humans cannot—and yet humans can eat all of them. If they are domesticated, humans can ensure that nearly

100 per cent of their meat goes to their human herders.

The option of domesticating large mammals was simply not available to the Australasians, for with the exception of a couple of species of kangaroos, they had entirely exterminated the mammal fauna in the size range of goat to horse. With such a limited surviving fauna, the chance of finding species suitable for domestication is extremely slim. Intriguingly, however, some Australasians were beginning to develop close associations with various species which could have led to domestication as long as 20 000 years ago. We know this because of finds made in archaeological sites on New Ireland to the north of New Guinea.

When they arrived on New Ireland some 30 000 or more years ago, the first humans encountered a relatively impoverished land-based fauna. This is because New Ireland really is a 'new island' which arose from the sea as little as a few million years ago. Thus it has not had time to be colonised by a great diversity of animal species. We know from examination of bones in archaeological sites that until around 20 000 years ago people ate snakes, lizards, bats, birds and rats.[49] These were the only land animals originally present on the island. In sediments dating to less than 20 000 years ago, however, the bones of a species of cuscus (a possum-like animal) suddenly appear, and shortly after predominate, making up around 90 per cent of the bone in the sediments by weight.

The cuscus is very much out of place on New Ireland. There is no doubt that it was carried there from New Britain by humans. Subsequently it spread as far afield as Timor in Nusa Tenggara, Sanana Island near Sulawesi and into the southern Solomon Islands. Throughout this vast area, it has been introduced by humans. Much of its spread may have occurred over the last 3500 years and may be related to the expansion of the Lapita people.

Just why these early efforts at animal translocation—which probably involved taming individual animals, and certainly caging, feeding and carrying them aboard canoes—did not evolve into full domestication is mysterious. It is made even more so by the fact that around 7000 years ago Melanesians undertook the same process of translocation with a species of wallaby, and more recently with a second species of cuscus.[49] Today, cuscus, wallabies and cassowaries are popular pets throughout Melanesia. Indeed, consideration has been given to farming cassowaries for sale at traditional feasts in Papua New Guinea. Despite these promising circumstances, in no case did animal associations in Melanesia lead to full domestication. Protein is one of the great deficiencies in the New Guinean diet, and it seems to defy common sense that New Guineans did not try to remedy this deficiency by intensifying management of their animal associates.

As it was, the Australasians had to wait until the Lapita people brought dogs and pigs before they could obtain domestic animals and enjoy the enormous economic and social benefits that go with them. This was a great pity, for the lack of animal domesticates, particularly large, powerful species, was to become a great tragedy for the Australasians. As Jared Diamond said:

> the extinctions [of Australian and American megafauna] may
> have virtually ensured that the descendants of those first settlers
> would be conquered over 10 000 years later by people from
> Eurasia and Africa, the continents that retained most of their
> large mammal species.[36]

The critical assets of a rich, weedy environment with a reliably seasonal if severe climate and retention of a variety of large mammal species were, I think, those assets that allowed the Europeans to become, as Diamond calls them, 'world conquerors'. By the late fifteenth century the Europeans had embarked upon an expansionary invasion that was to see them, albeit momentarily, establish colonies in virtually every corner of the earth. The fate of these colonies, and the fate of the people amongst whom they were established, was largely determined by biology and forms the basis of the coming chapters.

CHAPTER 28

AS IF WE HAD
BEEN OLD FRIENDS

O n 31 August 1699 William Dampier, Captain of the
Roebuck, met with a party of nine or 10 Aborigines while
searching for fresh water off the north-west coast of Western
Australia. The meeting was not a cordial one and ended with one
Aborigine being shot and possibly killed, while one of Dampier's
sailors suffered a severe blow to the cheek. Dampier was 'very sorry
for what had happened'[32] and avoided further contact.

Dampier's encounters with Aborigines on his two visits to Australia (in 1688 and 1699), were the first contacts between an
Englishman and the 'natives of New Holland' as Dampier called
them. The fatal shot fired by Dampier's party was the opening volley
in a conflict between Briton and Aborigine that was to remain unresolved for over three centuries.

On 22 December 1993, 305 years after the English and Aboriginal people first met, Australia's Prime Minister Paul Keating
celebrated the passing of legislation recognising the legality of native
title in Australia. With this Act of Parliament he abolished *terra nullius* from the law books (if not the minds) of the Australian people.
In effect, Keating's legislation was a declaration of peace; a recognition that Aboriginal tribal law has a place in Australian society. If the
indigenous people of Australia recognise that declaration also, Keating's historic legislation will indeed herald a new era in relations
between the people of Australia.

Despite this recent reconciliation, Australia's past cannot be
altered or denied. A violent war was fought, victims slain and the culture of the victors established by the dispossession of the vanquished.

In 1688 there were at least 300 000 people living in Australia. All but a tiny proportion of their genes (the exceptions possibly being some Macassan genes in the Arnhem Land area) were Australoid. In addition, the culture of the people had been shaped by at least 40 000 years of custodianship of the Australian continent. On 22 December 1993, Australia was home to nearly 18 million people, of whom only 265 000 claim Australoid descent. The vast majority of these people carry a mixed genetic heritage. Their culture and homeland has been dramatically transformed, and today they own just 13 per cent of the continent; much of it in the least fertile regions.

It is necessary to tell the story of Australia from 1788–1993 from a biological perspective. For without the understanding that this perspective brings, I fear that Australians will never be reconciled with their past.

Guess, if you can, where the following encounter took place.

Responding to our gestures of friendship one of them bounded down and was with us in an instant. He was a young man of twenty-two to twenty-four years, of robust constitution, and having no defect other than the characteristically thin legs and arms of his race. His facial appearance was neither stern nor fierce, his eyes were lively and intelligent and his manner expressed both goodwill and surprise...
What first interested him was the whiteness of our skins. Wishing to see if our bodies were of the same colour he opened our waistcoats and shirts, demonstrating his amazement by loud cries and by rapidly stamping his feet. His curiosity then switched to our boat. Disregarding the sailors it contained, he jumped in, totally absorbed by its construction. Silently and excitedly he closely examined the thickness of the ribs and timbers, the strength of its construction and the form of the rudder, oars, masts and sails...
It was not long before we saw the same family coming toward us along the beach. As soon as they saw us they gave shouts of joy and ran to join us. Their number was now increased to nine members by the addition of a young girl of sixteen, a small boy of four and a girl of about three years of age. They were returning from a successful fishing trip, being weighed down with shellfish

of the large type of 'sea ears' [abalone] found on these shores. The old man took de Freycinet by the hand and indicating that we should follow, led us to the dwelling which we had recently found...

A fire was quickly lit and after ordering several times 'Medi, Medi' (sit down, sit down), which we did, the natives squatted on their heels and prepared to enjoy the meal. Their method of cooking was simple, the shellfish being placed directly in the fire and baked in the shells. We tasted some and found them very tender and juicy... While these kind people were eating, we decided to treat them to some songs... At first the natives appeared more puzzled than surprised, but they began to listen attentively. Their meal was soon forgotten as they showed pleasure with such fantastic facial contortions and gestures that we could hardly refrain from laughing... The young man especially was beside himself, catching hold of his hair, rubbing his head with his hands and throwing himself all over the place, accompanied by loud screams...

The young Ouray Ouray, like her parents wholly naked, yet entirely unselfconscious, drew herself to our attention by subtle glances and an affectionate expression. De Freycinet, seated next to her, became the focus of her attention and smiles. Taking some charcoal in her hands she crushed it into a very fine powder. Holding this in her left hand, she then rubbed some of it with her right hand, first on her forehead and then her cheeks, seeming very pleased with the results...

Ouray Ouray alone carried an elegant rush bag of such unique construction that I was keen to have it. As the girl had previously given me a favouring eye, I took the risk of asking if she would give it to me. Immediately, and without hesitation, she put it in my hand, accompanying her present with a pleasant smile and some affectionate words which, sadly, I could not understand. In return, I gave her a handkerchief and an axe, the effectiveness of which astonished her brother and excited the whole family... Gradually they became accustomed to us, such that by the end they behaved as familiarly as if we had been old friends... We loaded them with presents but the only one that produced

pleasure was a plume of real feathers presented to Ouray Ouray.
She leapt with joy, called the attention of her father and brothers,
made loud shouts, laughed, and, in short, seemed transformed
with delight...
They all rose to accompany us, but after a few words from the
head man, the older woman, the baby and all the children except
the eldest, remained at the hut. De Freycinet gave his arm to
Ouray Ouray, while the head of the family was my mate. Charles
Lesueur was accompanied by the young man and the Midshipman
led the child. The undergrowth scratched their naked skins and
although Ouray Ouray was especially badly scratched, she did
not seem to notice. She was completely absorbed with chattering
with de Freycinet, accompanying her conversation with seductive
gestures and smiles, so gracious and expressive, that they were like
the most polished coquetry. As we pushed out two boats from the
shore their sadness was most moving. By means of signs, they
invited us to visit them again...
Thus ended our first meeting with the inhabitants of Van
Diemen's Land. Every detail of what I have related is minutely
exact.[136]

So, in January of 1802, did François Péron—one of the first men
ever to use the term anthropology—record with great humanity and
dignity his initial encounter with the Aborigines of the D'Entrecasteaux Channel, south-eastern Tasmania. This historic meeting
promised so much for both races, yet fate was to be extraordinarily
cruel to the participants. Within 30 years almost all were dead and
bloody war raged between European and Tasmanian.

Perhaps the most famous Tasmanian of all was Truganini—the
last of the Tasmanians who experienced life as Péron saw it. Her
death in 1876 brought to a disastrous close a chapter in the history of
human relations that began so promisingly in that summer of 1802.
She was born on Bruny Island in the D'Entrecasteaux Channel. We
do not know the name of her mother, but she must have been Ouray
Ouray's close relative, if not her daughter.

In August 1803 the first British settlers left the convict colony at
Port Jackson, intent upon establishing in Van Diemen's Land what

we would recognise today as a gulag or concentration camp. In time, this remote outpost of the British penal system was to become one of the most loathed of all places by the enemies of the English aristocracy. The convict system debased many of those who suffered within it. But it also debased the military who acted as warders and the administrators who supported its existence. For all had an intimate knowledge of 'the system' and all played their part in its perpetuation. Indeed, many thinkers of the time felt that the penal system tainted the entire society.

Macquarie Harbour, in the west of Tasmania, was little more than a death camp. The Lieutenant Governor of Tasmania described its purpose as follows:

> *Prisoners upon trial declared that they would rather suffer death than be sent back alive to Macquarie Harbour. It is the feeling that I am most anxious to be kept alive.*[66]

A place of extreme physical and mental torture, its inmates were reduced to a state of degradation repugnant to all Australians living today. There were few survivors. The inmates of Macquarie Harbour often cheered executions, for death was a merciful—and often the only—means of escape. On occasion, a man would murder a fellow prisoner simply to enjoy a week or two in a Hobart prison before his own inevitable death by hanging.

The penal colony at Macquarie Harbour was built by British in order to torture British. The atrocities carried out there were just a few distant salvos in an undeclared class war that racked Britain for a century or more. The ongoing war for the liberation of Ireland was also burning fiercely at the time and many Irish patriots were forcibly deprived of their status as indigenous people and sentenced to transportation. Too often they saw out their last days at places like Macquarie Harbour. As Robert Douglas—blinded and mutilated after six months unceasing torture by his prison guards—said of the convict system at his trial on Norfolk Island:

> *Let a man's heart be what it will when he comes here, his man's heart is taken away from him, and he is given the heart of a beast.*[66]

All too predictably, the brutal social system imported into Tasmania in August 1803 had a terrible impact upon the Tasmanians. Many of those set loose among the Tasmanians had no sense of common humanity or morality. A perplexed Governor Arthur showed an extraordinary blindness, which was perhaps universal in his class, when he summarised the conflict between black and white in Tasmania thus:

the natives are now visiting the injuries they have received, not on the actual offenders [the convicts], but on a different and totally innocent class.[66]

Arthur's 'innocent' class included the builders and administrators of Macquarie Harbour, the prosecutors of a vicious civil war, and the brutal colonisers of both Ireland and Tasmania.

What could young Ouray Ouray have expected from the people who built and suffered in such a system? Her fate is, perhaps mercifully, not recorded; but the destruction of her people is the best documented chapter in the history of Australia's war between black and white. The facts are almost too distressing to relate but, in a search for understanding, they must be told.

The first use of cannon to massacre Ouray Ouray's people took place on 3 May 1803, just eight months after European settlement. Mount Victory was named for just one of the many places where slaughters occurred. There, 30 people were murdered and their bodies thrown over a cliff. Elsewhere, police slaughtered 70 people and dashed out the brains of their children. Shepherds castrated men and Tasmanian women were tied to logs, burned with firebrands and forced to wear the heads of their freshly murdered husbands around their necks. In 1828 martial law was declared in Van Diemen's Land, and a five pound bounty was put on the head of each Tasmanian adult, two pounds being paid for each child.[37]

The Tasmanians did not take this persecution meekly, but fought back valiantly and persistently. A measure of their success is the statistic that one out of every four victims of the 'Tasmanian wars' was a European. The writer Louisa Anne Meredith provided a graphic account of a massacre by Tasmanians of Europeans:

One day, some persons went out to see Hooper, and were surprised at not finding him or any of the children about, or at work as usual, and proceeded towards the cottage where, lying all around, frightfully mangled and full of spears, were the dead bodies of Hooper, his wife, and all of their [eight] children. As usual with savages when not disturbed in their work of fiendish butchery, they had cruelly mutilated their helpless victims, hammering their bones in pieces, broken their fingers etc... A black woman some time after told the whole of their plans and schemes to achieve this terrible murder; she said that a party had for three days kept watch unseen on one of the rocky hills close to the cottage, intending to wait there until Hooper went to work without his gun.[39]

By 1830 only 300 Tasmanians survived and the government made an effort to round up these last resisters. Fewer than half survived the process. By 1847 all but 47 had died, the remainder living out their days on a reserve set up for them on Flinders Island. Truganini died in 1876. No European was ever tried for the murder of a Tasmanian.[37] And not one of the stooges of Macquarie Harbour was ever brought to justice.

The conflict between Tasmanians and Europeans is a vital part of Australian history, the significance and morality of which is still debated in our daily newspapers. It is not, however, often taught in our schools, nor is it often subject to rigorous analysis in a public forum.

The basic facts of the war are clear. Both conventional warfare and terrorism were prosecuted by a violent and debased people against a relatively defenceless one. The final tally, 200 Europeans and 800 Tasmanians killed, seems small by the standards of today's horrific conflicts, but is ghastly enough when one considers that the total Aboriginal population of Tasmania was only 5000.[115]

But these simple facts do not tell the entire story, for the war was also being fought on a biological front. Historian Geoffrey Blainey has recently commented in public on what others have long discussed in learned journals. That is, that Australia-wide, disease was a more important factor in the conflict between white and black than war. He has been accused of saying this in order to lessen the guilt felt by many Australians about their past. Clearly, the question has great social and political importance in the eyes of many people. This should not, however, prevent us from taking a dispassionate look at the facts.

After pushing his body to the limits in his search for scientific knowledge, François Péron finally succumbed to tuberculosis in 1826.[136] Today, in the town square of the small village of Cérilly in central France, there is a modest monument to the great man who was born and died there. The monument is decorated with a bronze plaque. On it is depicted a funerary monument of the Tasmanian people that Péron knew and loved.

It is almost certain that Péron or some member of his crew were carrying the tuberculosis bacterium with them when they landed at Port Cygnet in that summer of 1802. They also probably carried the common cold, measles, smallpox, gonorrhoea, syphilis and many other contagious diseases previously unknown in Tasmania. It is also clear that young Ouray Ouray and her people possessed as few defences against these diseases as they had against the bullets of the Europeans. What happens when such a deadly biological cocktail is lobbed into a microbially naive population? It is difficult to tell with people, for massacres and other hostilities often cloud the picture, and many deaths remain undocumented. But if we look at other species, we find analogies that show exactly how the situation works.

Christmas Island is an extraordinary place, for it was probably the last verdant place on Earth to remain a true wilderness. It lies just 320 kilometres south of Java, which is one of the world's most densely populated landmasses, yet paradoxically it remained undiscovered until 1615, when a passing European vessel sighted it. It was to remain a virgin land until 1888, when George Clunies Ross arrived and began mining the phosphate that had been deposited as droppings by the countless birds that used the island as a nesting place.

The fauna of Christmas Island was special because it had never been disturbed by people and had been isolated for a million years or more. It included two very strange and rather attractive species of rats which were unique to the island. They had presumably been derived from Javan ancestors a million years or more before. Maclear's rat

(*Rattus macleari*) was a handsome chestnut-brown species with a long, partly white tail. The burrowing rat (*Rattus nativitates*) was dark brown, short-tailed, and had a covering of fat two centimetres thick over the upper part of its body. Both were relatively large, perhaps weighing 300–500 grams.

The rats were studied by a Dr Andrews, who visited the island in 1897, He reported of Maclear's rat that:

> *In every part I visited it occurred in swarms. During the day nothing is seen of it, but soon after sunset numbers may be seen running in all directions, and the whole forest is filled with its peculiar querulous squeaking and the noise of frequent fights. These animals, like most of those found on the island, are almost completely devoid of fear, and in the bush if a lantern be held out they will approach to examine the new phenomenon. As may be imagined they are a great nuisance, entering the tents or shelters, running over the sleepers and upsetting everything in their search for food.*[44]

Unfortunately, the Christmas of 1897 was to be one of the very last that these splendid if troublesome creatures were to enjoy. Dr Andrews relates that in 1908:

> *In spite of a continual search, not a single specimen of either species could be found on the island.*[44]

The medical officer told Andrews (who had been absent for some years) what had happened. In 1902 or 1903 many rats had been seen in broad daylight, crawling about on the paths, apparently in a dying condition. Evidently, the black rat (*Rattus rattus*) had arrived on the island at this time. Its spread was slow, however, for Andrews found that even by 1908 it had not reached the remoter parts of the island. Diseases carried by it had evidently spread more quickly and had rapidly proved fatal to both species of native rat.

There is a remarkable parallel in the story of the Christmas Island rats and the spread of Europeans around the world. In both cases, the invading populations were drawn from great international entrepôts.

In the case of the English the streets of London provided many recruits, while in the case of the rats, London, Sydney, or Batavia may have provided them. These great entrepôts were extraordinary places. Every day, cargo, people and rats arrived from every corner of the earth, carrying their own tiny cargoes of microbes with them. Both human and murid inhabitants lived in squalid, crowded and unhygienic conditions ideal for encouraging the spread of disease. Moreover, most inhabitants were in a poor state of health, and the death rate appallingly high. Before the twentieth century, London's reputation as a consumer of excess labour was based more upon its high death rate than upon its labour market. Indeed, before the spread of modern hygiene and medicine, cities everywhere were population sinks.

The appalling death rate brought about by this situation causes extremely rapid selection for tremendous disease resistance, and individuals become carriers for a cocktail of diseases that would do the Pasteur Institute proud. Any one of these diseases could prove fatal to microbially naive individuals of the same or a similar species. The arrival of syphilis in Europe in the fifteenth century shows just how deadly newly introduced diseases can be until populations develop partial immunity to them. For the century after its arrival in Europe, syphilis was a terrifying disease. There were few survivors, for it often killed within weeks, infected individuals suffering terrible disfigurement during the short course of the sickness.

The catalogue of pestilences recorded for the Aborigines of the south-west of Western Australia between 1830 and 1894 are a good example. There, the population was visited by deadly epidemics of typhus (twice), whooping cough (twice), influenza (four times), chicken pox, smallpox (three times), measles (twice) and typhoid.[57] If, by good fortune, an individual happened to be immune to one, there were clearly many more diseases to challenge them. Each infection, even if not immediately fatal, would weaken or damage the person, leaving them ever more vulnerable to other diseases.

It is little wonder then that both native rat species on Christmas Island succumbed to black rat-borne diseases before most individuals had the chance to even meet an introduced black rat. Likewise, Tasmania's Aborigines were doomed to horrific pestilence through contact with the scum of Lyme Street. Tasmanian genes did survive of course, but only when joined with genes that offered some resistance to the Pandora's box of pathogens brought by the founders of Port Arthur and similar settlements.

CHAPTER 29

DIVERSE EXPERIENCES

The experiences of the Australasian peoples in the face of European colonisation have been extraordinarily diverse. Some groups, such as the Tasmanians, faced brutal war, pestilence and extermination. A few, such as the Wanggangurru of the northern Simpson Desert, left their land of their own accord. But their experience was unusual, being shared only by some groups of remote central and northern Australia.

While there were varied experiences, a common thread which runs through the great majority is that the usual contact situation was one of conflict. The 'war' as it is now sometimes called, began relatively well for the Aborigines of Port Jackson, or *Wéerong* as they knew it, in January of 1788. By December 1790 they had killed or wounded 17 of the new colonists, without the Europeans being able to inflict any reprisals.[131] Among the casualties was Australia's first Governor, Arthur Phillip. He suffered a severe spear wound to the right shoulder while he was visiting Aborigines at Manly, and was incapacitated for six weeks. Phillip ordered no immediate retaliation, possibly because of the precarious situation of the new colony and possibly because he could see no justice in it. It was only after the fatal and treacherous spearing of McEntire, the colony's 'game keeper' (in Aboriginal eyes a poacher of kangaroos), on 9 December 1790, by Pemelwey, a Botany Bay man, that Phillip ordered the troops into action. Their efforts, led by Watkin Tench, were a spectacular failure, for much to Tench's apparent relief they failed to even contact the Aborigines.

Governor Phillip promulgated some interesting laws which reveal something of his understanding of the situation of the new colony. He decreed, for example, that anyone who met with Aborigines while fishing on the harbour should hand over a portion of his catch to them. He did this in recognition of the economic importance of fish to the Sydney Aborigines and in the hopes of avoiding conflict. He also forbade the sale of Aboriginal artefacts in an attempt to stop the pilfering of vacant camps by convicts, which was a source of constant annoyance to the Aborigines.[131]

It is surprising to learn that, despite the enormous difficulties facing both black and white at *Wéerong* in 1788–90, some true friendships blossomed. Perhaps the most striking was that between Governor Phillip and Arabanoo, a man of the Manly tribe who was the first Aborigine to live with the Europeans. Arabanoo was kidnapped by the British in order to establish a dialogue with the Aborigines. Despite this unpromising start, Arabanoo and Phillip gradually came to know and respect each other. Tench records the distress that many at the new settlement felt at his death:

> *I feel assured, that I have no reader who will not join in regretting the premature loss of Arabanoo, who died of the small-pox on the 18th instant [May 1789]... During his sickness he reposed his entire confidence in us. Although a stranger to medicine, and nauseating the taste of it, he swallowed with submission innumerable drugs which the hope of relief induced us to administer him. The Governor, who particularly regarded him, caused him to be buried in his own garden, and attended the funeral in person. [This] regard was reciprocal. His excellency had been ill but a short time before, when Arabanoo had testified the utmost solicitude for his ease and recovery.*[131]

The friendship between Phillip and Arabanoo was not unique. Watkin Tench himself developed friendships with many Aborigines. He learned the Sydney language and had particular regard for Bennelong and Colbee, his companions on several long journeys. Perhaps these friendships flourished because Tench believed that in essence, 'man is the same in Pall Mall, as in the wilderness of New South Wales'.[131]

Such friendships were by no means restricted to the days of first settlement at Sydney Cove. The historian Neville Green records a particularly poignant example of such a friendship from Western Australia. It developed in the 1830s between Mokare, a Nyungar man and Alexander Collie, surgeon to the colony at Albany. These men held each other in high regard. When Mokare died, Collie, under the supervision of the Nyungar people, dug Mokare's grave and lowered him into the ground. Three years later, while on his death bed, Collie expressed a desire to be buried next to his friend. His wish was honoured.[57]

These fine friendships flourished in part because of the more equal balance of power between black and white while British establishments were in their infancy. But with a huge build up in the European population, and the establishment of agriculture, things were soon to change. Edward Curr, whom the historian Henry Reynolds suggests had the widest overview of white–Aboriginal relations of anyone in nineteenth century Australia, described the contact situation throughout Australia thus:

> *In the first place the meeting of the Aboriginal tribes of Australia and the white pioneer, results as a rule in war, which lasts from six months to ten years, according to the nature of the country, the amount of settlement which takes place in a neighbourhood, and the proclivities of the individuals concerned. When several squatters settle in proximity, the country they occupy is easy of access and without fastnesses to which the blacks can retreat, the period of warfare is usually short and the bloodshed not excessive. On the other hand in districts which are not easily travelled on horseback, in which the whites are few in numbers and food is procurable by the blacks in fastnesses, the term is usually prolonged and the slaughter more considerable.*[115]

In all, Reynolds calculates that, Australia-wide, some 20 000 Aboriginal people were killed as a result of these wars, and that some 2000 to 2500 Europeans and their associates lost their lives as well.[115] We must remember that these are rough estimates, but they give some idea of the magnitude of the wars that won most of Australia for the Europeans.

Between 1788 and 1890 the Aboriginal population of Australia had been reduced from at least 300 000 to 50 000. The numbers killed in war clearly fall far short of the numbers needed to account for this dramatic decline. Why did the Aboriginal population decline so precipitously, only to rebound in the mid-twentieth century? The diverse experiences and outcomes of European contact for various Aboriginal peoples give us a chance to examine these issues further.

Among the best understood and documented declines in population on the mainland of Australia took place in Victoria. The story of the Victorian Aborigines has become known largely through the efforts of Dr Diane Barwick, who has painstakingly accumulated a vast body of evidence on survivorship and mortality.[9] In 1830, before successful European settlement, the population stood at around 12 000–15 000. It may well have been higher a few decades earlier, as early settlers report seeing signs of smallpox everywhere and it is quite likely that this disease ravaged the Aborigines of Victoria during the late eighteenth and early nineteenth centuries.

The Aboriginal population declined dramatically between the late 1830s and 1863, when the first accurate census revealed that only 1907 Victorian Aboriginal people survived. In 1877 there were just over a thousand survivors, while by December 1967 only three people of pure Australoid descent survived.

A very unusual feature of the Aboriginal population through the period of greatest decline is the preponderance of adult males. They outnumbered adult females by around two to one, while children made up a much smaller proportion of the population than is normal. The large number of men has never been satisfactorily explained, for in traditional societies the sex ratios are equal or slightly in favour of women. The small number of children is more easily explained. With few women of child-bearing age the fertility of the population was reduced, but on top of this there was horrific infant mortality, with respiratory diseases being the most common cause of death.

Among adults, respiratory diseases were also by far the most common cause of death. Of these, it is likely that influenza, pleurisy and pneumonia would have accounted for many fatalities. But there were also large numbers of deaths from 'liver disease'. These may have been caused by the parasitic hydatid worm, which was probably introduced into Victoria with the European settlers. It is transmitted through contact with dog faeces or the consumption of lightly cooked meat.

A final informative aspect of the decline of the Victorian Aborigines concerns the fate of individuals of mixed Aboriginal and European descent. In 1886, when only 844 Victorian Aborigines survived, already half of the population was of mixed descent. These people had perhaps the most difficult time of all. In 1886 they were expelled from the Aboriginal reserves, and were left to feed and shelter themselves through the depression of the 1890s in a society where they were often blatantly despised. In addition, the intermarriage of what were then classified as half-caste and full blood was forbidden and the children of many families of mixed descent were taken from their parents. The feelings that such policies evoked in Victoria's Aborigines have been preserved, as Barwick touchingly recalls, in the:

Numerous letters and petitions in their own carefully elegant script [which] also survive to record their opinions, grateful or indignant, of the managers and missionaries, and of the Board and its policies.[9]

The terrible bitterness that the policies towards people of mixed descent caused are still evident today. As one survivor said:

We were too black to get work, or state relief, and too white to get help from the Board [for protection of Aborigines].[9]

Despite the extraordinarily difficult time that they had, it was these people who survived and whose descendants constitute the substantial and rapidly increasing Aboriginal population of Victoria today.

Elsewhere in Australia very different situations developed. An unusual situation concerning the Aboriginal people of northern South Australia allows some unique insights into the importance of events in the early contact period. The Wanggangurru and Yawarawarka people of South Australia's north-east deserts were culturally similar, both being adapted to living in the very driest and least hospitable region of Australia. But the stories of their contact with Europeans are about as different as any could be.

The Yawarawarka once occupied land in what is now the north-eastern corner of South Australia. Their contact history is a particularly savage one, although their experiences were not as atypical as most living Australians would care to admit. The anthropologist Louise Hercus wrote that the settlement of Yawarawarka land by Europeans followed exploration in tremendous haste, and amid a display of greed that has few parallels in Australian history.[62] J.W. Wylie, station owner at *Coongie* from 1881, fenced the Yawarawarka out of their own lands and shot or poisoned any who dared to return. They were also to suffer one of the most infamous massacres ever perpetrated by Europeans on the mainland. In the mid-1880s the Yawarawarka, Karangura, Dieri, Yarluyandi, and Ngamini had assembled at Cooncherie Point, a great ritual centre, to participate in important ceremonies. The entire group—men, women and children—were cut off at the narrow part of the peninsula and shot, their bodies burned. This was done because some Aborigines had killed a young bullock. Only three men survived the massacre.

The remnants of the Yawarawarka sought refuge far from their traditional land. Some came into Killalpaninna Mission, which had been established by German Lutheran missionaries in 1866, while others travelled as far as Innamincka, Tibooburra and Silverton. The Coongie area became known as *Ngura Warla* 'empty camp'. A few eventually made it back to their tribal home, where they found refuge on Lew Reese's station 'Minnie Downs'. Reese worked the Yawarawarka hard, but was not hostile to their traditional way of life. The very last *Mindiri* (Emu ceremony) was held at Minnie Downs in the 1920s.[62] Today, not a single Yawarawarka lives on their traditional land, and their culture is irreversibly altered.

The Wanggangurru were more fortunate in their contact history. They inhabited the most hostile part of the Simpson Desert, the vast dunefield and salt pan complex that stretches north of Lake Eyre. There, they eked out an existence around their only permanent source of water: nine wells, some of which had been dug as deep as seven metres in order to reach groundwater.

Theirs is a truly forbidding country. I stayed there during the spring of 1982 while searching for fossils around the small salt lakes that extend north from Lake Eyre. The main feature of the country is the parallel red sand dunes, running roughly north–south and stretching endlessly past the ever blue horizon. On them, clumps of cane grass make up the bulk of the vegetation, little of which stands taller than a man. Occasionally one finds, lying between the dunes, a glistening white salt pan. My most vivid memory of the land of the Wanggangurru is of waking up one morning to a strange, roaring sound. Lying still in my sleeping bag, I realised that the sound was not coming from outside, but that I was hearing the sound of my blood circulating through my head. Outside, in the first light of dawn, not a single bird sang, not an insect chirped. Everything was as frighteningly still as death.

It was only during years when sufficient rain fell that the Wanggangurru could leave their wells and travel widely, attending ceremonies and visiting distant parts of their country. Not surprisingly, their land was never taken up for pastoralism. Only one or two Europeans entered it while the Wanggangurru were still in residence. They learned of the arrival of the Europeans—and doubtless heard of the devastation they wrought—during trading visits to places such as Parachilna in the Flinders Ranges, which they visited to obtain ochre. They also learned, probably from the courageous explorer David Lindsay, who encountered them in 1886, that Europeans were not all bad and that there were good things to be had, such as abundant food and water, around station houses.

The story of what happened to the Wanggangurru is known to us primarily from the memories of Irinjili, a man born in the desert, who told his story to anthropologist Louise Hercus in the 1970s. Irinjili was a boy when he last experienced life beside the wells upon which his people depended. He recalled life there as follows:

We knew nothing of weeks, or months, or years; we had no idea of how long we had been there. We just stayed there one night after another.[62]

Of their food he said:

*We had meat and we were satisfied. We weren't really
worrying about food, not like today when people eat every
five minutes.*[62]

Finally, the attraction of a constant supply of European food and
plentiful water became too much of a temptation for the Wanggangurru. Irinjili's family were the second last group to leave the desert,
entirely of their own accord, in the summer of 1899–1900. His
father and mother, paternal grandmother, two uncles and his infant
sister made the trek to Poonarunna outstation. The walk was too
much for Irinjili's grandmother, who died on the way, despite being
carried by Irinjili's father. After leaving Balcoora Well they carried
water in the skins of hare-wallabies and rat-kangaroos. On the way,
Irinjili remembers eating hare-wallabies, bilbies (both now long
extinct in the area), carpet snakes, goannas and frill-necked lizards.

Most Wanggangurru eventually settled at Killalpaninna Mission,
and although individual groups occasionally returned to their wells
for brief visits, their homeland has remained uninhabited to this day.
Perhaps because of their voluntary presence at Killalpaninna, the
Wanggangurru were the most persistent of tribal groups in retaining
their language and culture. But this advantage seems to have counted
for little in the long term. The great influenza epidemic of 1919
killed many Aborigines living at Killalpaninna, but even before this,
populations were in dramatic decline from a number of diseases. By
the mid-1970s only three Wanggangurru who had been born in the
desert survived. Now they are all gone. To judge solely from the present fortunes of the Wanggangurru and Yawarawarka, it would be
difficult to decide which group had been dispossessed and which had
voluntarily relinquished living on their land.

The land of the Wanggangurru is the most hostile country in Australia. Making a living there would always have been difficult and
fraught with uncertainty. The fact that the Wanggangurru never permanently returned to their land, even though the possibility was
always open to them, may seem to us extraordinary, for the strength
of the bond between Aborigines and their land is profound. Hercus
experienced the strength of that bond in her meetings with Murtee,
the last survivor of the Strzelecki Desert Jandruwanta. She recalls:

He was nearly a hundred years old when I saw him for the last
time. He was blind and crippled, and was being well cared for at
the Wami Kata Home in Port Augusta. Yet he would always pack
and repack his suitcase, saying 'I want to go back to my own
country, my country with the red sandhills.'[62]

In order to understand what happened to the Wanggangurru, we must consider the unique ecology of the Aboriginal people. Like all Aboriginal people, the Wanggangurru were accustomed to moving over the landscape, settling down to take advantage of a temporary abundance of food wherever it occurred. During such temporary abundances it was usual to move into a neighbouring group's land and join them at the feast. The Wanggangurru probably thought of Killalpaninna as just another such food source, albeit one that never dried up. Thus, mentally, the Wanggangurru probably never 'left' their land, but simply moved away temporarily to harvest a new resource. It was probably only in their twilight years that many Aborigines, such as Murtee, realised that they would never see the country of their birth again.

Throughout Australia Aboriginal groups came to camp around outstations, towns and missions, 'going walkabout' whenever they felt the need to revisit their land. This suggests that the strategy employed by the Wanggangurru for exploiting resources brought by the Europeans may have been more widespread among Aborigines than is generally recognised.

In 1836 Charles Darwin visited Wallerawang Station—now drowned by an artificial lake built for the coal-fired power station of the same name—then a pastoral property at the western foot of the Blue Mountains. He observed the interaction of Aborigines and pastoralists carefully:

The Aborigines are always anxious to borrow the dogs from the
farm-houses: the use of them, the offal when an animal is killed,
and some milk from the cows, are the peace offerings of the
settlers, who push further and further into the interior. The
thoughtless Aboriginal, blinded by these trifling advantages,
is delighted at the approach of the white man, who seems

predestined to inherit the country of his children.[33]

The outcome of coexistence of black with white in New South Wales was not always as peaceful as Darwin depicted it. In 1838, a mere two years after Darwin's visit, a party of Europeans captured some Aborigines near Myall Creek. They tied them together, shot them, then burned their bodies. Although 11 Europeans were brought to trial, all were initially acquitted. In what was then an unusual act of impartiality, the Governor of New South Wales ordered a retrial, as a result of which seven Europeans were convicted of murder and hanged. Some elements of the European population were, needless to say, in uproar at this turn of events.

In other parts of Australia, warfare between black and white resulted in the recruitment of Aborigines to fight other Aborigines. The Native Police were young Aboriginal men who were armed, then sent into the territory of enemy or distant Aboriginal groups to massacre, terrify or displace them. Mainland Aborigines were sent to Tasmania to help kill Tasmanians, while throughout Victoria and New South Wales, but especially Queensland, Native Police were used to help control other Aborigines living on the frontier. Similar policies have been used throughout history to eliminate enemies. They are highly successful, for no-one knows the habits of a people quite like their near neighbours. Unfortunately, it is all too often the case that near neighbours hate each other, nursing grudges down the generations.

Perhaps more importantly, the Native Police were young men who were given a gun and a position of authority in the new European society. The authority that they could wield would only have been won after many years building relationships in their own societies. In effect, they had all-power but no responsibility. Not surprisingly, Aboriginal accounts of the conflict suggest that the Native Police were the most feared of all enemies.[115]

Some of the most fascinating and important histories of Australia's people concern those of Arnhem Land. Their history is particularly important in the biological context because the Arnhem Landers have experienced a long period of contact with people from outside. As a result, they have succeeded extraordinarily well in repelling invaders and maintaining control over their land and destiny.

Asian visitors have been arriving on the coast of northern Australia for a very long time. The importation of the dingo some 3500 years ago, and the subsequent export of dingos from Australia back to Asia, argue for long and early contacts between the peoples of these two lands. But more recently another group of Asian people have made regular journeys to Arnhem Land.

These are the Macassans—otherwise known as Bugis—who are still the greatest maritime traders in the entire Indonesian archipelago. Today they undertake annual voyages lasting many months, being blown before the Trades in their wooden praus which are often raised high at stern and bow in diminutive imitation of the archetypal Errol Flynn pirate ship. Their great white sails billow in the wind, powering them across the ocean from Malacca to Irian Jaya—a distance comparable to that from Perth to Sydney.

Today, the Bugis trade principally in clothing and other goods bought cheap in the markets of Ujung Pandang or Jakarta, and sold less cheaply in the far-flung outposts of Indonesia. They are perhaps the spiritual successors of the Lapita people, for the sea is their true home.

In 1991, I encountered a Bugis fleet anchored off remote Obi Island in the north Moluccas. The local people I was travelling with were greatly disturbed at being caught at a remote location near the Bugis. They explained that they never camped near the beach at night in the 'Bugis season', for fear that they would be robbed and killed in their sleep. Their fear was perhaps not totally unfounded, for the piratical deeds of nineteenth century Bugis may have given rise in part to the English term 'Bogey Man', which was used so widely by English-speaking parents to mollify their wayward children.

For centuries the Macassans had been arriving annually in northern Australia with the north-east winds in order to harvest bêche-de-mer. These great black sea cucumbers are related to starfish and sea urchins, although they look like so many fat black turds lying on the bottom of a sandy lagoon. They are prized by Chinese gourmands, who pay huge prices for their dried bodies. The warm, shallow waters of northern Australia are literally crawling with them. Until the Macassan voyages were outlawed by the Australian government in

1907, each year they drew a Macassan fleet southwards from Ujung Pandang.

The impact of the Macassans upon the Aboriginal populations of Arnhem Land were profound. For they brought with them new goods, new diseases, new genes and a new language. As a result, the Aborigines of Arnhem Land called the first Europeans they saw *Balanda*, a Bahasa Indonesia term for Europeans which is derived from 'Hollander', as the Dutch were once known.

Relations between Macassan and Aborigine were not always peaceful, and Arnhem Landers living today can recall at least one massacre that their people inflicted upon the Macassans. The revenge raid, in which 10 praus full of armed Macassans participated, resulted in the deaths of many Aborigines and the kidnapping of others.[142] Despite these incidents, the trading relationship between the two groups seems to have been generally peaceful. Macassar trade goods, including tobacco, calico, metal axes, dugout canoes, glass and knives were rapidly absorbed by the local Aboriginal populations. In exchange the Macassans received turtle shell and help in harvesting bêche-de-mer.

There is no doubt that Macassan trade goods had an important impact upon the culture and ecology of the Arnhem Land Aborigines. But I suspect that the most important Macassan imports were the hidden cargoes of genes and microbes that they introduced into Australia's north. And there is no doubt that both germs and germ lines found ready entry into Aboriginal society.

As early as 1803, François Péron detected, among the Aborigines of northern Australia, the presence of people who were clearly carrying Macassan genes.[136] Genetic interchange is thus venerable and it persisted into the twentieth century. One elderly Aboriginal woman fondly remembered her youth in Macassan times:

Of course the Macassar men liked the young women here. You know we didn't wear any clothes then, just a sheet of paperbark and some stringy bark to cover us in front. The girls still showed their breasts.[142]

Other Aboriginal women explained that when the praus were in, it was just normal for people to talk and flirt and become attracted to each other. The impact of Macassan genes in some Arnhem Land Aboriginal populations is plainly evident in the faces of people even today. Interestingly, the flow of genes was not entirely one-way, for some Aboriginal men and women travelled north to Sulawesi with the returning fleet. There, they married and brought up families. Some even returned to Australia at a later date.[142] As a result, Australoid features can still be seen in the crowded streets of Ujung Pandang.

Today, the impact of the diseases brought to Arnhem Land by the Macassans is not clearly evident, although it was highly visible in times past. In 1827 a visiting British doctor noted smallpox scars on Arnhem Land Aborigines.[142] This was long before the establishment of the British settlement at Port Essington. The disease was almost certainly introduced to Arnhem Land by Macassans, its progress to the south perhaps impeded by the thinly populated regions of the southern Northern Territory. Tuberculosis, leprosy and malaria may also have been first carried to Australia aboard a Macassan prau.

As devastating as these diseases undoubtedly were when they were first introduced, they were probably introduced a few at a time over hundreds of years. The Aboriginal population thus had time to recover from the initial impact of one disease before being assaulted by another. This gradual inoculation, along with the Macassan genes which carried some resistance to the diseases, were to be long-term boons to the Arnhem Landers.

By the time the British settlement of the Top End was underway, the Aboriginal population of the region was large and relatively inured to several major diseases which decimated populations of indigenous people elsewhere. They were also culturally sophisticated, for they had several centuries of experience and acclimation to novelties as diverse as trade with outsiders, tobacco, alcohol and new cultural practices. No longer biologically or culturally naive, they—of all Aboriginal groups—were in a strong position to contest ownership of the land with the invading Europeans.

The diseases and genes introduced by Macassans not only provided a good defence for the Arnhem Landers, but it armed them with offensive weapons as well. For they possessed in their microbial armoury diseases to which the Europeans were exquisitely sensitive.

In biological terms, in the Top End it was the Aborigines who had the 'biggest guns'. And the biggest and best of them all was malaria. This disease decimated would-be European colonists throughout the Top End during the nineteenth century.

In all, Europeans made three attempts to settle the Top End during the first half of the nineteenth century; and all failed spectacularly. The British established their first outpost in 1824. Fort Dundas was a garrison town established on Melville Island, some 40 kilometres distant from the mainland. Aboriginal resistance was fierce; both the doctor and storekeeper of the fledgling settlement falling to well-aimed Tiwi spears, as did much of the settlement's livestock. On 31 March, just four-and-a-half years after landing, the remaining settlers abandoned the north. Then, in 1827 Fort Wellington was established on the Coburg Peninsula. A mere two years after, its last malaria-struck residents were evacuated. Finally, in 1839 the British established a settlement on the Arnhem Land coast at Port Essington. The hope was that it would grow into a great entrepôt, dealing with a vibrant Asian trade. But trade did not eventuate and by 1849 the last of the British withdrew to the south. Lack of an ability to make a living, and malaria and other tropical diseases for which they had no immunity, had vanquished them. Not until Europeans were armed with the medicines and technologies of the twentieth century could they settle the Top End in substantial numbers. Even with such advantages, they have been unable to displace the Aboriginal owners.

The success of European settlement in the 'new' lands was initially dictated in part by the range of domestic plants and animals they possessed. European animal and plant domesticates, and indeed even their pests, are entirely temperate climate species. The tropical climate of northern Australia was to prove a fatal obstacle to Europeans, their crops, foxes and rabbits alike until new, tropical-adapted animals and crops could be borrowed from South-East Asia.

In New Zealand, the Maori were also fighting a battle for survival with nineteenth century European colonisers. But there the decisive factors were quite different. The Maori were an agricultural people, but their derivation from tropical ancestors disadvantaged them enormously in New Zealand. They had relatively few crops to start with, and none were entirely suitable to the cool New Zealand climate. Their most highly prized crop, *kumara* or sweet potato, was cultivated 1000 kilometres further south in New Zealand than it could be cultivated in its native South America. In New Zealand it became an annual which had to be harvested and the tubers protected before frost killed them in winter. Even when it yielded, its tubers were pathetically small. This lack of temperate-adapted crops meant that the South Island was extremely thinly populated, being entirely unsuitable for the tropical crops which the ancestral Maori brought with them. The few South Island Maori had to rely upon a few indigenous species, such as bracken fern and cordyline, and could not utilise at all the vast grasslands of their island.

The Maori were also unfortunate in their lack of domestic animals. Their only domesticated species was the dog, which was not suited for use as a pack animal or for large-scale meat production. Furthermore, being a carnivore, it was not an efficient transformer of plant matter, such as root crops, into protein. It had to be fed things that people could consume directly. Unfortunately, it could not utilise at all the abundant grasses that European domesticates such as sheep thrived on.

In contrast, the Europeans had a plethora of plant and animal species preadapted to the temperate conditions of the cooler parts of New Zealand. This gave them an enormous tactical advantage, for with little effort they could generate huge wealth from the grasslands of the South Island. The potential wealth was so great that by the 1840s a rush of new settlers began to arrive, drawn by a standard of living and affluence that their British stay-at-home relatives could only dream of.

The struggle for land meant that it was not long before open warfare had erupted between some Maori on one side, and Europeans and their Maori allies on the other. The wars of the 1840s finished inconclusively, neither side gaining a decisive advantage. But by the 1860s, conflict erupted again, and these wars, which became known as the Settler Wars, were particularly violent. The New Zealand historian Rob Steven writes of them:

The 'White Man's Anger' unleashed upon the Maori People a

savagery which is incomprehensible unless one understands how
settler societies are driven to destroy indigenous peoples.
Eventually even British troops, although brutalised by imperialist
exploits all over the world, left the prosecution of the war to the
settlers, in disgust at the brutalities the latter were inflicting.[127]

Disease and these wars were to take a dreadful toll of the Maori. In 1800 the Maori population stood at between 100 000 and 200 000. Population declined precipitously throughout the nineteenth century, so that by 1893 it had reached its nadir, and only 42 000 Maori survived. Over the same period, the European population had grown from zero to over 600 000. During the worst years, the Maori population declined by around 1.9 per cent per annum; a figure horrific enough, but nothing compared with the 10–15 per cent annual decline suffered by the Tasmanians. Despite this sad and difficult history, the twentieth century has seen a rapid recovery in the Maori population, so that by 1991, 511 947 New Zealanders claimed Maori descent. This is double the population of pre-European times.

New Guinea and nearby Melanesian islands have perhaps the most interesting colonial histories of all. The colonial history of New Caledonia presents a remarkable parallel with that of Australia, yet the eventual outcome as regards population mix was entirely different. For over 60 years after Captain James Cook discovered New Caledonia in 1774, the island was left to its Kanak (Melanesian) inhabitants. But then in 1853 it was annexed by France. In 1864 a penal colony was established there. As at Port Jackson, the penal settlement was to form the beachhead of a larger European invasion. Transportation of France's undesirables continued until 1897, but by that time free settlers had arrived and set up homes.[126]

The last half of the nineteenth century saw a bitter struggle develop between the land-hungry French colonists and the 27 000 indigenous Kanaks of New Caledonia. Beginning in 1856, mobile columns of French troops were sent against the rebellious Kanaks. Gradually, Kanak villages were uprooted and re-established in small, infertile reserves. Kanak resistance remained strong, and in 1878 a great rebellion broke out against the French. The European settlements on the west coast, between Bouloupari and Poya, were all destroyed. They would not recover for 20 years.[126]

Even with the Kanak population forced into reserves, the French oppression was not complete, for two-thirds of the land originally set aside for Kanaks was taken away by the Governor on the basis that it was not being properly used. Native regulations were introduced that dictated compulsory labour and restricted the movements of the Kanaks to their shrinking reserves. As a result, a second Kanak rebellion occurred in 1917 resulting in many casualties, and by 1921 the Kanaks had reached their population nadir, having decreased from around 27 000 to 18 600. The downturn however was brief, for by 1939 the Kanaks had again reached their pre-European population level.

New Caledonia and the east coast of Australia share much in common. Both were made known to Europeans by James Cook, who was struck with the similarity of the two lands. Both were settled as penal colonies, and in both places a bloody war of repression was fought against the indigenous peoples by European settlers. Why is it that the Aborigines, particularly in southern Australia, were all but eliminated, while Kanaks still make up 44 per cent of the population of New Caledonia?

In the face of the military ferocity hurled against them and their almost complete dispossession from their land, the relatively small dip in the Kanak population, followed by rapid recovery, is remarkable. It suggests that European diseases, which were such a decisive factor in the decline of Australian Aborigines, were not such a problem for Kanaks. It seems possible that their relatively short period of isolation (3500 years) may account for this. For their ancestors, like the Maori, may have been exposed to European and Asian diseases, thus granting them some immunity. Unfortunately for the Kanaks, malaria is absent in New Caledonia. Had it been present, it would almost certainly have devastated the European population, and the balance of races, which is so finely set at present (44:56 per cent), may have come down decisively in their favour.

While some factors favoured the Kanaks, there were certainly others which disadvantaged Europeans in New Caledonia. I suspect that the most important of these were New Caledonia's appallingly poor soils. The first division of land for European agriculture—825 hectares near Noumea in 1859—yielded results so poor that it had to be abandoned.[126] The introduction of cattle met with better success, for it allowed Europeans to make something of the rank pasture that grows on much of drier parts of the island. Even so, Europeans found it extremely difficult to make a living, and by 1946 their population stood at a paltry 18 700. It was only with the massive nickel boom of the 1960s–1970s, and its concomitant immigration from France, French Polynesia and other French ex-colonies, that the numbers of new settlers finally exceeded those of the Kanaks. Today, the Caledoche (as the older French settlers call themselves) survive on the proceeds of nickel mining (New Caledonia is the world's third-largest producer) and on generous subsidies from France. Agriculture has proved so unsuccessful that New Caledonia must import food.[126]

The colonial history of Papua New Guinea was very different from that of the other 'new' lands, as was its outcome. Although Europeans had known about New Guinea since the sixteenth century, they were very slow to colonise it. Even when they did annex the land and pour funds and people into the area, their success was limited and transitory. Interestingly, their major settlements, at Port Moresby, Lae and Jayapura, are or were set in small areas of savanna grassland caused by a rain shadow and poor soils. It seems likely that this habitat was the only one in New Guinea which was even marginally suitable for Europeans. Curiously, the people who speak Austronesian languages in New Guinea have a broadly similar distribution, suggesting that the Lapita people may also have used the savanna habitats to gain a foothold in Melanesia.

The reasons for the European failure to colonise New Guinea are not difficult to see. Europeans did not possess a single crop and very few domesticates that will thrive under the conditions that prevail in Melanesia. If they attempted to grow local crops, they faced competition with local people, who have 10 000 years of experience in tropical agriculture behind them. This long history of agriculture has also allowed the New Guineans to build up very large and dense populations. Thus when Europeans entered the New Guinean highlands in 1930, they found an area that would have been suitable for them, but the fertile, malaria-free valleys were already overcrowded with people.

Without doubt the most potent factor working against Europeans was their exquisite vulnerability to the Pandora's box of tropical diseases that have coexisted with the Melanesians for countless millennia. An extreme of what happens to Europeans who try to settle in a traditional European manner in Melanesia, is seen in the settlement of Port Breton, on southern New Ireland.

In 1872 the Marquis de Rays proclaimed personal sovereignty over 'La Nouvelle France' an empire that included the eastern half of New Guinea. Between 1879 and 1882 four shiploads of European colonists were sent to Port Breton, a small harbour situated in the wettest part of southern New Ireland. It is a place singularly unsuited to European settlement. Poorly supplied, and with the situation about as inimical to European survival as it could be, all but 70 of the 1000 colonists died of starvation, malaria or other diseases before the failed settlement was abandoned in 1882.[116] Dutch settlements in what is now south-western Irian Jaya met the same fate. Settlements were more successful, however, in the northern parts of New Ireland and elsewhere. There, European plantation owners made a living for some decades selling copra. But even today the lives of Europeans who visit New Guinea are endangered unless they carry modern medicines. Many succumb to malaria.

Melanesia is now the last great stronghold of the original Australasian people. The Indonesian province of Irian Jaya still has a largely Melanesian population, although increasing numbers of transmigrants from western Indonesia are finding a living in towns and, less successfully, in transmigrant settlements reliant upon rice agriculture. To the east, Papua New Guinea, the Solomon Islands and Vanuatu are the world's only Australoid nations. For them, malaria has been a more important weapon of defence than rifles.

Despite the enormous variety of outcomes of colonisation for the original inhabitants of the 'new' lands, there are some underlying similarities that link experiences and explain results. Everywhere, the early contact period was characterised by a population decline among the indigenous peoples as new diseases took their toll. In some places, such as Papua New Guinea and New Caledonia, the decline was small; but in others, such as in southern Australia, it was much greater. Following this initial decline, with the acquisition of immunity and new genes, indigenous people in all areas have increased in numbers rapidly.

The pre-European inhabitants have suffered most where they have been longest isolated and where conditions have suited the specialised European ecology. These conditions coincide with eastern Australia's temperate areas, particularly where soil fertility is greatest. In these places the Aboriginal population has been all but exterminated. These patterns have prevailed no matter what the nature of race relations in an area has been. They are determined by tens of thousands of years of adaptation to very different ecological conditions, to the prehistoric loss of resources and most of all to the germ warfare that humans unwittingly carry on with each other every day of their lives.

On the broad scale, the results of this great invasion of Europeans into the 'new' lands can be measured by the genetic contribution made by the various groups to the existing populations. Today, Australia has a population of over 17 million, of whom around 265 000 claim Aboriginal or Torres Strait Islander descent. This is only slightly less than the number that existed at the time of first European contact, but the vast majority of the existing population carry some admixture of European genes that confer defence against disease and conditions such as diabetes, which arise from dietary change. Aborigines and Torres Strait Islanders own only around 13 per cent of Australia, much of it the least productive regions. Despite this, the lands that they control may eventually turn out to be a source of enormous wealth, for much is rich in minerals and has enormous tourist potential.

In 1992, New Zealand supported a population of 3.5 million, of whom just over half a million claim Maori descent. This is some 15 per cent of the total population, a considerable increase on the 9.5 per cent that they constituted in 1982. But the Maori have won a political victory also, for many New Zealanders now recognise the Maori, their culture and their laws, as coequals in Aotearoa.

In New Caledonia there are some 54 000 Kanaks in a population of around 145 000 (44 per cent), and Kanaks essentially control the provincial governments that administer the northern part of the island. In Papua New Guinea there are nearly four million people.

Almost all are Melanesian in origin, and they own virtually 100 per cent of the land.

The mosaic of peoples that inhabit the 'new' lands today is far from stable, for changes are occurring all the time. As the last of the Europeans are leaving their final toeholds in Papua New Guinea, Melanesians are finding their monopoly challenged from the west. Wallace's Line has marked the limits of advance of the Mongoloid people for millennia. To the east, poor soils, ENSO and dense rainforest have provided a barrier that the rice-growing peoples of the east have been unable to overcome. But the twentieth century has brought new technologies and political alignments which challenge these ancient defences. It remains to be seen how successful the push of the rice-eaters into Melanesia will be.

Presently, some 25 per cent of Irian's population of 1.2 million have migrated from western Indonesia or are descended from people who have done so. As with the establishment of convict settlements in Australia and New Caledonia, this settlement is government subsidised. The Indonesian government has plans to continue to increase the number of immigrants into Irian. Some reports suggest an ultimate population target of four million for the province. It is doubtful if this will be achieved, for the new settlers continue to struggle with old problems. Irian's soils are far less fertile than the soils of Java and ENSO brings a highly unpredictable climate. Many immigrants find the challenge too great. They end up drifting into towns where they find jobs as vendors or in government offices. It seems probable that for many transmigrants, their future lies in such occupations.

CHAPTER 30

LIKE PLANTATIONS IN A
GENTLEMAN'S PARK

The European history of the colonisation of Australia has followed the same pattern as has the history of all of the colonists of the 'new' lands. All have arrived at what they are convinced is a virgin land. All have found resources that have never before been tapped, and all have experienced a short period of tremendous boom, when people were bigger and better than before, and when resources seemed so limitless that there was no need to fight for them. Because there was enough for everyone, egalitarian, carefree societies with the leisure to achieve great things, have prospered. There was a period of optimism, when people imagined great futures for their nations. Inevitably however, each group has found that the resource base is not limitless. Each has experienced a period when the competition for shrinking resources becomes sharper. The struggle between people increases, whether it be a class struggle or a struggle between tribes. If people survive long enough, they eventually come into equilibrium with their newly impoverished land—and their lifestyles are ultimately dictated by the number of renewable resources that their ancestors have left them.

The history of European colonisation of the 'new' lands is on a shorter time scale than that of either the Aborigine or Maori. As a result, it documents that most fascinating phase in the biological process of colonisation—the initial boom as the standing resources of a naive land are consumed. It is thus of extraordinary interest to the biologist.

The way that colonisation of the various 'new' lands was financed and arranged was very different from one to another. This is because the way colonisation proceeded was driven by the resource opportunities that presented themselves to the new European settlers. In New Zealand, the enormous, evident and immediately exploitable wealth of the land meant that free settlement could occur from the very start. Once the Treaty of Waitangi had been signed, and the problem of conflict with the Maori decided, New Zealand could offer a lifestyle to its new settlers that was the envy of the English-speaking world. Living and working conditions there were, from the beginning, of a far higher quality than prevailed in Britain. When wool production began in the 1840s and 1850s, it provided an easy means for people to make vast sums of money.[127] All that was needed was land; in particular the open grasslands of the South Island which, until then, had provided few resources for anyone.

The situation in resource-poor Australia and New Caledonia was quite different. There, poor soils and climatic unpredictability made free settlement, at least in the initial stages, a high-risk and largely uneconomical venture. In both cases, a solution to this difficulty was found through the penal system, for both France and Britain chose to transport convicts to found their new colonies. The expense of this government-funded form of settlement was large, but was acceptable to the public as it offered a solution to a severe social problem at home.

In Australia, transportation of convicts began in 1788, and ended in 1868. In New Caledonia it began in 1864 and ended in 1897. In both instances, the prisons and administrative infrastructure provided a large, immediately accessible market for local produce. This market made free settlement an economic proposition. As a result, privately owned farms became established and gradually, once a transport infrastructure and local agricultural expertise developed, the free settlers began to dominate the economic life of the colony. But it seems unlikely that settlement in either case would have been so successful were not enormous amounts of government funds poured into the infant colonies through the socially acceptable medium of the system of convict transportation.

The forces that led to European expansion in the Pacific had very little to do with conditions in Australia and her neighbours, and much to do with conditions in Europe at the time. That this was the case is perhaps most eloquently borne out by the fact that Britain sent no ships to reconnoitre the Australian east coast, where the first European outposts in the region were to be established. This was because the government of William Pitt, which was footing the bill, had other priorities than the long-term prospects of the settlement to contend with.[66]

Pitt's problem, as Robert Hughes, author of *The Fatal Shore*[66] has pointed out, was that the undeclared civil war between the British landed classes and their underlings had brought the British penal system to its knees. With the loss of its American colonies, vast numbers of convicts were confined in intolerable circumstances in prisons and hulks. There, they were a visible reminder to all of how terrible, unjust and barbaric the undeclared civil war was. Its victims, the starving masses in the poorhouses and hulks, were also potential sources of contagion, and a constant drain on the economy.

The enormous prison population of Britain may seem anomalous considering that literally hundreds of crimes, including such relatively innocuous acts as walking on a road with a blackened face, carried the death penalty. For convicts are much more economically stored in paupers' graves than in hulks. Although public hangings were major events, the application of the death penalty on the massive scale allowed by the law would rightly have been seen by many as mass, government-sanctioned murder. Despite the barbarism of the time, dramatically increasing the number of executions was not really an option for the Pitt government. As Hughes points out, where would they find juries willing to convict? And if such juries could be found, the social situation was so critical that by increasing hangings they risked full-blown civil war. As it was, vast numbers of people, at the connivance of judge, prosecutor and defence alike, were convicted of stealing 39 shillings worth of goods regardless of the real amount, for the theft of 40 shillings carried the death penalty, and few had the stomach to do such murderous work day in, day out.[66]

In an attempt to dispose of the victims of the 'system', the British government investigated several sites in Africa to establish a new depot for convicts, but they were all so dismal that plans to settle them had to be abandoned. By 1786 the situation had become so desperate that the entire British prison system was facing meltdown.

There was simply no time to send out more scouting parties. Thus, on 6 January 1787 the first convicts were loaded from the hulks onto the ships of the First Fleet. By 26 January 1788 they had arrived in what was to be their new home.

It is fascinating to read the opinions of conditions that were submitted to the Pitt government by various people who saw advantage in establishing a settlement at Botany Bay. Foremost amongst those who ventured opinions were Joseph Banks, by then one of the beacons of the European scientific community, and James Matra, a man who held no official position in England but who had seen Botany Bay when he sailed as a midshipmen under James Cook.

Banks gave evidence to a British parliamentary committee, and his opinions were of considerable importance, for he was well connected and respected, as well as being the most highly trained botanist to have visited the place. In contrast, Matra was a petitioner for a commercial scheme which he hoped would land him a job. He opined that, with the help of Chinese slave labour, Botany Bay might produce tea, silk, spices, tobacco and coffee. He also pointed to the flax of New Zealand, and the pines of Norfolk Island, which he imagined could be used in the construction of a vast British naval fleet. The fantasy of these views were not lost upon the Pitt government, for neither Chinese slave labour nor the supplies necessary to begin Matra's industries were sent from England with the convict fleets.[66]

Matra's plans were based upon optimism and a short, first glimpse of a new land. Although extreme, one finds them echoed in the first glimpses of many other English visitors, particularly those not trained in the biological sciences. A common theme of these first impressions is that Australian ecosystems are described as if they were European ones, with the same potential for productivity, and the same responses to European agriculture and animal husbandry practices. These early observers also often saw the Australian continent as being so vast that the resources of the land were thought of as limitless. Both illusions were quickly shattered as experience with the land grew. Despite their transience, these first impressions are worth recording, for they reveal the initial approach to the land by the new European settlers.

In 1770 the barque *Endeavour* was sailing northwards along the coast of the Illawarra region of New South Wales. Her commander, James Cook, described the scene before him:

trees, quite free from underwood, appeared like plantations in a gentleman's park.[27]

He was not alone in forming this impression, for several decades later Sir Thomas Mitchell said of western Victoria that the country resembled 'a nobleman's park on a gigantic scale', which 'gladdened every heart'. In the south-west of Western Australia, in 1831 Lieutenant Preston described the countryside around what is now Busselton as like 'a fine park in England'; while in Tasmania, Lieutenant John Bowen said that the land surrounding the Derwent River near Hobart, with its 'copses and meadows', reminded him also of a nobleman's park.

Within a few months of arrival in most places, there was no more such talk from the new settlers. Instead, their delight turned quickly to disgust as their inability to unlock the riches of the land became apparent. By 1789 Robert Ross, Lieutenant Governor of the new colony at Port Jackson, wrote:

in the whole world there is not a worse country. All that is contiguous to us is so very barren and forbidding that it may be with truth said that here nature is reversed; and if not so, she is nearly worn out... If the minister has a true and just description given him of it he will surely not think of sending any more people here.[66]

For the first few years, the situation for convicts and their overseers were indeed desperate. Starvation was ubiquitous, with several convicts actually dying from want of food. It was not until February 1791 that James Ruse coaxed a fair crop from the truculent soil of New South Wales. Ruse, along with the others who were experimenting with agriculture, complained of a problem from which Australian soils had been suffering for over 35 000 years, for they could find no animal dung with which to fertilise. In a moment of genius, Ruse came up with a solution that was a remarkable parallel with Aboriginal firestick farming. He decided to burn off the timber on his acre of land and dig the ashes into the soil. He then turned the sod over to compost with the weeds and grass. ENSO was relatively kind and despite a drought, by December 1791 Ruse was supporting his family from his own land. After 1792 the infant colony was growing enough grain to support itself—in most years.

Despite their success, the new settlers were still enormously vulnerable to ENSO, for all of their agricultural lands lay along the flood-prone Hawkesbury. In 1806 this led to disaster when an enormous ENSO-spawned flood swelled the river, covering more than 14 000 hectares and destroying all standing crops, along with the colony's tools, livestock and seed reserves. Sydney and Parramatta could not feed themselves, while fledgling Hobart, established only three years before and still dependent upon Sydney for food, faced starvation.

Disillusionment was inevitable, for those first impressions of Australian landscapes were mistaken. Australia's woodlands, which so impressed European explorers as being like an English gentleman's park, bear only the most superficial similarity to English woodlands. English woodlands grow in a seasonal climate on fertile soils. They are the result of generations of shaping. The trees are planted, and are kept in their place by intense grazing pressure exerted by a vast biomass of large herbivorous mammals. The grazers fertilise the grasses in a nutrient cycle that allows for tremendous productivity.

Australian woodlands, in contrast, are shaped by infertility and fire. Anyone who has walked through an Australian woodland and an English park will realise that even the apparent similarities vanish when examined at close quarters. English pasture is so soft, rich and dense that it often entirely obscures the soil. In an Australian woodland, the grasses consist of clumps of long, usually brownish stems, between which much soil is exposed. If grazing animals such as sheep and cattle interrupt the cycle of the firestick, Australia's open woodlands turn into a dense thicket of woody shrubs. This is because fire suppresses woody shrubs in favour of grass. If cattle remove the grass, which is the major fuel for fires, the shrubs sprout. Those that are unpalatable to cattle and sheep have an enormous advantage, and thrive, smothering the remaining pasture.

The small extent and poor quality of arable land available on the eastern side of the Great Dividing Range in New South Wales drove explorers to attempt to cross the mountains to see if suitable pasture existed to the west. In 1813 Blaxland, Lawson and Wentworth succeeded in finding a way across, and the vast, if fragile west was opened to European exploitation.

Over the next century Australia was to experience an agricultural expansion that, by the beginning of the twentieth century, would bring her inhabitants the highest standard of living in the world. She looked set to become the brightest star in the southern firmament, if not one of the brightest of all. But even at this early stage there were signs that not all was well.

Charles Darwin, perhaps the greatest biologist of all time and universally acknowledged as an astute observer of nature, visited Australia in January of 1836 after a voyage of four years that had taken him half way around the world. Here is his summary of Australia's then and future prospects:

The rapid prosperity and future prospects of this colony are to me, not understanding these subjects, very puzzling. The two main exports are wool and whale oil, and to both of these productions there is a limit... Pasture everywhere is so thin that settlers have already pushed far into the interior: moreover the country further inland becomes extremely poor. Agriculture, on account of the droughts, can never succeed on an extended scale: therefore, so far as I can see, Australia must ultimately depend upon being the centre of commerce for the southern hemisphere, and perhaps on her future manufactories. Possessing coal, she always has the moving power at hand. From the habitable country extending along the coast, and from her English extraction, she is sure to be a maritime nation. I formerly imagined that Australia would rise to be as grand and powerful a country as North America, but now it appears to me that such future grandeur is rather problematical.[33]

One must remember that Darwin was an extremely modest man. For it is clear that he understood 'prosperity and future prospects' only too well. Academics have argued that, after four years at sea,

Darwin was intellectually exhausted by the time that he reached Australia. Others have suggested that he deeply disliked Australian society. There is no doubt that he detested the convict system, for the sight of men and women utterly deprived of dignity and slaving in chains on road gangs was abhorrent to him. There is also no doubt that he was horrified by the destruction of the Tasmanian Aborigines. He was well aware of the appalling circumstances surrounding their demise, for just before his visit the last few resisters had been transported to Flinders Island.

Perhaps Darwin's dislike was so strong because he saw in Australia the 'flip side' of his peaceful and enjoyable life in rural England. Australia was, after all, the repository for much of the ugliness and suffering upon which a tranquil British life depended. It was also the quarry from which raw materials such as wool and whale oil were extracted. Whatever the case, Darwin's famous valedictory on Australia was not a kind one:

> *Farewell, Australia! you are a rising child, and doubtless some day will reign a great princess in the south: but you are too great and ambitious for affection, yet not great enough for respect. I leave your shores without sorrow or regret.*[33]

It is probably a reflection of the general disregard in which the biological sciences are held that Darwin's analysis of the fledgling Australia's prospects has been largely forgotten, while those fleeting impressions of Matra have driven the nation on. Even today many Australians see their nation as having the potential to be a second North America, with the ability to support hundreds of millions of people. In a sense, it is this great illusion that has been the tragedy of the European Australians. It is the illusion that has pushed them towards limitless growth in population and expenditure. It has encouraged them to be profligate with their few resources and to imagine a grander future than their land would ever have permitted.

The other feature of the Australian environment that settlers failed to understand is its profoundly erratic nature, due largely to ENSO. This lack of understanding had its greatest impact in the semi-arid regions. After a good year the semi-arid country of inland

Australia looks wonderful, with palatable saltbush, grass and other fodder in abundance. I have had the pleasure of driving through such country in a good season. The small trees look resplendent in their flush of new growth, while each sand dune is covered in bright yellow daisies and each swale clothed in purple and white flowers. At night, when all is still, the scent of the flowers is delicious. If there happens to be a full moon, a wondrous sight is seen, for all across the silver-crusted landscape the flowerheads track the moon's brightness.

Eighteen sixty-three was one such year in South Australia. Pioneers flooded north into the magic country, and the breathtakingly beautiful Flinders Ranges area looked set for a wonderful future. Even though pastoralists had pushed further north than ever before, virtually all were making a fortune. The area was carrying more stock than it had ever done before, and wool production was so great that the track south to Port Augusta was almost crowded with bullock teams heaving high piles of wool bales to the ships, which were waiting to carry the bounty to the mills of England. To add to the wealth of the area, copper had been discovered and was being mined, producing a small fortune for those involved in the non-pastoral sector.[133]

Then the inevitable drought came. It lasted from 1864 until 1866. The alarmed government of South Australia sent a commission to investigate in 1865, and they found that from Port Augusta to Yudnamutana in the northern Flinders Ranges, nearly all vegetation fit for pasture was gone, and that the invaluable salt bush was almost entirely destroyed. Many settlers quite literally walked off their land, even their horses having succumbed to the ghastly conditions. Almost unbelievably, the magnitude of the tragedy was forgotten with the onset of a few more good years, and the entire farce was played out again in the droughts of 1888, 1896–1902, and so on.

This flogging of the land was to have dire effects. Aboriginal culture was one victim. Mr Hughes of Yudnamutana wrote to *The Register* in 1864:

> *In ordinary years the natives have ample supply of food in the numerous animals indigenous to the country. This year the terrible drought has been as fatal to those animals as it has been to the sheep and cattle of the squatters... They no longer afford the*

natives the supply of animal food on which they heretofore depended. As blackfellows have pathetically told me, 'Euro no food now—big one tumble down...'.
The natural severity of the drought is greatly aggravated by the flocks and herds of the squatters which have utterly consumed or trodden out every vestige of grass or feed within miles of water. The Aborigines have therefore two resources—they are compelled to crowd around the dwellings of the squatters and beg for food, or to ... prey upon the flocks and herds.[133]

A second casualty was the native fauna, for the 1864–66 drought, along with those of the 1890s had—in concert with the devastation wrought by introduced stock, the rabbit (which arrived in 1878) and the fox (widespread by 1910)—caused the local extinction of most of the native mammals of the region.

But perhaps the greatest tragedy was the destruction of the soil itself, upon which everything depended for existence. By the early twentieth century, soil erosion had become a problem of such proportions that it could no longer be ignored, and by the 1930s most state governments were taking some action to assess the situation. In the 1930s, the CSIR (forerunner to the CSIRO) appointed the British scientist Francis Ratcliffe to investigate the problem in northern South Australia. He arrived when much of the state had had no or little useable rain for 12 years. But, as he astutely observed, the drought and its associated soil erosion was not solely a consequence of this lack of rain. As he travelled north, Ratcliffe found a courageous but deluded people struggling to maintain their dignity and economic viability in a land which had died under them.

One of his hostesses told him that they had experienced seven dust storms during the month he arrived. She elucidated:

I don't mean just days with dust in the air, but bad enough for me to have to clean the house out. Three of the seven were bad; and by that I mean that I got about half a kerosene tin of dust and sand out of every room![113]

On one property, the sheep had been left with a strip of unshorn wool 20 centimetres wide down the back, for it was so full of grit that it would ruin the shears if it were shorn. On others, hundreds of sheep were buried alive in the blowing sand, the windscreens of vehicles were frosted by the blowing particles, and the paddocks everywhere were completely stripped of vegetation. In the midst of all this human-caused misery Ratcliffe records a small miracle:

> I was once shown a little corner, a long way from the nearest water, which had managed to survive in something like its virgin state. It was a sight for sore eyes, and a very useful indication of the extent of the changes which had taken place since the white man settled the land. There was actually grass about, and the foliage of the shrubs grew down to the very ground; and I saw little bushes here which had practically vanished from the general landscape. There were also scores of kangaroos, which can always be trusted to discover every spot where good feed is growing.[113]

The philosophy of the people who had cost the land so much was recorded by Ratcliffe on several occasions, but never so strikingly as during his visit to Yudnapinna, north-west of Port Augusta. Mr Patton, the station manager there, bade Ratcliffe to look at the pictures hanging on the walls of the station house, saying:

> 'They are symbolic. On this side you see how we make our money, and on this how we spend it.' Round the right hand walls were photographs of stud rams and ewes... [while] on the left hand wall were pictures of racehorses.[113]

I quiver with rage when I think that it was for racehorses that the Yawarawarka and Dieri people of South Australia were barbarically slaughtered and their cultures destroyed. It was only for this that future Australians were deprived of 23 species of native mammals, all driven to extinction in our dry country. And it was for nothing more than this that the very soil of northern South Australia itself was lost, never to be replaced. The deserted and desolated wastes of this pastoral country, much of it now relinquished by leaseholders, is a pathetic monument to extraordinary folly.

To us, the foolishness of the Europeans in nineteenth century Australia seems to have known no bounds, but it is not fair to damn them by the standards of today. In reality they knew nothing of Australia's past, could not see her natural riches, and were only acting in accordance with the principles that their European environments had inculcated in their ancestors. Although they appear foolish in hindsight, they were in reality only terribly maladapted.

The question of adapting to Australian conditions has only recently taxed the minds of many Australians. For throughout the nineteenth century, and well into the twentieth, Australians thought instead of ways to make Australia adapt to them. In its most extreme form, this view saw people attempting to create a second Britain in Australia.

The story of what was done to achieve this almost inconceivably arrogant goal is one of the saddest chapters in the history of our continent. For Australia's ecology floundered in the attempt. People found that a second Britain could not be established, but that old Australia could be all too easily destroyed. The feelings that motivated people to do this were not evil—indeed they may have been such noble sentiments as love of home and childhood memories. But the entire episode shows that such emotions, when combined with misunderstanding, can be as environmentally disastrous as greed and rapine.

The mid-nineteenth century saw groups of landowners, academics and others come together to form acclimatisation societies, whose aim was to introduce European or Eurasian animal and plant species into the southern lands. Eric Rolls has said of them in *They All Ran Wild*, his definitive study of Australia's introduced pests:

> *There was never a body of eminent men so foolishly, so vigorously, and so disastrously wrong. The words of colonial governors were as wild as any orator's in the Sydney Domain, and professors of zoology were unbelievable. The Englishman's dislike of change was translated into a clamour for a homogeneous world—the English nightingale was to sing in English elms in Australian paddocks, and when a lucky colonial could return home, with his fortune made out of wool, the Australian magpie was to sing to him in an Australian eucalypt in an English field.*[117]

Among the greatest of all acclimators was Edward Wilson of Victoria. He once said of species suitable for acclimatisation 'if it lives, we want it'. A vast number of species were tried. Some members spoke in favour of monkeys, so that they:

would delight [the traveller] as he lay under some gum tree in a forest on a sultry day.[117]

Others favoured the boa constrictor, and yet others the South American agouti—lauded by one member as an 'inestimable acquisition'.

Despite their great ambitions, the acclimatisation societies were not a great success, for the greater number of successful feral animal species were not introduced through their actions, but by ordinary Europeans pining for the familiar fauna of 'home', or through accidental escapes of domesticated animals.

Today, as a result of such actions, Australia is home to a large variety of Eurasian plants, birds, small mammals and 'megafauna'. Cattle, water buffalo, horses, donkeys and camels are among the largest of the surviving Eurasian fauna. Today, all roam free in Australia, as do goats, deer and pigs. In all, some 21 species of placental mammals have established themselves there. Interestingly, it was not the larger species which caused the greatest damage, but the smaller ones—rabbits and foxes. Just why this is, and what can be done with Australia's feral animals will be discussed later.

CHAPTER 31

UNBOUNDED OPTIMISM

By the middle of the nineteenth century, Europeans living in Australia had begun to find ways to make European agriculture work, albeit temporarily, under Australian conditions. The 1870s saw the invention of the stump-jump plough, and the stripping and harvesting machines which would revolutionise wheat production the world over. The plant breeding genius William Farrer experimented in breeding various strains of wheat, some of which were rust-resistant, and others of which thrived under the dry Australian conditions. Irrigation, initially introduced to Australia from the USA in 1887, had an astounding effect, turning 'a Sahara of hissing hot winds and red driving sand'[77] in the Mallee, into a garden.

The waters of Australia's first irrigation scheme were driven by great pumps, each 4.5 metres in diameter, that were locally designed and built by George and William Chaffey. Ironically, this great 'improvement', came about in part as a result of a dubious land scam, under which the brothers were given outright 50 000 acres of land, the right to purchase another 200 000 at one pound per acre and the right to whatever water they could pump, which could be sold in turn to settlers who purchased their land. Here indeed was a licence to print money!

Along with gold, it was technological advances and development schemes such as this which led to Australia's great early prosperity; and to the great depression of the 1890s. As a result of these changes, by 1900 the country had largely come of age and progress towards federation was well advanced. The 3 773 801 Australians of European descent enjoyed one of the highest standards of living in the world. Australia had entered what Manning Clark called 'The age of the optimists'.[24]

The Chaffey brothers, George (left) and William Benjamin, 1888. They made their reputations by introducing irrigation to the Mallee country. Their massive irrigation scheme was combined with a dubious land speculation project. Ecologically unsustainable in the long-term, the scheme was designed to deliver profits of 5 or 6 million pounds to the brothers. (*Australasian Sketcher*, 24 March 1887. *Melbourne Punch*, 1888)

This period in Australian history spawned a fervent wave of nationalism. By 1901, 82 per cent of the European population was Australian-born, the highest proportion ever reached until then, and indeed a higher proportion than is the case today. These native-born people saw themselves as Australians first and foremost, although most still felt a firm attachment to Britain. Their nationalism was perhaps partly engendered by a desire to throw off of the epithet 'colonial', which was still flung at Australians whenever they visited 'home', as they referred to the British Isles. It was also in part a reaction to the enormous and growing wealth and influence enjoyed by the United States of America, a nation seen by many Australians as pointing the way to Australia's future. It was a tempting comparison indeed, for the similarity in size, ethnic origin and recency of European colonisation of both nations seemed to beg Australians to see their land as a fledgling, southern hemisphere United States, and to envisage as grand a future for their new country. Here are a few predictions made between 1883 and 1923, concerning the future that Australia might enjoy:

*There is every reasonable probability that in 1988 Australia will
be a Federal Republic, peopled by 50 millions of English-speaking
men, who, sprung from the same races as the American of the
Union, will have developed a separate recognisable type.*
The Spectator, 28 January 1888.[111]

*There exists today circumstances, similar to those which created
the United States of America, that are going to assist in the future
creation of a United States of Australia. [For] a well conceived
system of settlement and development will assuredly lead to results
like those exhibited in the New World.*[95]

*If the United States has grown in the last century from five mil-
lions to a hundred millions, there is no reason why, in the coming
century, we should not grow to a population of two hundred or
three hundred millions of white people in the Empire...*
Leopold Emery, Under Secretary for Colonies, 1923.[111]

Popular notions of the time concerning the wealth that might be
extracted from Australia were quite extraordinary. Most writers
seemed to believe that Australia could support a population of
between 100 and 500 million, the upper figure being quite eagerly
accepted by the populace as entirely reasonable. One of the most
extraordinary and influential books written on these matters was
Edwin Brady's *Australia Unlimited,* first published in 1918. It is a
magnificent, gold-embossed, lavishly bound and illustrated, two-vol-
ume set. It clearly sought to impress through appearance, for its
written contents are laughable. Its distortion of undeniable reality
seems infantile today, but at the time Brady's views represented mainstream thought. He wrote that the absence of flowing rivers in
Central Australia had led to the development of a 'desert myth', and
that the lack of rivers was in fact a great boon, for nature had provi-
dently stored all of its water underground. There it was hoarded, just
waiting for the wells of pastoralists to transform the mythical desert
into a land of milk and honey, 'destined one day to pulse with life'. In
his summing-up of virtually every region of Australia, he states that it
is 'the best agricultural land in the world', 'the best grazing land in

the world' or some other 'best in the world'. It is difficult to read his work today without baulking at the ingenuousness of his arguments.

Along with the early success and prosperity of the Australians, the very vastness of the continent itself seemed to be responsible for this outrageous optimism. As a result, estimates of the size of resources were often vastly inflated. It was estimated in the late nineteenth century, for example, that half of New South Wales and 92 per cent of Victoria was timbered country. Such gross overestimates led to extreme overcutting of forest resources throughout Australia.

In 1847 Commissioner Fry of the New South Wales public service informed a commission of inquiry that the Big Scrub of northern New South Wales could not be cleared within five or six centuries. Clearance of the Big Scrub began in earnest in the 1880s, and by 1900 it was all gone. As Eric Rolls has pointed out, timber that today would be worth hundreds of millions of dollars was destroyed for nothing, exposing a deep red soil which was soon exhausted, and which today supports a dairy industry worth a fraction of the value of the now-vanished timbers.

Indeed, the attitude of Australian pioneers towards timber seems to have been rather similar to that of the early Maori towards moa. For in both cases their use of the resource was typified by extravagant waste. It seems clear that neither people ever thought of their precious moa or timber as finite. The botanist Leonard Webb has summarised the folly of Australia's early timber industry:

> *Whether motivated by greed or irresponsibility, the early tenants of Australia's forested land had one thing in common—a supreme faith in the inexhaustibility of the timber resources of the country.*[137]

During the nineteenth century such views ensured that there was no concern at all about wastage of timber. Because of their knowledge about the sustainability of timber resources, foresters became Australia's first hands-on conservationists, a position they occupied for a century or more until the various national parks services were established. Indeed, it is only recently that foresters, including such fine conservationists as Webb, have had to bear the opprobrium of the community as forest exploiters. The best of them have expressed dismay—even incredulity—at the rapid shift of public opinion regarding their role in this area.

Webb makes it clear that during the nineteenth and early twentieth centuries timber was wasted on an enormous scale. In Victoria an official suggested that at a minimum, just one-tenth of the useable timber cut was ever used. On the goldfields of New South Wales prime timber trees were destroyed for their bark, which was used to roof huts. Instead of felling a tree and removing all the useable bark, only one sheet was taken from the base of a standing tree, which ringbarked it. In Queensland, a group of cedar getters felled three million super feet of red cedar, arguably Australia's finest timber tree, into the Barron River in order to float it to the coast. They had not counted on the power of the Barron Falls, which, at 300 metres high, smashed the logs to pulp.

Yet these depredations were not the worst, for throughout Australia settlement was spurred on by those who wished to obtain valuable timber. As a result, vast areas of Australia which are entirely unsuitable for agriculture were cleared and left to erode. Politicians in particular seemed to love schemes which promoted this. Webb says that:

> *Damnable as all these instances of forest spoliation under the guise of land development may seem, they are equalled if not excelled by the failure of the much-advertised Group Settlement Scheme in West Australia after the First World War. This scheme was proposed by Sir James ('Moo-cow') Mitchell, Premier of Western Australia at the time... [It] was an expensive experiment promoted by ignorance, inexcusable in terms of past experience, and crowned by the fallacy 'that in Australia in the twentieth century a plantation of an imitation European peasant community could have proved successful'.*[137]

Under the scheme 75 000 subsidised immigrants were to clear five million hectares, including some of the finest karri and jarrah forests in existence, in order to set up dairy farms. After enormous financial loss, human hardship and forest destruction, Moo-cow

Mitchell's scheme was exposed as being as foolish, gross and palpable as the man himself. But he was not alone in his foolishness. The New South Wales Forestry Department was initially situated in the Department of Mines. Clearly, it was seen as an extractive industry!

It is hard for people today, who are used to seeing at least some scientific analysis applied to development issues of such enormous importance, to imagine subjective and entirely unjustified assessments of the economic potential of Australian lands being taken seriously. But taken seriously they were, and politicians in particular parroted them *ad nauseam* to the delighted multitudes. So blinded were some people by them, that they simply refused to recognise the crises of resource overexploitation that were unfolding before their eyes. One such was Albert Dunstan, Premier of Victoria in 1935 when Victoria's Mallee country was becoming so degraded that sand drifts blocked railways, roads and irrigation channels, and bare dunes moved over fertile country once covered with sheoaks and blackwood. After taking a tour of the stricken country he said 'I saw no erosion there', and was promptly dubbed 'Albert the Ostrich' by journalists, because they figured the only way that he could have missed seeing the sand was by burying his head in it. Clearly, Australians have long delighted in injuring the dignity of politicians by bestowing appropriate if ignoble nicknames on them!

Australia was not alone in its unbounded optimism at future prospects in the early twentieth century. In 1900 the *Union Agricole Calédonienne* described New Caledonia thus:

> *With a marvellous climate and abundant rainfall, endowed with fertile lands alternating with mineral outcrops that turn certain parts of the land into blocks of iron, nickel and copper (the principal minerals of a long list), New Caledonia is perhaps one of the finest possessions—in terms of its resources rather than its size—of our colonial empire.*[125]

New Caledonia presents some valuable lessons to Australia, for ecologically, and in some senses economically, New Caledonia is like an Australia writ small. Despite the fine hopes of the *Union Agricole*, agriculture on New Caledonia has, by and large, been a failure. Between 1956 and the present, the workforce engaged in agriculture has dropped from between 50 per cent to less than 20 per cent, and New Caledonia now has one of the smallest agricultural workforces in the Pacific. As a result, the population is becoming increasingly urbanised. Today, agriculture plays a minor role in the New Caledonian economy, and New Caledonia imports large amounts of agricultural produce. The main pillars of the New Caledonian economy are mining (particularly nickel) and tourism. New Caledonia could not survive however, without the lavish handouts that France provides, for more than half of New Caledonia's financial needs are provided as subsidies and grants from France.

In Australia, in 1911, in the midst of this orgy of jingoistic nationalism, a geographer at Sydney University made the then extraordinary prediction that by the year 2000, Australia might support a population of 19 million people. His name was Griffith Taylor and his prediction—then viewed as being as outrageous as it was insulting—may have been entirely ignored were he not one of the greatest and most courageous scientists that Australia has ever produced.

Taylor was thrust into international stardom quite fortuitously, and while still quite young. A student of geologist Professor (later Sir) Edgeworth David, he was appointed Weather Service Physiographer on Captain Scott's tragic *Terra Nova* expedition to the South Pole. Public interest in the fate of Scott and his expedition was insatiable, and as senior geologist and leader of the successful Western parties, Taylor was in particular demand as a public speaker and darling of the wealthy and influential. Professor Powell of Monash University, in a brilliant partial biography, recently noted that Taylor, upon his return to Australia, was regarded then as a homecoming sporting hero who has achieved a great triumph 'overseas', would be today.[111] Perhaps only the return of the heroes of the 1983 America's Cup, or victorious Olympic athletes, would provoke a similar display of public adulation among contemporary Australians.

An honourary DSc was bestowed upon Taylor by Sydney University for his work in the Antarctic, and he obtained a prestigious position in Australia's then fledgling Weather Service. His great passion in life was, however, as he wrote to Edgeworth David in 1909:

*a careful and scientific study of the physical controls governing life
and industry in Australia.*[111]

He was a founding member of the Advisory Council of Science
and Industry and recommended the compilation of a national resource atlas for Australia. This task was left undone for decades. It
was clearly thought of as unnecessary as it would have been unpopular among those who believed in 'Australia Unlimited'.

Taylor's other great strengths were his ability to write in a popular
style, and his eagerness to write for the popular press. These were
highly unusual attributes among the scientists of his day, and they
made him enormously vulnerable, for his academic colleagues hated
and envied him for them; while the popular press was free to unfairly
pillory and parody his views.

If Taylor were unpopular enough for being a golden boy who told
his adoring populace the truth about the future of their nation,
unpopularity turned to outrage as his non-racist views became
known. For not only did the nation that spawned the White Australia
policy imagine itself bloating to a size of half a billion or so in a few
centuries, but it was determined that the bloat should be white. Taylor suggested that Asians were better suited to making a living in
tropical conditions such as those of northern Australia and that, if
development were to go ahead, Australia would be best served by
allowing coloured immigration. He also saw nothing wrong with
interracial marriages and even said that real benefit could be derived
from them.

The reaction was not long in coming. The press framed him as a
'croaking pessimist', a 'council for the yellow streak', his work 'the
half frozen cry of a scientific gentleman with mental chilblains'. But
worse, he was seen as a traitor to the class that, as the great survivor
of Scott's polar expedition, had once adored him. Anthropologist
Daisy Bates summarised the feeling of British Australians when she
wrote:

> *Are there no young British pioneers with the courage and the spir-
> it to venture out and blaze new tracks for those of their kin yet to
> come? Are our young men so frightened of the bogies raised by our*

WILL SOMEBODY TELL HIM?

Professor Griffith Taylor (Sydney University) asks: Why are we so horror-stricken at any suggestion of marriage with Mongolians?

Griffith Taylor lampooned by the popular press. *Daily Telegraph*, 25 June 1923.

*stay at homes that they will not do their part in giving the lie
direct to the defamers of our heritage? ... The central portion and
the north of our continent will be taken up in God's good time by
British pioneers, and developed by them, for Australia is going to
be forever British.*[111]

Taylor's colleagues provided him no support in this great and
public debate. As one of his academic friends put it:

*Many scientific men know perfectly well that you are right, but
will not say so, and so you have to fight against the hordes of
ignorance practically alone.*[111]

There is no doubt that academic jealousy of his early success pro-
voked some of this lack of support, as did his natural reticence at
having his work or statements publicly examined.

Despite (or perhaps because of) Taylor's strong publishing record
and a growing student following, he was steadfastly refused a full pro-
fessorial chair at Sydney University. Finally, vilified on all sides by a
vituperative press, betrayed by his fellow academics, yet firmly convinced of the correctness and importance of his views, Taylor could
do no more for his country. In 1928 he resigned from his post at
Sydney University and took up an appointment at the University of
Chicago, whose geography department was then one of the world's
finest.

The debate that Taylor engendered seems to have damaged the
reputation of geography in Australia, for no chair of geography was
created at Sydney University for a quarter of a century after he left.
Furthermore, despite the correctness of his views, Taylor did not win
the debate. Indeed, in the 65 years since he left his country, Australia
has seen a phase of population growth unparalleled in its 60 000 year
human history. Even today, echoes of the wild claims made by his
opponents such as Brady are commonly heard, while careful, biological analyses of the potential of Australia to support a human
population of a particular size are a rarity in the popular press.

In 1993 the CSIRO organised a seminar at the Canberra Press
Club to discuss the topic *Australia's carrying capacity: how many people*

to the acre? Mr Des Moore, Senior Fellow with the Institute of Public Affairs, argued that the Australian continent was, despite popular impressions, water-rich and had by far the highest availability of crop land per capita in the world. He said that Australia, like the United Kingdom and Japan, could even become a net importer of food to cope with a population of more than 100 million. Mr Moore is no member of a lunatic fringe. He is a respected economist who reflects one of Australia's mainstream bodies of thought. One even hears of plans from his camp to populate Australia's north with half a billion or so people. We might be able to regard these ideas as laughable were it not for their dangerousness and their persistence in the Australian psyche. I greatly fear that should Australia ever decide to try to attain a population of 100 million, and fail, our economy, ecology and people will be destroyed.

What then is a desirable population size for Australia? Estimates have varied widely over the years. Taylor himself produced a number of estimates. While he predicted that Australia would have a population of 19 million by the year 2000, he also suggested that the maximum population for a people, living US or European living standards, lay between 20 and 65 million. Less scientific estimates of the maximum population have also varied widely, from 10–15 million to 480 million. The most recent and carefully thought out estimates are those of Professor Henry Nix of the Australian National University. He has based his estimates on the amount and quality of arable land available in Australia. His earlier computer-based estimates, made in the 1960s, suggested that Australia had just under 125 million hectares of arable land. When he revised his figures in 1976, taking into account better knowledge about soil types and climate, he came up with an estimate of only 77 million hectares. He predicts that, when the data are examined again in the light of new satellite information, the estimate will be again lower.[103]

Currently, 22 million hectares of arable land is being used in Australia. Much of this land would be considered marginal agricultural land on other continents. Yet it is by far the best of our arable land. The rest is decidedly marginal even by Australian standards, and is largely untested. Already, after less than 200 years of use, 70 per cent of that 22 million hectares is degraded and in need of soil restoration programs. Much degraded land will have to be taken out of productivity if current intensive use continues. Even if it could be restored (an enormous task) and national parks, forest and urban sprawl abolished from potential arable land, the capability of Australia's arable land will always remain relatively small. Nix estimates that Australia is currently capable of feeding around 50 million people, but that much of this food is exported. Given current technology and investment, we can take this figure as a carefully estimated, if perhaps trifle optimistic, maximum population for Australia—unless we, currently one of the world's half dozen or so reliable food exporting regions, decide to start importing food.

But what of an optimum population? This is much harder to estimate, for much depends upon non-biological factors such as mineral resources and commodity prices. But from a purely biological perspective a few comments can be made. Given the desire of Australians to reserve some potential arable land for purposes other than agriculture, particularly national parks and forests—and given the enormous challenge presented by soil degradation—a more realistic maximum population for Australia may be 20–30 million. A population of this size might also give Australians a chance to earn some money from food exports.

This, however, is not the whole story. By examining those societies which live in a sustainable manner, a few important principles of sustainability for humanity can be discerned. Most hunter–gatherer societies are ecologically sustainable, the basic lifestyle having been in existence for 100 000 years or more. Virtually all hunter–gatherer societies seem to possess a 'golden rule' of population. This is, that in 'normal' times, the human population of a given area rarely exceeds 20–30 per cent of the carrying capacity of the land. This occurs because people are long-lived and usually reproduce slowly. Their population takes a long time to recover from disasters such as droughts, but their food species recover much more quickly. This means that, for most of the time, hunter–gatherers lead a leisurely lifestyle. They spend a few hours each day finding their food, and use the rest in recreation or ritual. In general, they are healthier than agricultural people from non-industrialised countries.

Australia, the land of ENSO, is no stranger to natural disaster and it also has a fragile natural environment. Thus, it makes enormous good sense to observe the 'golden rule' of population. If this were done, Australians might decide upon an optimum, long-term population target of 6–12 million.[45]

This raises the question as to whether Australia is already overpopulated. In 1992 Australia's population stood at 17 482 600. For humans, impact on their environment results from the interplay of three factors: population, affluence (that is, demand + consumption) and technology. If the interplay of these three factors results in environmental degradation, then we can say that a region is overpopulated. It has proved to be extraordinarily difficult to quantify how much environmental degradation results from population increase, and how much from affluence, but Harrison in his book *The Third Revolution* has recently estimated that in developed countries between 1961 and 1985, 46 per cent of the growth in arable land, and 59 per cent of the growth in livestock numbers, has been due to population increase.

Many people look to technological advances to negate the effects of both growing population and affluence. There is no doubt that technology has had some impact, but thus far it has been insufficient to reverse environmental degradation. Although it may well happen, it is of little value to assume that technology will solve these problems. This is because if we make assumptions about future change in such areas, we create a hypothetical species; one that might exist *if only* we could change our habits. We must deal with people as they have acted in the past, and as they exist now, in order to make progress in determining future action.

There is no doubt that since 1788 European population, affluence and technology have led to enormous environmental degradation in Australia. True, much damage has also been done by introduced species such as rabbits and foxes, but much more is clearly attributable directly to human activity. To this extent, and to the extent that these problems are ongoing, Australia can be defined as being already overpopulated.

Because Australia is part of a world economy, there are ways that we could continue to fuel growth if we really so desired. We could sell our minerals, labour or tourism at an even cheaper and faster rate in order to buy food to support a population of more than 50 million. Many economists argue that Australia's population must reach a 'critical mass' so that our manufacturing industry has a sufficiently large domestic market to be competitive. Defence experts also argue for a larger population. Both of these arguments have recently lost much of their validity. The lowering of trade barriers is rapidly changing the nature of world trade, and many Australian businesses have realised this and now seek their major markets overseas, regardless of the local situation. The defence argument likewise suffered a blow, when the outcome of the Gulf War demonstrated the ability of state-of-the-art technology to neutralise a vast superiority in numbers.

Furthermore, a quick look at the state of the world suggests that building Australia's population beyond the carrying capacity of the land would be very foolish. Australia is one of only half a dozen reliable food-exporting regions worldwide. Presently, world population stands at 5.4 billion, and is increasing at the rate of approximately 100 million per year. At the same time the world's soils, forests and oceans are all rapidly becoming exhausted. Any nation that counts upon buying massive amounts of food far into the next century is likely to face severe difficulties and increased costs.

So what are the implications of continued population growth, or a population reduction, for Australia? During the Hawke era of the late 1980s, high levels of immigration increased our rate of growth to one of the fastest in the developed world—as fast as that of Indonesia. Australia's population is still growing rapidly, and at the present rate it will continue to grow indefinitely. It will probably reach 30 million or more by 2060, the vast majority of whom will inhabit cities of six million people or more.

It is almost certain that the social inequality that has increasingly begun to characterise Australian society will grow as a result. This is because large cities seem to spawn such inequality, and because in times of resource shortage the privileged cling to their resources, while those who are more vulnerable are sacrificed. Initially this may come through the imposition of more 'user pays' amenities such as tollways, beaches and national parks. Eventually it may lead to direct exclusion. The current high rate of unemployment, which is unlikely to change significantly for the next decade, also reflects this trend.

There is no doubt that all of this will lead to a much diminished quality of life for most Australians. Ultimately, continued population growth has no future. But the point at which it ceases determines how finely the wealth must be divided and how much future eating must occur in order to sustain it.

But what of the option of ceasing population growth now? Many people have great concerns about living in a society with a decreasing population. One of the most frequently cited anxieties concern living with an ageing population. But except in exceptional circumstances, this is not a major problem. Australia already has a surplus of labour, as our high unemployment rate shows. Relatively few people are needed to earn the foreign exchange necessary to support the nation. Were Australia's population to age slightly, although spending on the aged would increase, savings would be made in the fields of education, child-care and possibly immigration-related costs. Compulsory retirement is already largely a thing of the past, and as medical care improves, people can be expected to prolong their working lives, even if only on a part-time basis.

In any case, an ageing population is inevitable in Australia. The baby-boom generation is now middle-aged. Within 20 years they will swell the numbers of the aged, who will then comprise a large percentage of the population. No conceivable immigration or other social program could alter the demographic trends that will bring this about.

There is no doubt that, were Australia's population size to stabilise or decrease, the pace of 'development' would slow. The urban sprawl that presently consumes Sydney's natural environment at an unprecedented pace would cease. Likewise the development of freeways, bridges, tunnels and the like would slow as the need for them, which is driven by an ever-increasing population, diminished. Investment in property, which is presently a disastrous national obsession, would become less attractive, encouraging diversification of investment into more profitable ventures. But some kinds of development would not slow. Large tourist developments would be unaffected, as would spending on airports, hotels and other things related to tourism.

There would be real rewards for Australians once a sustainable population size was reached. Because of the structure of our economy, a population of 6–12 million would give Australians enormous flexibility in dealing with environmental and other problems. Australians are still dependent upon exports of fossil fuel, mineral and agricultural commodities and, to a lesser extent, tourism for foreign dollars. There is no prospect of this changing in the short- to medium-term. Because both mining and agriculture are highly mechanised industries which require small workforces, Australia can manipulate population size without compromising earnings.

This means that if Australia's population were smaller, we could afford to do many things differently. The argument over logging of old growth forest, for example, would be less intense, for the housing construction industry, which uses much of the hardwood timber produced, would be scaled down. Likewise, increased levels of affluence would mean that the dollars earned through woodchip exports would not be so vital. It would also give us much more choice when it came to the preservation of other areas, for the cash earned through their exploitation would not be as vital as it is now.

There is a further aspect to this argument which is worth examining here. Until the early 1980s agricultural products were the single most important income earners for Australia. Since then, a rapid growth in mineral exports has superseded agriculture, so that today mining earns Australia some 29 billion dollars, while agriculture earns only 16 billion dollars. All our other export earnings (including all manufacturing) earns around 11 billion dollars.

Australia's increased reliance upon non-renewable mineral resources to earn export income suggests that the capacity of the nation's biota to support her population has been exceeded. This is worrying enough, but even more problematic is the way in which the earnings from non-renewable sources are spent. The royalties earned by the Australian government from mining go towards the general budget. A percentage of this is spent on fuelling further population growth. Fortunately, the profits returned to stockholders are often reinvested in industries that may provide a future for the nation. Given its nonrenewable nature, Australia's economy can be buoyed by mineral wealth for only a short time. It seems to me that government should use that one-off bonanza to achieve long-term goals.

As noted earlier, the old 'populate or perish' argument that drove successive governments to increase Australia's population size is no longer valid, for an affluent, prepared Australia of 10 million would be far better able to defend itself against outside attack than a poorer, divided nation of 30 million. But there are other kinds of threats to Australia's security to consider. Despite its rapidly growing economy, Indonesia, Australia's nearest neighbour faces severe problems in bringing a reasonable standard of living to its population of almost 200 million. It has an excellent family planning program, but despite this, it will add another 100 million people to its population over the next 20 years. Indonesian agriculture is presently running at near capacity, and it seems unlikely that sufficient food can be grown in time to feed its additional mouths. Food imports will probably become necessary. Although many may see danger in the proximity of an empty Australia and an Indonesia which is bursting at the seams, this may not necessarily be the case. Australia's best defence against both refugees and aggression may be having the means to help other people at home. And that means having a well-focussed aid program and the ability to export food.

Australians must decide how many of them there will be. The Hawke government caused Australia to support the most rapidly growing population in the developed world. Yet that government had no population policy. Because of the nature of demographic change, decisions made during the Hawke era, and even more so today, will shape Australia in 2020 and beyond. It is no longer sufficient to blunder on, blind to long-term consequences. If there ever was a critical time for a population policy, it is now.

Australia is quite literally at the crossroads as far as population growth is concerned. Professor Lincoln Day of the Australian National University has calculated the following possible future populations for Australia. If survival ratios and age-specific fertility rates continue as at present—and there is no immigration—then the population of Australia in 2067 will be 17 312 800. This is after reaching a peak of 19 451 800 in 2017. If however, these same conditions apply, but immigration is held at 1988 levels, Australia's population will increase to 40 706 000 by 2067. A slight increase in fertility on top of this (to replacement levels by 2002, and to 10 per cent above by 2037) would see population burgeon to over 53 million by 2067.[35]

Australia therefore has a very stark choice to make. By cutting immigration to very low levels (say 30 000 or so to match emigration) and keeping other factors as at present, Australia could probably create a stable population. By increasing immigration we could rapidly create cities of 10 million or more in our most populous states.

I think it is clear that the single most important factor in restoring biological stability to Australasia is consideration of the total human population that the various lands can support. In the cases of Australia and New Caledonia, this issue is of even greater importance, for these lands are particularly impoverished, and the European populations of both have been subject to delusions of grandeur, believing that the carrying capacity of the land is far greater than in reality it is. The discovery of non-renewable mineral wealth has in both cases fostered and fed such delusions. In light of the enormous subsidy from France to New Caledonia, and the declining living standards and increasing reliance upon minerals by Australians, it appears that the optimum population of both of the mineral-rich lands has been exceeded. New Zealand, which has never possessed such illusion-creating wealth, is now settling into population stability. Over the 15 years from 1976 to 1991 its population grew by only 300 000, and seems destined to stabilise near its present 3 434 950. Although this population might be too great to deliver the comfortable lifestyle enjoyed by past generations of New Zealanders, population stabilisation does give them a chance to enjoy some of the benefits outlined above.

The last of the 'new' lands where increasing population is playing an important role in shaping its future is Papua New Guinea, which is a relatively large and fertile land. Its population, while standing at about 3.5 million, is growing rapidly. If population growth can be stemmed, I feel that Papua New Guinea has the brightest future of any of the 'new' lands. Already largely self-sufficient in food, it has sufficient high-quality agricultural land to become a major exporter of some foodstuffs. In addition, its recently discovered mineral wealth is breathtaking. Some of the world's largest reserves of gold have recently been located there, while its enormous copper reserves are already well-known. Furthermore, substantial reserves of oil and gas are already being tapped, and the possibility of more discoveries is great. Its reserves of rainforest timbers are also vast. By far the most lavishly endowed of the 'new' lands in a wide variety of products, Papua New

Guinea now awaits social change in order to flourish. Its government needs an unwavering commitment to educate its people, a clear national population policy, sharply focussed social policies and a strong commitment to using its one-off mineral wealth wisely. With this achieved, Papua New Guinea promises to become a significant power in our region.

But to return to Australia. Perhaps the most fundamental philosophical adjustment that Australians need to make relates to growth. Earlier, our national image, born and fostered on a frontier, was that of a young nation with boundless potential for growth. Growth was seen not only as positive but inevitable. In reviewing our attitude towards growth we must differentiate clearly between the various kinds of growth. Clearly, further population growth in Australia is pernicious. Recognition of this will affect our attitudes to such fundamental issues as motherhood, immigration and a society and economy obsessed with construction and property speculation. All of these must change if we are to survive; yet one can hardly imagine a more difficult, yet more necessary, adjustment. But not all growth need be viewed as deleterious. Economic growth that does not demand increased resource utilisation (for example, that originating from new technology) could be positively beneficial.

I recently spoke to a Rotary meeting in the northern Sydney suburbs. The Returned Services League Club where the meeting was held was filled with middle-aged businessmen and their wives, all beneficently looked down upon by a large portrait of the Queen. The Rotary members had grown up in a culture that extolled growth, and the very nature of their livelihoods ensured that they saw growth as good. I spent the best part of an hour outlining the limits of the Australian environment in supporting a large human population. Most were stunned when confronted with the facts. At the end of the meeting we sang the national anthem. When we came to

For those who've come across the seas, we've boundless plains to share

some voices faltered. Perhaps hope for change is encapsulated in that faltering.

CHAPTER 32

RIDING THE RED STEER—FIRE

AND BIODIVERSITY CONSERVATION

IN AUSTRALASIA

A ustralian graziers have long referred to fire as 'the red steer'. The phrase is an apt one, for it discerns the ability of both fire and large herbivores to shape ecosystems. In Australia, fire is such a critical factor in the conservation of biodiversity, and has such profound effects on productivity, landscapes and people, that its role must be discussed at length. It cannot be examined in isolation, but should be seen in the context of the struggle to conserve biodiversity in Australasia as a whole.

A most important part of this examination concerns the present management of reserved lands in Australia, and why it is failing to produce the outcomes that Australians wish for. Land reserved for the primary purpose of conservation comprises approximately 5.2 per cent of Australia, a figure which has grown from 1.2 per cent in 1968. The land has been reserved at various times by all State as well as Federal governments under some 50 different categories. The percentage of land reserved for conservation in Australia is in accord with that reserved by similar nations, for approximately the same amount is reserved in the USA and Canada.

The superficial overview presented by these figures gives many Australians some comfort, and a feeling that biodiversity conservation is proceeding well in Australia. For the amount of reserved land has obviously grown rapidly, and there is a feeling that the National Parks system really is working to protect Australia's biodiversity. But the truth of the matter is far different, for the reserved lands system is widely acknowledged at the political and professional levels as being inadequate for the protection of biodiversity.

In part, this is because Australia's system of reserved lands has grown haphazardly. Land has often been chosen for reservation for reasons other than its importance for conserving biodiversity. Often, reserves were created in lands that were perceived to have little economic use. This has resulted in a configuration of reserved lands that is not ideal for conservation purposes in size or location.

The problem of conserving Australia's biodiversity within reserved lands is enormous, for Australia is a 'megadiverse' region, with many species occupying small ranges scattered across the continent. To protect all biodiversity within reserved lands would require a huge increase in the reserved lands system. Given the economic reality of Australia's dependence on agriculture and mining—and the current policy of single use strategies for reserved lands—it is almost certainly not possible to expand Australia's reserved lands system far enough to protect the biodiversity of the continent.

Even if the reserved lands could be extended, the cost of managing the land would be astronomical. Thus far government funding for National Parks, particularly at the State level, has been appallingly niggardly. It has often been so bad as to leave staff unable to carry out basic management. Thus they are left open to the criticism that they lack either the will or expertise to manage land. The demoralisation that this has brought about has drained the various State services of their finest people.

But there are other problems in attempting to carry out conservation solely within reserved lands. Climate change brought about by the greenhouse effect may mean that reserved lands may become unsuitable for various species over the next century. Clearly other ways must be found of protecting biodiversity. But having said this, there is no question but that Australia's system of reserved lands, as flawed as it is, is the major tool presently available to protect biodiversity.

Although we are far from devising a system for the protection of biodiversity on privately owned lands, legislation protecting endangered species (presently only some vertebrates and plants) has been passed by the Federal and various State governments, and is a step in this direction. There are also encouragements for farmers to reserve parts of their land, and other schemes which help contribute to this goal. Legislators must tread a very fine line when dealing with these matters, for badly drafted legislation can do more damage than good to endangered species.

The historical problems concerning the distribution and extent of reserved lands show a dimension of the problem which is difficult to solve. Other, more approachable problems, concern current management. One of the most important, yet more easily solved problems of the reserved lands of Australia is the numerous categories under which they are currently administered. Indeed, there are nearly 50 different designations for reserved lands.[74] The International Union for the Conservation of Nature (IUCN) recognises only six categories of protected areas, and Australia's reserved lands system would benefit enormously were the IUCN categories (or a variant of them) to be applied. The categories are:

1 Scientific Reserve and Wilderness Area
   Strict protection, managed mainly for science and wilderness protection.
2 National Park
   Managed mainly for ecosystem conservation and specific natural features.
3 Natural Monument
   Managed mainly for conservation of specific natural features.
4 Habitat and Wildlife Management Area
   Managed for conservation through active intervention.
5 Protected Landscape
   Managed for landscape conservation and recreation.
6 Managed Resource Protected Area
   Managed for sustainable use of natural ecosystems.

All of these categories have clearly defined goals within which management strategies can be framed. This can be said for few of the 50 existing Australian categories. However, one of the most problematic IUCN categories of land reservation is that of 'Wilderness Area'. The definition of such land as given by the IUCN and recognised by the Federal government is:

Large areas of unmodified or slightly modified land, or land and water, retaining their natural character and influence, without permanent or significant habitation, which are protected and managed so as to preserve their natural condition.[74]

I find this definition quite extraordinary in an Australian context. The newly proclaimed Mabo legislation may have wiped the concept of *terra nullius* from Australia's law books, but it is still alive and well in the minds of those who promote such a concept of wilderness. I hope that this book has demonstrated that wilderness as defined by the IUCN simply does not exist in Australia. For the entire continent has been actively and extensively managed for 60 000 years by its Aboriginal occupants. To leave it untouched will be to create something new, and less diverse, than that which went before.

Some of the best examples of the effects of lack of management on biodiversity comes from Northern Australia. Dr David Bowman of the Northern Territory Conservation Commission has described large parts of the Arnhem Land plateau as 'the new wilderness'. For millennia the plateau was managed and burned by its Aboriginal inhabitants. Now, they no longer visit many parts of it, and a new, 'natural' fire regime is developing. It consists of occasional vast wildfires which kill fire-sensitive species such as *Callitris* pines, and which threaten the small vine thicket patches that survive in the moister areas. If this lack of management is allowed to continue, there is no doubt but that the 'new wilderness' of the Arnhem Land plateau will be poorer in species, and less stable, than the system which was shaped by humans over the millennia.

The history of Australia's environment makes its continued management a particularly difficult task. Until around 60 000 years ago, Australia's ecosystems were fully self-sustaining. Then, vast extinctions devastated the entire continent. Following this, for 60 000 years Aborigines managed the crippled ecosystems, preventing them from degenerating further. Now, Europeans have arrived and forced the discontinuance of that management. These changes beg the question of what we should aim for in our management of reserved lands.

Should we aim to keep them as they are today, as they were 200 years ago, or as they were 60 000 years ago, when they functioned without the interference of humans?

The answers that Australians give to these questions will have a profound impact upon the nature of our reserves. To give an idea of how profound that impact is, it is worth considering one or two areas whose history is well-known. Captain Cook's landing place at Kurnell is a good example.

Sixty thousand years ago the area would have supported a diverse flora, including most of the species currently growing there as well as many plant species which we associate with rainforests today. Likewise, the fauna would have been much more diverse, for in addition to the great majority of native species historically known from the area, there would have been great grazing and browsing marsupials, which were probably preyed upon by a gigantic goanna, a relative of the Komodo dragon. The scraps from any kill would have been wolfed down by thylacines and Tasmanian devils.

It is likely that following the extinction of the giant marsupials, the region changed rapidly, and began to resemble the landscape described by Joseph Banks in April of 1770. He said that the area was 'a very barren place'. The vegetation then occupying the site was clearly very open, for he remarks that he saw Aborigines retreating up the hill from Cook's landing place, and that a musket was discharged at them by the landing party from such a distance that the shot was harmless. Banks also supplies evidence that there was an open, grassy understorey in the vicinity. This all suggests that the area was occupied by a very open woodland with a grassy understorey. This is a habitat type that seems to have dominated the Sydney and Botany Bay areas in the early contact period, yet it has all but vanished from the region today. As I have shown above, it was clearly burned regularly, and fuel loads would rarely have been allowed to exceed a couple of tonnes per hectare. Fires of this type are cool and rarely widespread.

At present, a vastly denser vegetation type covers the site, and grass is a very minor part of the natural ecosystem. Indeed, the forest is presently so dense that it is not possible to see someone once they have entered a few metres into it from the shore. This dense forest consists primarily of fire-promoting species which periodically burn in a vast conflagration. These fires are exceptionally hot, for fuel loads regularly exceed 10–15 tonnes per hectare. The hot fires that this fuel feeds are also often widespread. They can easily drive fire-sensitive species, such as rainforest plants, to extinction in these small reserves.

Partly as a consequence of this changed fire regime, most of the mammals present in the area 60 000, or even 200 years ago, are now gone. Historic records indicate that grey kangaroos, eastern quolls, short-nosed bandicoots, white-footed rabbit-rats, bettongs and many other mammals were present in the Botany Bay region 200 years ago. All of these are locally, and some are entirely, extinct. In their place are a few introduced species which inhabit the more disturbed habitats. The dense bush supports a very limited native mammal fauna.

It is not difficult to see why these changes have occurred. Botany Bay National Park, is, like many of Australia's reserved lands, a small, isolated fragment. The Park was by chance spared on 6–8 January 1994. But the nearby Royal National Park was ablaze in what has since become known as the Sydney bushfires. Over 90 per cent of the park was burned, including invaluable rainforest which will never fully recover during our lifetimes. This catastrophic event means that, if not fire-managed, the entire fire-prone vegetation in the park will, for the foreseeable future, be dominated by a relatively uniform-aged stand of growth. This uniformity of vegetation structure is a disaster for biodiversity. Many animal species need old growth areas for shelter, and patches of young growth for food. A mosaic of vegetation age structures is thus critical to the maintenance of biodiversity in Australia.

A further, even greater tragedy is that, within the next 30 years the vegetation of the Royal will again be ready to carry an equally vast blaze which may banish yet more species from the park. With a pattern such as this in our reserved lands, there seems little hope of preserving viable, complex and self-regulating ecosystems.

Another option for managing such reserves is to try to recreate the conditions that existed there during the 60 000 years of Aboriginal occupation. This would clearly involve the implementation of a fire regime that would be regarded as sacrilegious by many conservationists, for it would effectively banish forest from much of the Botany Bay National Park and similar areas, replacing it with very open, fire-maintained woodland.

Were this accomplished, we would need to reintroduce the many species that play a role in such ecosystems. Unfortunately several of these, such as the white-footed rabbit-rat are extinct and therefore cannot be replaced. Of those that can be introduced from elsewhere, there is no guarantee of survival, for foxes and cats have been introduced. The woodland species would find few secure hiding places in their newly created open habitats.

Even if these problems could be overcome, there remains the issue of funding the enormous, ongoing commitment to management that such a plan entails. Even the current, less onerous plans of management which are being implemented in many reserved lands are putting considerable financial strain on the administering bodies. With their small tax base, and enormous biodiversity to maintain, Australians may simply not be able to afford such management.

One other, at first unlikely looking option, exists. It is clear that today's ecosystem, in places like the Royal and Botany Bay National Parks, is the poorest in terms of species diversity, that has existed since humans first arrived. The richest biota that the region has supported existed some 60 000 years ago. Then, rainforest plants, giant marsupials and gigantic goannas coexisted with most of the species native to the region today. But is it really feasible to try to recreate such an environment?

Such a task is much more difficult in Australia than it would be on other continents. If North Americans, for example, ever wanted to recreate their Pleistocene megafauna, they could import horses, zebras, elephants and lions from Asia and Africa, and spectacled bears, peccaries, tapirs and camelids from South America. This is because although many species or distinctive populations of large mammals became extinct in North America 11 000 years ago, a large number of similar species survive elsewhere. The situation in Australia is quite different. We have no chance of finding a single one of the large marsupial herbivores present in the past, for all are extinct.

Many imbalances and vacant niches exist in Australian ecosystems. True, Australians have introduced a dozen or so species of large herbivores, such as cattle, buffalo, horses and deer, which may in part fill the vacant ecological niches of the large marsupials. Yet imbalances still clearly occur between plants which are being grazed by new and unfamiliar herbivores; between herbivores and the invertebrates that process their dung or live in association with them; and between herbivores and the carnivore guilds that they would normally support.

A major imbalance has recently been remedied. This was the lack of large, dung-eating insects. On every other continent these insects, predominantly dung beetles, break up and bury dung, hastening the return of nutrients to the soil. Australia's large dung beetles presumably followed their food source to extinction 60 000 years ago. After the introduction of large placental herbivores, other insects, predominantly bush flies, proliferated by feeding on the exposed, watery dung. This created many problems, the most important of which is probably that flies do not hasten the recycling of nutrients as do dung beetles. The recent introduction of dung beetles to Australia has partially rectified this imbalance.

One of the most important remaining imbalances in the Australian ecosystem is the lack of large carnivores. Without carnivores, herbivore populations can go through cycles of boom and bust, causing major problems of soil degradation resulting from periods of overstocking. Smaller carnivores can also become superabundant. Today, the largest Australian carnivore is the dingo (*Canis familiaris*). Although capable of killing animals as large as pigs, its overall impact on the introduced herbivores is rather limited. It is certainly not as effective as a suite of large carnivores would be.

Humans of course can still play a substantial role in regulating herbivore numbers, particularly, as is the case with the water buffalo (*Bubalus bubalis*), where a species has a limited distribution. Australia is, however a vast and thinly populated land, and it is doubtful that, without massive market encouragement, humans could ever entirely fill the large carnivore niche.

While many of the large herbivores introduced to Australia have not been entirely suitable to replace the long-extinct species, the situation for the carnivores and insectivores is better. It is true that the gigantic snake *Wonambi*, *Quinkana* the land crocodile, and the marsupial lion are long-extinct and can never be replaced, for all belong to extinct families. A long-beaked echidna survives today in New Guinea—but only just, for the few remaining populations are highly endangered. Were it to be introduced into Australia it may be able to fill a similar ecological niche to the now-extinct long-beaked echidnas of Australia's Pleistocene. Likewise, the Tasmanian devil still survives in Tasmania, and could be reintroduced to the mainland, from which it has been absent for only 3000 years or less.

The introduction of the Tasmanian devil is likely to be successful despite the fact that it had became extinct on the Australian mainland by 500 years ago. Its extinction, as with the earlier extinction of the thylacine on the mainland, was probably due to the introduction of the dingo, possibly exacerbated by human hunting. Today, both dingo and intense human hunting have been banished from large parts of the continent.

Importantly, there may be some real advantages to its reintroduction, for it has been suggested that Tasmanian devils may affect fox numbers by preying upon cubs. In this regard it is interesting to note that there have been several attempts to introduce foxes into Tasmania. All have failed. If fox numbers could be limited in part through reintroduction of the Tasmanian devil, the re-establishment of vulnerable, medium-sized native species (such as the eastern quoll, which became extinct on the mainland in the mid-1960s) may be possible. Failing this, the more widespread reintroduction of the dingo may be considered, for there are strong indications that it has a powerful impact on both fox and cat numbers.

Seeking replacement species for Australia's lost large carnivores is fraught with difficulties. Ideally, the carnivore should be one that the existing marsupial fauna is preadapted to, yet it must be large and efficient enough to play a role in limiting herbivore numbers. Placental carnivores are, by and large, unsuitable. This is because there has never been the equivalent of lions and tigers in Australia, probably because there is insufficient energy in Australian ecosystems to support a population long-term.

In the past, gigantic goannas were found throughout Australia, the islands of Nusa Tenggara (Indonesia) and on New Caledonia. Today, small populations of a single species (*Varanus komodoensis*) survive on

Komodo and nearby islands, as well as on western Flores, all in Nusa Tenggara. These relict populations are presently endangered. At around 100 kilograms in weight, the Komodo dragon is much smaller than the extinct Australian *Megalania*. It is nonetheless a near relative of the Australian varanids, as biochemical studies by Dr Peter Baverstock (personal communication) have shown.

There is no doubt that the Komodo dragon is, in ecological terms, the closest living species to any of Australia's lost reptilian carnivores. There is also every reason to suspect that Australia's surviving marsupials are better adapted to predation by it than any other exotic species. Australian marsupials have, after all, evolved and proliferated over millions of years in the presence of *Megalania*, the largest goanna of all time.

Because of their ecology (being cold-blooded and large), large goannas such as the Komodo dragon may not spread into the more temperate parts of Australia. They nonetheless have the potential to bring greater ecological stability to Australia's tropical regions. This stability could be achieved by helping to recreate a guild of carnivores that would regulate herbivore numbers enough to smooth out the cycle of boom and bust that currently characterises Australia's introduced placental megafauna. The Komodo dragon may also have the benefit of reducing the numbers of introduced carnivores, either through competition for food or by preying upon them.

Such an introduction should of course be approached with extreme caution. A long period of experimentation and data collection would be necessary. Initial experiments may involve the introduction of single sex individuals into fenced areas, followed if appropriate, by single sex releases into unfenced suitable areas.

The problem of predation by large varanids upon humans should not be understated, but the risks posed are probably small when compared with the risk of falling prey to the salt-water crocodile (*Crocodylus porosus*). Australians are willing to extend protection to the 'saltie' and are rapidly adapting their behaviour so as to minimise that risk. There is no reason to imagine that they could not also adapt to the smaller risk posed by large goannas.

Were it possible to re-establish a stable community of large herbivores and carnivores in Australia, it is just possible that the fire balance could be again tipped in favour of fire-sensitive species in some environments. This is because much vegetation can be removed by grazers and browsers before it burns. Studies have shown that this has indeed happened in some parts of Australia.

A well-documented case concerns the rainforest–woodland boundary in heavily disturbed and grazed areas of north-eastern Queensland. As a result of fire suppression through the removal of grass by cattle and soil disturbance associated with tin-mining, the rainforest has been able to invade deep into the woodland, re-establishing itself after an absence of 60 000 years or more. The effect of water buffalo in Kakadu National Park has also been substantial. Until culling began, there were thousands of buffalo in the park. Each ate about 40 kilograms of grass daily, resulting in relatively cool fires on the floodplains in the dry season. Now the buffalo have been removed, the intensity of the dry season burns has increased dramatically.

Changes in vegetation type could be encouraged in many areas where fire is a particular problem today. The urban fringe, for example, could be planted with a mix of rainforest plant species, which, with suitable buffers and management, could protect property from fire.

Such proposals may seem extreme, or too much like tinkering with nature, to some. Yet the time scale over which they can be examined, toyed with, and abandoned or carried out, is a long one. Many Australian trees live for a thousand years or more. For them, the time scale for change is long indeed.

Slattery's Sago Saga is the last, unfinished, yet perhaps greatest work of the Irish writer Flann O'Brien. It is a comic work in which the chief protagonist blames all of Ireland's woes upon the potato, claiming that it has made the Irish base, grovelling and backward. This indeed is not far from the truth, for when the potato was introduced in the sixteenth century, it allowed Ireland's population to burgeon far beyond its safe carrying capacity, resulting in widespread poverty and suffering. The subsequent potato blight killed over a million people, and forced even more into exile. Today, the country still carries the scars of those great social traumas.

In *Slattery's Sago Saga* a wealthy American widow of Irish ancestry proposes that Ireland can be converted into a paradise through vast plantings of the sago palm. The palm will feed the multitudes and it will soften the climate, she says. Monkeys and exotic birds will frolic in the treetops. The Irish, weaned from their beastly tuber, will become a noble and civilised people.

Perhaps the sago palm could never soften the climate of Ireland, but as we have seen, Australia's climate may have been modified and its lakes dried or filled, depending upon the kinds of trees that predominated. O'Brien died before his saga could be finished. I do not expect that anyone reading this book will live to see the outcome of Australia's great, ongoing saga of fire, rainforest and fauna.

While prospects for Australians to return stability to their ecosystems seem at present remote, they are far better than those facing the people of New Zealand. There, many of the irreplaceable land bird species are extinct, as are many amphibians, reptiles and one of the three bats. Of the survivors, many eke out an existence on tiny islets. Where will New Zealand find a suitable ecological replacement for the great adze-bill, or its moa, or the majestic Haast's eagle?

At present, conservationists in New Zealand have not had time to ask such questions, for they are involved in a desperate struggle to preserve critically endangered remnant populations of species such as the kakapo. In light of the extremely limited funding available to them for wildlife conservation, emergency procedures must take precedence over strategic planning.

Two other factors make returning stability to New Zealand ecosystems very difficult. First is the fact that the extinctions are so recent; so recent indeed that still-living trees once provided food and shade to moa. Some of these trees will live for hundreds, or perhaps even a thousand years more. In such a situation the dramatic 'cascade' effects of extinctions have a long way to go before they are played out. Another factor is the vast number of alien species which have been introduced. From starlings to possums and moose, each of the hundred or more species presents its own problems to the natural environment.

But one factor in New Zealand's favour is that fire seems to play a different role there than in Australia. This is because the natural sources of ignition are few. Thus, if people are prevented from lighting them, fires will not necessarily prevent the regeneration of fire-sensitive forest.

Compared with New Zealand and Australia, the problems of New Guinea and, to a lesser extent, New Caledonia, appear to be small. The vast forests of Papua New Guinea are largely intact, and its fauna has suffered few extinctions. Indeed, only 10 mammal species from a fauna of 217 are known to have become extinct on the island of New Guinea since humans arrived. Introductions are likewise few, with only a couple of rats, the dog and pig having spread widely. Many of these are restricted to, or are most abundant in, disturbed or marginal habitats.

At present, it is difficult to say much with certainty about New Caledonia. Several unique and irreplaceable keystone species, such as the *Sylviornis* and the Caledonian crocodile, are already extinct. The unique kagou is clearly severely endangered, but the conservation status of many other species is undocumented. What is clear is that many of its animal and plant species are rare and little known.

I feel that the conservation challenge in the 'new' lands is, in general, far more difficult than most Australasians previously suspected. At present we lack even the basic data with which to assess a direction in which we should head. I have introduced some radical and provocative views principally because I believe that, given our present understanding, they are the right way to begin. Even if they are eventually discarded, the knowledge gained in investigating them would be invaluable as a base from which to make a beginning.

CHAPTER 33

ADAPTING CULTURE TO
BIOLOGICAL REALITY

Culture means many things to different people, but one of its key elements is the embodiment, in beliefs and customs, of actions that help people survive in their particular environment. This is not to say that all cultures are in tune with their ecology, indeed it could be argued that the great majority of the world's cultures are not. This is because cultures that we can call 'ecologically attuned', are the result of many thousands of years of experiencing and learning about a particular ecosystem.

The only one of the 'new' lands in which the majority of the population is presently 'ecologically attuned' is New Guinea. There, most of the population live in their ancestral homelands and are embedded in ancient, indigenous cultures. This gives people an extraordinary understanding of their land, which has been refined and handed down over thousands of years. But today, even with these advantages, the ecological equilibrium is threatened in New Guinea. As we saw above, this is because of rapid technological and social change. Despite this disruption, the deep understanding of their environment possessed by most New Guineans gives them an enormous advantage in working through these problems.

The cultures possessed by the great majority of inhabitants of the other 'new' lands are clearly not 'attuned'. The issue of 'cultural maladaptation' is a critical one for these people. In many instances, their maladapted cultures are dramatically incompatible with the environment they find themselves in, and it may take a very long time for them to adjust.

The problem of cultural maladaptation seems to be particularly acute in Australia. For it has the highest number of new settlers of any of the 'new' lands, and it has an extremely difficult and unusual ecology. Perhaps this accounts for what outsiders perceive as the obsession Australians have with defining themselves. But to Australians, that obsession makes perfect sense. It arises from a frustration borne of the long-felt inability to live in harmony with the land. It comes from the dismay one feels when seeing the extraordinary beauty and complexity of unique environments wither—even from an apparently gentle touch by a European hand—and from the floods and bushfires that constantly remind Australians that the land does not hold them comfortably. Finally, and most importantly to many, it arises from the great gulf of culture and understanding that exists between Aborigines and other Australians.

As a result of these feelings, Australians have long struggled with the issue of national identity; yet they have done so without really trying to understand the nuts and bolts workings of their land. It is, I think, now clear that any lasting notion of Australian nationhood must arise from an intimate understanding of Australian ecosystems.

Throughout the 1980s and '90s the struggle of Australians to define their sense of nationhood has intensified. Until recently, the image of the suntanned stockman—laconic, self-reliant, but a dependable mate—has been the role model for Australians; their way of defining themselves. But is the stockman a true Australian? Do the ideals he encapsulates help us survive in modern Australia? The same question, I argue, must be asked of our other cultural icons: our flag, oaths, anthems; and policies such as multiculturalism, a haven for large-scale immigration, and the relatively new idea of becoming part of Asia.

The stereotype of the stockman was born on the Australian frontier at a time of rapid expansion into a hostile land. Pioneers had to be tough and self-reliant and had little time for recreation. But most importantly they had to stick together, for hard times could see all perish if the community was not cohesive. Thus, mateship was esteemed above almost any other value.

The great Australian writer Henry Lawson saw mateship as a central tenet of Australian culture and through his writings extolled its importance in Australian life.[25] His sense of its importance arose in part from his own experience of various crises—particularly droughts—while he travelled through the Australian bush in his youth. His works are full of references to the vulnerability of Australia's pioneers to these disasters. His poem *Bourke* is a classic example:

> *No sign that green grass ever grew in scrubs that blazed beneath the sun;*
> *The plains were dust in Ninety-two and hard as bricks in Ninety-one.*
> *On glaring iron-roofs of Bourke, the scorching, blinding sand-storm blew,*
> *No hint of beauty lingered there in Ninety-one and Ninety-two.*

It is easy for contemporary, urban Australians to forget the importance of the social bonds inherent in the ideal of mateship. After all, most of the time people are busy trying to live their lives away from the scrutiny of their near neighbours and to avoid the petty conflicts that arise as a result of living in such close proximity. But the Australian environment will not permit even urban Australians to escape indefinitely from the difficulties faced by the pioneers. The Sydney bushfires of January 1994 are a good example, but ENSO-spawned flood and fire have, at one time or another, devastated parts of virtually every major Australian city. In such circumstances, mateship suddenly re-emerges *en masse* in the suburban wilderness and people do extraordinary things to help those whose lives have been affected.

During the Sydney bushfires I witnessed one such event. Real estate agents are not often known for their sense of morality, but on the morning following the fire I saw an agent locate a new flat to rent, for a single mother who had lost everything, and give her $1300 in open cheques to cover her initial costs. The Commonwealth Bank—another institution not usually known for its generosity—opened its doors over the weekend, suspending housing repayments, arranging emergency funding and acting as a depot for donations. Neighbours, many of whom rarely spoke to each other, were working side by side to help clean up. Others were discreetly handing over cash to those more drastically affected than themselves. So many opened their homes to those whose houses had been burned that there was no shortage of accommodation. As a result—in contrast to the aftermath of the Los Angeles earthquake of 19 January 1994—there was no tent city, indeed not a single tent in Sydney following the fires. For the people of places like Jannali, the prospect of a tent city housing their neighbours would have been a deep insult to their sense of mateship. They would have done anything in order to avoid it.

I find it both intriguing and heartening that European Australians should seize upon the ideal of mateship so quickly after colonising their new home. That it should persist for so long in an environment as adverse as the sprawling Australian suburbs is little short of a miracle. Such caring is the very sweetest of the uses of adversity. Perhaps there is something quite fundamental about such social obligations that makes them indispensable in the Australian environment. There is no doubt that Aboriginal culture, particularly as it existed in the more hostile environments, had this sense of sharing. Aborigines had, and indeed still have, social obligations which link people over thousands of kilometres. In times of crisis, these social obligations could see people sharing their few resources with visitors from even more severely affected areas. It is perhaps a tribute to the harshness of Australian environments that these two human groups, which are so different in many other ways, should both develop and maintain such an onerous system of social obligation and sharing.

To return to the stockman, the key thing about him is that he is an image from the frontier. The frontier was created when people introduced cattle to grasslands that had lacked large herbivores since the diprotodons vanished at least 35 000 years ago. The uneaten grass of their rangelands was an enormous resource which could not be fully exploited by humans in the absence of large herbivores. For only herbivores could turn its energy into meat and hides. Following the extinction of the diprotodon and in New Zealand the moa, the resource lay essentially untapped, or at least vastly underutilised, by humans in Australasia—until the coming of the Europeans. In New Zealand, it was the shepherd with his flocks that turned the grasses into gold. In Australia, the stockman was foremost, although shepherds, particularly in the nineteenth and early twentieth centuries, played a vital role as well.

Contrary to popular opinion, the stockman is not uniquely Australian. Give him a moustache and *maté*, broaden the brim of his hat and you have an Argentinian gaucho. Give him fringed leggings and a six shooter and you have a North American cowboy. It is no coincidence that stockman, cowboy and gaucho all come from newly settled continents. For just as the Australian grasslands were cleared of their diprotodons by the first invaders, the prairie and the pampas lost their native elephant, horses, camels and sloths when the first Indians arrived 11 000 years ago. Thus, the grasslands of all three continents presented a bounty that had not been reaped for millennia. The stockman/gaucho/cowboy arose to take advantage of a particular, short-lived ecological niche which resulted from this situation.

Today, the first bounty of the open range has long been exhausted, rendering the ecological niche untenable over vast regions. More intensive land uses have now taken over in these areas. In addition, changes in Australian society, such as increasing urbanisation, feminism, multiculturalism and the declining role of agriculture in the Australian economy, are all acting to render the image of the stockman anachronistic. Yet despite this—perhaps as a result of the inherent appeal of his independence and mateship—the image of the stockman still has immense emotional value to many Australians.

If the stockman arose out of a special and short-lived window of economic opportunity and was not entirely a uniquely Australian phenomenon, at least it arose as a response to local conditions. But there are many aspects of Australian culture that developed in the enormously productive, highly seasonal and 'weedy' ecosystems of Europe or east Asia which have, to our detriment, thrived in Australia. Given the differences in environment, it is remarkable how tenacious these European traditions and views have been in the new land.

At the beginning of this book I described a Christmas ritual as I experienced it in Melbourne in the early 1960s. Christmas was originally a pagan feast which gave comfort to subsistence communities in the depths of the European winter. Despite its Christianisation, its essential form has not changed for millennia. I suppose it is not surprising that if the Christian Church, with all its power, did not substantially change the form of the mid-winter feast, then little else would be able to do so. Australia's reversed seasons and alien environments have certainly had little impact. The ritual of Christmas has not changed for a century or more in many families. My own family, of largely Irish descent, has endured over a hundred insufferably torrid Christmas days, perspiring in front of a hot roast turkey and plum pudding. They may well endure a hundred more.

As well as a love of tradition, there is just a touch of the 'cultural cringe' in the stubbornness with which many Australians cling to the most unsuitable aspects of the European Christmas ritual. Certainly, the cultural cringe is alive and well in Australia today, although it takes on a different aspect from that which it did in the 1960s when Australian films such as *The Adventures of Barry MacKenzie* parodied the lack of sophistication of Australians and reinforced the message that young Australians should undertake their pilgrimage 'home' (to Britain) before settling down.

Science has not been exempt from the cultural cringe and this has already cost Australians dearly. I have a number of colleagues who graduated from Australian universities during the late 1940s and early '50s, following war service. It was at this time that the very last of the bridled nailtail wallabies, pig-footed bandicoots, lesser bilbies and many other Australian mammals were becoming extinct. I have asked several of my older colleagues why they did not devote their studies to preserving these unique elements of Australia's natural heritage. The answer given has always been the same. They were, they said, absolutely rivetted by the developments occurring in biology overseas. One must not forget that this was the time when Watson and Crick were unravelling the secrets of DNA, when major advances were being made in evolutionary studies and when the names Oxford and Cambridge reigned supreme in academia. Thus, despite repeated warnings that the Australian fauna was in dire trouble, our promising young graduates left the country one by one to pursue the 'big questions' in biology. With them left the last hope of preserving Australia's unique and vanishing fauna.

Today, the cultural cringe seems to be most evident in the reticence of Australians to develop a strong, vibrant and unique culture of their own. Instead, the government produces policies such as multiculturalism, or we talk of becoming a part of Asia. Neither of these options will help Australians to live comfortably in their own land.

Indeed, pursued to their logical ends, these policies could destroy the slow process of adaptation between humans and their environment which is currently occurring in Australia.

No-one seems to have defined what it means for Australia to 'become part of Asia', yet many Australians accept it as an imperative. It is clear that Australia can never be like Asia in an ecological sense. Therefore, our economy and culture must always differ fundamentally from that of the Asian nations. There are certainly excellent arguments for Australia increasing its trade, cultural, scientific and other links with its Asian neighbours. Indeed, provided our overall population does not grow as a result, there is no disadvantage in having a large proportion of the Australian population derived from people of Asian ancestry.

Even if this were achieved, Australia will always be startlingly different. For ENSO, soils and biology will begin to change people of Asiatic origin into Australians just as surely as it has done with Europeans. It is clearly a great challenge to forge links with Asia. But the greatest challenge will be to fit Australia's uniquely shaped bit of the world economic jig-saw into the puzzle as a whole. In order to do that, we must know our shape. An unthinking push to 'become part of Asia' will not help us with that.

And what of the government policy of multiculturalism? Although widely adopted and believed in by many Australians, the community concept of multiculturalism is fuzzy. Many people confuse it with immigration. But these concepts are in fact quite separate. Australia's population is so diverse today that a cessation of immigration would have no effect on multiculturalism.

In fact, the government policy of multiculturalism simply means that it is a national goal that people with diverse cultural practices should live side by side in Australia without obliterating each other's culture. From a biological perspective there is absolutely no reason why this should not happen, as long as all cultures modify the few of their practices that can adversely impact on the Australian environment. These, in general, relate only to family size, some agricultural practices, the harvesting of wild resources and attitudes towards the bush.

In the face of the challenges outlined above, Australians have begun to search more deeply for symbols of their identity. Increasingly, they are looking towards 'the bush' to provide them. From a biological perspective this is a good thing, for hopefully it will build strong links based upon understanding between Australians and their environment. Such links are vitally important, for Australians are caretakers of a disproportionately large share of the world's biological riches and Australia is a land that tolerates few mistakes. If Australians do not possess a culture that values these things, they will be lost to the world.

Despite the potential good to come out of this, some strange amalgams are arising out of the process. Few, perhaps, are as strange as the 'Easter bilby'. The Easter bilby is a chocolate gift which is being touted as a replacement for the chocolate Easter bunny. The rabbit is reviled in Australia as a great environmental destroyer and many Australians have long felt equivocal about its role in our Easter festival. Yet in some ways the rabbit is a supremely appropriate symbol of Easter. Despite its Christian veneer Easter is a pagan festival which celebrates *Eostre*, the European goddess of fertility who loaned her name to oestrogen and the oestrous cycle. The rabbit and egg were chosen by the ancestors of the Europeans as symbols of fertility. This perhaps should have warned nineteenth-century Australians concerning the suitability of the rabbit as an introduced species.

Admittedly, the bilby does resemble a rabbit with its long ears, but in many other ways it is decidedly non-rabbit like. This applies in particular to its powers of generation, for the bilby reproduces slowly and is one of Australia's most endangered species. It is hardly the symbol of fertility needed for our Easter celebration. I have suggested that the long-haired rat (*Rattus villosissimus*, otherwise known as the plague rat) is more suitable, for it is arguably our most fertile native mammal. Perhaps it lacks the bilby's charm!

Although this might seem to be an entirely frivolous example of the struggle of Australians in coming to grips with their new home, it is nonetheless an interesting one. It may stem from the same failure of understanding that has some people view the kangaroo—symbol of Australia—as some sort of holy cow, which is not to be used in any way, but just to be celebrated and loved from afar.

Australians are not alone in facing the challenge of adapting to their environment, for within their region many New Zealanders and

New Caledonians face a similar challenge. Yet it is a striking fact that neither New Zealanders nor Caledoche seem to be as obsessed with their search for a national identity as the Australians are. A good example of this concerns the recent announcement that New Zealand may become a republic. The public response was lukewarm indeed, with few caring to argue on either side. In Australia, the debate quickly became frenetic, with many rightly seeing it as a chance to further the cause of becoming better-adapted Australians, rather than remaining British in a strange land.

I must say that I do not understand the tepid response of many New Zealanders to the great issues of cultural adaptation. It is possible that the ecology of New Zealand is more amenable to European lifestyles and agriculture than Australia. Perhaps, in the lack of reminders by fire and flood New Zealanders forget that they do not fit comfortably; that they are still strangers in a strange land.

The case of the Caledoche appears to be quite different. Few Europeans living in New Caledonia are long-term settlers, and many French are temporary visitors who intend to return to France when their employment terminates. As a whole, they appear to live a largely unchanged French lifestyle on the strangest island on earth. They may be insulated from the reality of the New Caledonian environment by the wealth of nickel and the enormous subsidy to the island's economy from France.

Further afield, many other parts of the world are peopled by migrants who have little understanding of the ecology of their new home. Such people often suffer from cultural maladaptation. A most striking example was brought to my attention on a recent expedition in eastern Indonesia. In this region, recently arrived Muslim people of Asian origin live side by side with long-established animist or newly Christian Melanesians. There is great cultural diversity in both groups, but gardening practices and plant cultivars are very similar among all the people of the region.

Despite this, a fundamental difference in animal husbandry practice exists between the groups, for the Muslims do not keep pigs, while they are the principal livestock of the Christians and animists. Laws prohibiting the consumption of pigs are ancient, and arose in the semi-arid Middle East, where they may well have conferred considerable benefits upon their proclaimers. This is because pigs need lots of water, and indeed they often foul their water sources. Thus it may be counter-productive to keep them in more arid regions. Unfortunately, pigs are the only large domestic animals that thrive in the wet, tropical conditions of the Moluccas, and there are large feral as well as domestic populations. These populations are not genetically separate, for feral boars are often relied upon to inseminate domestic sows, and captured feral piglets are constantly being added to the domestic stock. The animists hunt wild pigs avidly for food and hunting pressure can be so intense that wild pigs can be eliminated locally, although they will occasionally leave a single wild boar to continue the supply of piglets. In this way the Melanesians gain valuable protein while at the same time keeping pigs from damaging gardens. Muslims, on the other hand, abhor pigs and, paradoxically, pig numbers are often extremely high in Muslim areas. Consequently, pig damage to gardens is frequently catastrophic.

The ecological implications of these two cultural traditions, living side by side just to the north of Australia, are obvious to the western visitor. But Australians are blind to even more startling 'cultural handicaps' in their own country.

The production and consumption of meat is a notoriously conservative area of human behaviour and provides many major examples of cultural maladaptation. Indeed, so conservative is this area of human behaviour that modern Australians are almost entirely dependent for meat from mammals upon the three species (cattle, sheep and pig) that were initially domesticated in the late stone age of western Asia some 8000–10 000 years ago. I find it remarkable that despite a shift in continents and the space of 200 years, with some minor exceptions, Australians have not added a single new source of meat to their limited diet.

Despite this conservatism, Australian primary producers support, albeit inadvertently, a vast array of edible animals on their grazing properties. A typical grazing property in western New South Wales supports large populations of three or four species of kangaroos, sheep, goats, cattle, feral pigs, rabbits, dogs, horses and emus. All of these species are highly edible, and all are esteemed by one or more cultural groups somewhere in the world. Yet the traditional Australian attitude to them is strange and highly wasteful, for most Australians consider that only sheep and cattle are fit for human consumption. All the rest, which often constitutes a substantial part of the meat produced, is ignored, shot, or poisoned and left to rot where it dies, breeding blowflies which further reduce the productivity of sheep.

One of the most striking of all European maladaptations concerns our management of fire. As was discussed earlier, the pioneers quickly wrested control of the firebrand from the Aborigines and with it control of productivity in Australian environments. Graziers have long controlled fire. In many areas they burn more frequently than is good for the ecosystem. In these cases hollow logs and other shelter needed by wildlife is destroyed and soil nutrients lost.

In forests and urban areas Europeans initially (and indeed even today) often had no fire policy, resulting in a massive build up of fuel and the ignition of fire storms that scorched, with monotonous regularity, millions of hectares, resulting in massive losses of life and property. Such devastating fires climaxed in south-eastern Australia in the 1920s and 1930s, and it was then that a policy of fire exclusion was implemented in many areas, combined with protective patch strip burning. Although initially successful at containing wildfires, the policy resulted in fuel building up to unprecedented levels. Bushfires of extraordinary ferocity followed.

Slowly, Europeans learned that throughout much of Australia, one must fight fire with fire. Prescribed burning was thus used extensively from the 1950s onwards. But even this policy has often been applied lackadaisically, with little thought being given to overall fire management. Indeed, the Sydney bushfires of January 1994 show how fragile the Australian control of fire is; and not surprisingly they have kick-started development of a better fire policy. It is to be hoped that, after 200 years of ignorance, modern European fire management techniques are coming to resemble Aboriginal firestick farming, or even (as discussed above) reconstituting the older, less fire-prone Australia of pre-Aboriginal times.

The highly urbanised nature of contemporary Australia is another example of maladaptation. The problems of catering for continued growth in Australian cities seems almost insurmountable. The only solution offered so far is to increase urban density. This, of course, results in a dramatic decline in urban wildlife, and in generations of young Australians growing up in concrete jungles without any opportunity to learn first-hand about or experience their unique wildlife heritage. Without that, there is little hope for the future indeed.

But the cities present even more severe problems in terms of nutrient and energy transfer. At present they act as vast nutrient and energy sinks for the surrounding countryside. The nutrients and minerals are quite literally stripped from the hinterland and are freighted *en masse* to the cities or overseas for export. The nutrients, once accumulated in the urban areas, become an enormous problem in the form of sewage and other pollutants, which cannot be easily dealt with given current funds and technology. A.D. Hope, Australia's Poet Laureate, saw this with remarkable prescience when he wrote *Australia* in 1955:

> *And her five cities, like five teeming sores*
> *Each drains her: a vast parasite robber state*
> *Where second-hand Europeans pullulate*
> *Timidly on the edge of alien shores.*

The list of examples of cultural maladaptation is almost endless, but Australia's beach-oriented culture (with its coincident appallingly high rate of melanoma), its taxation and banking practices as applied to rural areas (with its potential for financial and ecological ruin) and its economic addiction to home ownership and construction (which adds pressure to continue with population growth) are a few. Doubtless many, many more will become apparent as the process of cultural adaptation continues.

Given that it is essential that Australians evolve a culture that will help them survive long-term on their continent, two major questions arise. What laws and values should Australians encourage; and what artefacts and ceremonies should Australian society adopt in order to symbolise these goals?

The critical values that a truly adapted Australian culture must enshrine are dictated by the impoverished nature of Australian ecosystems. These conditions demand the following features from such large, warm-blooded creatures as humans:

1 That population remains small.
   Attitude to growth should therefore be revised so that rapid growth and 'big' ventures are very carefully assessed.
2 We need to retain flexibility in our decisions relating to the environment.
   Because resources are so limited, and our demands so great, painful choices must sometimes be made.
3 The Australian environment is complex and difficult to understand.
   We must do all in our power to ensure that Australians have intimate contact with, and a working knowledge of, their ecosystems.

If these basic values are not enshrined in our culture, then any sense of Australian nationhood is bound to conflict with environmental sustainability.

After population, perhaps the most important of these elements is a new sense of flexibility. The achievement of this will be as difficult as it is essential, for it will encounter opposition from special interest groups at every turn. At present there is no sign of increasing flexibility towards environmental matters in political or social spheres. Indeed, the trend has recently been towards more inflexible approaches.

The clearest need for such flexibility concerns attitudes towards our natural resources. At present the various elements of the Australian biota and landscape are roughly divided into two groups: those that can be pillaged by whoever has the means; and elements which, whether they need it or not, are rigorously protected from any human interference.

Into the first category fall Australian fisheries, feral animals, occupied arable lands and mineral resources. Into the second fall national parks, most indigenous vertebrate species (except fish) and rainforests.

The resources that are presently free to be exploited have been discussed above. In the case of Australian fisheries, the problem is in most part one of failure of management. For arable lands, it is largely an economic problem and one of lack of understanding. But in the case of our mineral wealth, which I have not discussed previously, I think that we are simply asking the wrong questions.

Non-renewable resources such as minerals and fossil fuels must, by their very nature, be treated differently from renewable resources. This is because they represent a one-off bonanza, not a source of continuing wealth. The present argument about mining in Australia is concerned almost solely about where and whether mining should occur. From a long-term environmental perspective, this seems to be a strange focus. This is because although mining can have a locally significant impact, it affects a relatively small area and is presently well-regulated. Thus, under existing conditions, it presents a very minor threat to Australia's biodiversity. But remarkably, no-one presently seems concerned about the utilisation of our one-off mineral wealth. The real issue, from a long-term environmental perspective, concerns how the wealth gained from mining is to be used.

It is now time to discuss the situation as regards the elements of the Australian biota that are considered as sacrosanct by many Australians. The acceptance that whales, kangaroos and certain environments such as rainforests are inviolable 'holy cows' prevents us from utilising our few renewable resources in the least destructive way. This means that we must often rely upon other resources far more heavily than good management would allow. There is a parallel here with Aboriginal resource utilisation. The Aborigines utilised an extraordinarily wide array of resources, from insects to marine resources, plants and all kinds of vertebrate animals. With the exception of fish in Tasmania and a few other examples, taboo species seem to have been relatively few. This utilisation of an exceptionally wide array of species allowed Aborigines to 'tread softly' in Australian environments. For if fewer resources were utilised, those targeted would have been quickly exhausted and destroyed.

I feel that a far better situation for conservation in Australia would result from a policy which allows exploitation of *all* of our biotic heritage, provided that it all be done *in a sustainable manner*. It may seem shocking to some conservationists that anyone should advocate the sustainable utilisation of endangered species and rainforests. But if it is possible to harvest for example, 10 mountain pygmy-possums (*Burramys parvus*) or 10 southern right whales (*Balaena glacialis*) per year, why should we not do it? The economic gain made from such utilisation may allow us to ask less of critically over-exploited resources. Is it more moral to kill and consume a whale, without cost to the environment, than to live as a vegetarian in Australia, destroying seven kilograms of irreplaceable soil, upon which everything depends, for each kilogram of bread we consume? I fear that Australian environments are now in such crisis, our population so large, and our affluence so dearly protected, that it is only by carefully utilising all of our renewable resources that we can hope to avoid further environmental damage.

The financial cost of such a policy would be considerable, for management plans would have to be developed and rigidly adhered to for every species. This would mean vast investment in regulatory bodies and, above all, knowledge. For our science would have to be unimpeachable, our regulation enforceable and our response to changing conditions instantaneous.

A quite different example of the inflexibility of our views on the environment concerns the impact that protected species legislation has on the interaction of people and fauna. This also reveals something of the third of the points mentioned above—the need for better understanding of Australian environments.

A few generations ago a large proportion of young Australians kept indigenous species for pets and lived in semi-rural or rural areas. From my own 1950s childhood, I vividly remember a family friend's pet magpie and my grandmother's cockatoo. I myself kept goannas, snakes, blue-tongue lizards and a wide variety of frogs. Great benefits in terms of familiarity and fondness for wildlife developed from such interactions. Today, many such associations are illegal, unless specifically licensed by the relevant government authority. Even where they remain legal, there is a general community perception that it is somehow wrong to have native animals as pets.

This great legal fence that divides ordinary Australians from their fauna is, I believe, highly destructive. Today, many young Australians may *like* their fauna, but few *understand* it as their grandparents did. Unfortunately, continued urbanisation and urban consolidation is forcing further alienation of people from their environment. Urban consolidation is removing bushland and even gardens from much of our immediate habitat. These areas give most young Australians their first chance to learn about their environment. As the larger trees, lizards, frogs and birds gradually vanish from the urban areas, the alienation of the great majority of Australians from their land will become complete.

A further aspect of protected species legislation is the impact that it has had on organised crime and wildlife smuggling. Prohibition of trade in many common species, some of which are destroyed as pests under licence, has inflated the value of such species to specialist collectors overseas. To cater for this trade a black market has grown up, which has tainted many areas of society, including some charged with wildlife protection. With the connections and power that easily won cash from sale of common species has brought, illegal wildlife traders have been able to risk the capture and sale of even highly endangered species. Clearly, this legislation has been far more destructive than a well-regulated legal trade would ever have been.

Side by side with growing legislative protection of native fauna has come an increase in animal liberation. Biologists frequently observe in nature that what is good for the individual is harmful to the species. Because of this, animal liberationists are often pitted head-on with those interested in ecosystem conservation. With animal rights increasingly being incorporated into the law, it becomes ever more difficult to manage ecosystems as a whole. Today, the Royal Society for the Prevention of Cruelty to Animals feels quite justified in telling National Parks authorities how to run their management programs. This is a truly appalling state of affairs which cannot last if Australian ecosystems are to survive intact.

Viewed this way, the period from the 1950s through to the end of the 1980s has seen a backwards march from the goal of developing an environmentally attuned Australian culture. My grandfather ate more native animal foods than I do, for he loved his mutton birds more dearly than chicken. My grandmother and her family knew more about native birds than I, for she lived in the countryside and kept an assemblage of cockatoos, magpies and other birds about the house. Most of my predecessors had a better feel for nature as children than my children do, for few grew up in such a densely urbanised and highly regulated environment as my son and daughter.

I find it sad that so much of the energy of the debate about what it means to be Australian is misdirected. Many Australians would consider that the person eating the meat pie is more 'Australian' than the one eating the souvlaki. Yet no-one cares to ask whether the meat has come from kangaroo or cattle, or how much soil was lost in the production of the wheat products used to wrap the meat. Likewise, many people worry about whether immigrants come from Asia or Europe, yet never for a moment think of the effect of Australia's total immigration intake on population growth.

It is ignorance of the past that dooms each new wave of immigrants to the 'new' lands to be future eaters. So certain are they of their superiority; so sure of their ability, that they do not think to learn from those who have gone before them, nor do they take the time to read the signs of the land until disaster has overtaken them. We, perhaps, are the first generation of future eaters who have looked over our shoulder at the past, but we have done so quite late in the process of environmental destruction. If we can change our ways before we have consumed all of the future that we are capable of, then we will have achieved something very precious.

It is true that the most recent immigrants are slowly being shaped by their new homes, but like Aborigine and Maori before them, the process is costing dearly in terms of biodiversity and sustainability. This, in turn, is severely limiting the future for their children. More than 60 years ago, Sir Keith Hancock wrote in his great work *Australia*:

> *When it suits them, men may take control and play fine tricks*
> *and hustle Nature. Yet we may believe that Australia, quietly and*
> *imperceptibly ..., is experimenting on the men... She will be*
> *satisfied at long last, and when she is satisfied an Australian*
> *nation will in truth exist.*[61]

If Australasians are happy to progress towards adaptation at the imperceptible pace of change at which evolution proceeds, it will be a long time indeed before an Australian or a New Zealand nation will in truth exist. But humans are different from all other creatures. We can think, understand, and act to make our lives better. Yet despite all of our advantages—our technology and our intellect—we seem to have made as disastrous a series of mistakes as any other species. Even now, we seem to be seeing only the very first glimmer of the change that Sir Keith foretold. For the moment, I hope that Australasians one and all will begin to ask the right questions. For this is the first, necessarily wobbly step on the road to discovering what it means to be custodians of the wonderful and enigmatic 'new' lands.

POSTSCRIPT

I am writing this perched in a tiny flat high above *Wéerong*. Below me spreads the inexpressibly beautiful Sydney harbour, its complex blue waters interweaving with the green of parks and suburbs. I have seen, each morning, the ships and sailors of the Australian Navy bustling at the docks, just as the sailors of the First Fleet must have done over 200 years before. Yet there is much that is hidden to the casual observer in this charming scene. Somewhere near me lies the grave of Arabanoo, who was buried with so much care and regard by Governor Phillip all those years ago. I cannot visit his grave, for it is unmarked, the exact location now being lost. Somewhere behind me—far to the west—great machines drone on, converting forest into yet more suburbs or cropland. Like Arabanoo's grave, I cannot see them. But I know that they are there, for I can see, from my eagle's eyrie atop the 'vast parasite state' where the spoils go.

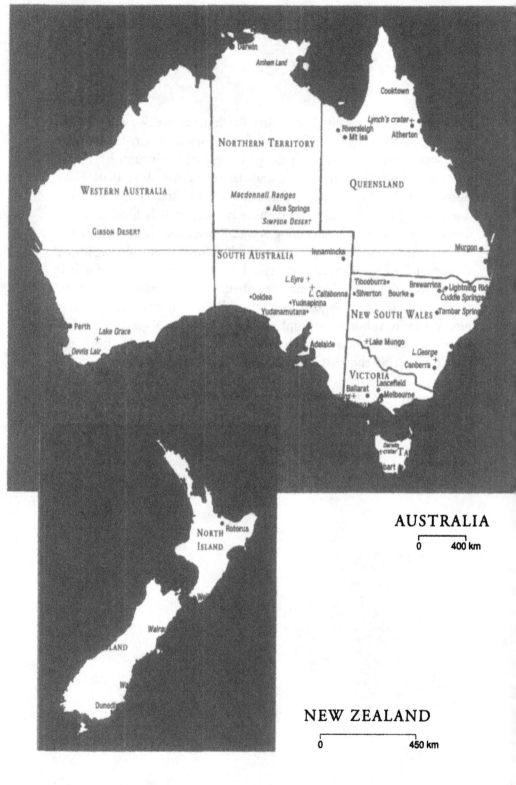

AUSTRALIA

0   400 km

NEW ZEALAND

0   450 km

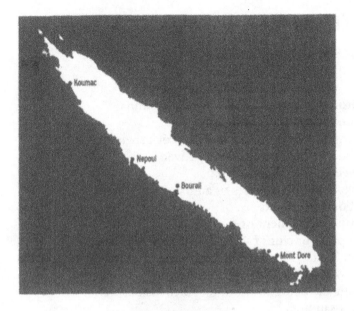

NEW CALEDONIA

```
0          70 km
```

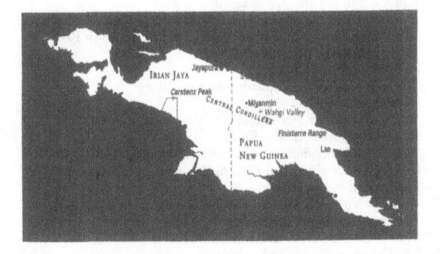

NEW GUINEA

```
0          700 km
```

List of Photographs

The skeleton of a diprotodon
A reconstruction of diprotodon
Lake Callabonna, northern South Australia
A diprotodon trackway on Lake Callabonna
A reconstruction of a large moa
Drawing of a moa on the wall of a rock shelter
Moa butchering site at Waitaki River
A moa egg from Wairau Bar
The skull and neck bones of a moa
Moa bone fish hooks
A fortified settlement or *pa*
Maori patus or clubs carved from wood and bone
Tamati Waka (Thomas Walker) Nene, chief of the Ngati-Hoa
Maori woman wearing huia feathers
Drawing of Maori attack upon Abel Tasman's crew
A Macassan prau
Probasso, captain of the Macassan trepang fleet
A Macassan trepang processing station
View of Malay Bay from Probasso's island
Part of the Tahitian naval fleet
Easter Island as James Cook saw it
Drawing of inhabitants of New Ireland made during Abel Tasman's visit
The *Endeavour* careened at the mouth of the Endeavour River
Early representation of New Guineans
A New Guinean hunter
Elephant seals of Sea Elephant Bay, King Island
A Tasmanian family
Southern Victorian Aboriginal family
A man and woman of Van Diemen's Land
A group of Aborigines at Coranderrk Aboriginal Station
Victorian Aborigines with possum-skin cloaks and bark canoe
Queensland Native Police
Group of Victorian Aborigines participating in the Jerail ceremony
Destruction of forest at Beechmont
Timber-getting in the Atherton district
ENSO-induced floods in Brisbane
Anglican church near Ipswich

Photographic Acknowledgments

For supplying photographs thanks to the following institutions:

American Museum of Natural History, Department of Library Services, New York; Alexander Turnbull Library, Wellington; Auckland City Art Gallery, Auckland; The Australian Museum Trust (*Australian Natural History* magazine); The British Library, London; The British Museum, London; Fryer Library, The University of Queensland, Brisbane; John Oxley Library, Brisbane; Mitchell Library, Sydney; Museum of New Zealand, Wellington; National Gallery of Victoria, Melbourne; National Library of Australia, Canberra; National Maritime Museum, Greenwich; South Australian Museum Anthropology Archives, Adelaide.

And the following individuals:

Richard Cassels, Otago Museum, Dunedin; John Mulvaney, Australian Academy of the Humanities, Canberra; Peter Murray, Museums and Art Galleries of the Northern Territory, Darwin; Neville Pledge, South Australian Museum, Adelaide; Michael Trotter, Canterbury Museum, Christchurch. While every effort has been made to determine the source of reproduced material, in some cases copyright proved to be untraceable.

REFERENCES

1 Anderson, A. (1989). *Prodigious Birds.* Cambridge University Press.

2 Anderson, A. (1991). The chronology of colonisation in New Zealand. *Antiquity* 65:767-795.

3 Anderson, A. and McGlone, M. (1992). Living on the edge—prehistoric land and people in New Zealand. Pp 199-241 *In* J. Dodson (ed). *The Naive Lands,* Longman Cheshire.

4 Archer, M., Flannery, T.F., Ritchie, A. and Molnar, R.E.(1985). First Mesozoic mammal from Australia—an early Cretaceous monotreme. *Nature* 318:363-366.

5 Bahn, P. and Flenley, J. (1992). *Easter Island, Earth Island.* Thames and Hudson, London.

6 Bain, R. (1985). Fisheries administration in the AFZ. *Infofish Marketing Digest* (4)85:17-22.

7 Balouet, J.C. (1991). The fossil vertebrate record of New Caledonia. Pp 1383-1409 *In* P.V. Rich, J.M. Monaghan, R.F. Baird and T.H.V. Rich (eds). *The Vertebrate Palaeontology of Australasia.* Pioneer Design Studio, Melbourne.

8 Barboza, P.S., Hume, I.D. and Nolan, J.V. (1993). Nitrogen metabolism and requirements of nitrogen and energy in wombats (Marsupialia: Vombatidae). *Physiological Zoology* 86:807-828.

9 Barwick, D.E. (1971). Changes in the Aboriginal population of Victoria, 1863-1966. Chapter 21 *In* D. Mulvaney and J. Golson (eds). *Aboriginal Man and Environment in Australia.* Australian National University Press, Canberra.

10 Bates, D., (1985). *The native tribes of Western Australia.* National Library of Australia, Canberra.

11 Bauer, A.M. (1990). Phylogenetic systematics and biogeography of the Carphodactylini (Reptilia: Gekkonidae). *Bonner Zoologische Monographien* 30.

12 Bauer, A.M. and Sadlier, R.A. (1993). Systematics, biogeography and conservation of the lizards of New Caledonia. *Biodiversity Letters* 1:107-122

13 Beaglehole, J.C. (ed.) (1962). *The* Endeavour *Journal of Joseph Banks. 1768-1771.* Angus and Robertson, Sydney.

14 Bellwood, P.S. (1989). The Colonisation of the Pacific: some current hypotheses. Pp 1-59 *In* A.V.S. Hill and S.W. Serjeantson (eds). *The Colonisation of the Pacific, a genetic trail.* Clarendon-Oxford Press.

15 Bellwood, P.S. (1992). Southeast Asia before history. Chapter 2 *In* N. Tarling (ed.). *The Cambridge History of Southeast Asia.* Cambridge University Press.

16 Berndt, R.M. and Berndt, C.H. (1988). *The world of the first Australians.* Aboriginal Studies Press, Canberra.

17 Birdsell, J.B. (1953). Some environmental and cultural factors influencing the structure of Australian Aboriginal populations. *American Naturalist* 87: 171-207

18 Boulton, G. (1992). *Spoils and Spoilers. A History of Australians Shaping Their Environment.* Allen and Unwin, Sydney.

19 Brady, E.J. (1918). *Australia Unlimited.* George Robertson, Melbourne.

20 Burenhalt, G. (general editor) (1993). *The Illustrated Encyclopaedia of Humankind.* Vols. 1 + 2. Harper, San Francisco.

21 Burroughs, W.J. (1991). *Watching the World's Weather.* Cambridge University Press, Cambridge.

22 Butler, P. (ed.) (1980). *Life & times of Te Rauparaha. By his son Tamihana Te*

Rauparaha. Alister Taylor, Waiura, Martinborough.

23 Challies, C.N. (1990). Red Deer. Pp 436-457 *In* C.M. King (ed.). *The Handbook of New Zealand Mammals.* Oxford University Press, Auckland.

24 Clark, M. (1963). *A Short History of Australia.* Mead and Beckett, Sydney.

25 Clark, M. (1978). *Henry Lawson. The Man and the Legend.* Sun Books, Melbourne.

26 Colhoun, E.A. (compiler) (1988). *Cainozoic vegetation of Tasmania.* Special Paper. Department of Geography, University of Newcastle, New South Wales, Australia.

27 Cook, J. and King, J. (1785). *A voyage to the South Pacific.* Vol 3. London. pp 124-5.

28 Cooper, A., Atkinson, I.A.E., Lee, W.G. & Worthy, T.H. (1993). Evolution of the moas and their effect on the New Zealand flora. *Trends in Ecology and Evolution* 8:433-437.

29 Cooper, R.A. and Millener, P. (1993). The New Zealand biota: historical background and new research. *Trends in Ecology and Evolution* 8:429-432.

30 *CSIRO Annual Report,* 1990-1991. Canberra.

31 Cumpston, J.S. (1970). *Kangaroo Island 1800–1836.* Roebuck Society, Canberra.

32 Dampier, W.C. (1729). *The Voyage To New Holland.* Republished 1981, Alan Sutton, Britain.

33 Darwin, C. (1845). *The Voyage of the Beagle.* Republished Edito-Service, S.A., Geneva.

34 Daugherty, C.H., Gobbs, G.W. & Hitchmough, R.A. (1993). Mega-Island or Micro-Continent? New Zealand and its fauna. *Trends in Ecology and Evolution* 8:437-441.

35 Day, L., (in Day, L. and Young, I.) (1994). *Australia's Demographic future: determi nants of our population.* Draft paper given at Population 2040: Australia's Choice. Australian Academy of Sciences, Canberra.

36 Diamond, J. (1992). *The Rise and Fall of the Third Chimpanzee.* Radius.

37 Diamond, J. (1993). Ten Thousand Years of Solitude. *Discover* March 1993:48-57.

38 Dodson, J., Fullagar, R., Furby, J., Jones, R. and Prosser, I. (1993). Humans and megafauna in a late Pleistocene environment from Cuddie Springs, north western New South Wales. *Archaeology in Oceania* 28:94-99.

39 Ellis, V.R. (1981). *Trucanini. Queen or Traitor?* Australian Institute of Aboriginal Studies, Canberra.

40 Finlayson, B.L. and McMahon, T.A. (in press). *Runoff Variability in Australia: causes and environmental consequences.*

41 Flannery, T.F. (1981). *A review of the genus* Macropus *(Marsupialia: Macropodidae), the living grey kangaroos and their fossil allies.* M.Sc. Earth Sciences, Monash University.

42 Flannery, T.F. (1989). Who Killed Kirlilpi? *Australian Natural History* 23:234-241.

43 Flannery, T.F. (1990). Pleistocene faunal loss: implications of the aftershock for Australia's past and future. *Archaeology in Oceania* 25:45-67.

44 Flannery, T.F. (1990). The rats of Christmas past. *Australian Natural History* 23(5):394-400

45 Flannery, T.F. (1991). Australia: overpopulated or last frontier? *Australian Natural History* 23: 769-775.

46 Flannery, T.F. (1992). A revision of the *Thylogale brunii* complex (Macropodidae: Marsupialia) in Melanesia, with description of a new species. *Australian Mammalogy* 15:7-24

47 Flannery, T.F. and Gott, B. (1985). The Spring Creek locality: a late Pleistocene megafaunal site from southwestern Victoria. *Australian Zoology* 21:385-422

48 Flannery, T.F. and Seri, L. (1990). The mammals of southern West Sepik Province, Papua New Guinea; their distribution, abundance, human use and zoogeography.

Records of the Australian Museum 42:173-208.

49 Flannery, T.F. and White, J.P. (1991). The zoogeography of New Ireland mammals, including evidence for the earliest deliberate human translocation of fauna. *National Geographic Research and Exploration* 7:96-113

50 George, A.S. (1981). The background to the flora of Australia. Pp 3-25 *In* Robertson *et al.* (eds). *Flora of Australia.* Australian Government Publishing Service, Canberra. Republished 1979. Doubleday, Lane Cove.

51 Gibson, D.F. (1986). The Tanami Desert: Research on Aboriginal Land. *Australian Natural History* 21:544-546.

52 Giles, E. (1889). *Australia Twice Traversed.* London. Republished 1979. Doubleday, Lane Cove.

53 Gillespie, R., Horton, D.R., Ladd, P., Macumber, P.G., Rich, T.H., Thorne, R., and Wright, R.S.V. (1978). Lancefield Swamp and the extinction of the Australian megafauna. *Science* 200:1044-1048.

54 Godthelp, H., Archer, M., Ciselli, R., Hand, S.J. and Glikeson, C.F. (1992). Earliest known Australian Tertiary mammal fauna. *Nature* 356:514-516.

55 Gould, J. (1843-1862). *The mammals of Australia.* London.

56 Gould, R.A. (1991). Arid-land foraging as seen from Australia: adaptive models and behavioural realities. *Oceania* 62:12-33.

57 Green, N. (1984). *Broken Spears. Aboriginals and Europeans in the southwest of Australia.* Focus Education Services. Cottesloe.

58 Groves, C.P. (1993). The Neanderthals. Pp 68-73 *In* G. Burenhult (ed.) *The Illustrated Encyclopaedia of Humankind.* Vol 1. Harper, San Francisco.

59 Groube, L., Chappell, J., Muke, J. and Price, D. (1986). A 40,000 year old human occupation site at Huon Peninsula, Papua New Guinea. *Nature* 324:353-355.

60 Halliday, T. (1978). *Vanishing Birds.* Hutchinson, Australia.

61 Hancock, K. (1930). *Australia.* Ernest Benn, London.

62 Hercus, L. (1990). Aboriginal People. Pp 149-160 *In* Tyler, M.J., Twidale, C.R., Davies, M. and Wells, C.B. (eds). *Natural History of the north east Deserts.* Royal Society of South Australia.

63 Hiatt, L. (1986). *Rom* in Arnhem Land. Pp 5-14 *In* S.A. Wild (ed.). Rom, *an Aboriginal Ritual of Diplomacy.* Institute of Aboriginal Studies, Canberra.

64 Higgins, L.V. (1993). The Unusual sea-lions of Kangaroo Island. *Australian Natural History* 24(9):30-37.

65 Holdaway, R.N. (1989). Terror of the forests. *New Zealand Geographic* 10:56-65.

66 Hughes, R. (1987). *The Fatal Shore. A history of the transportation of convicts to Australia 1788-1868.* Collins Harville. London.

67 Hunter, J. (1793). *Historical Journal of the Transactions at Port Jackson and at Norfolk Island.* Republished 1968. Library Board of South Australia.

68 Hunter, J. (1989). *The Hunter Sketchbook: birds, flowers of NSW drawn on the spot in 1788, 89 & 90.* J. Calaby (general editor). National Library of Australia, Canberra.

69 Hurditch, W.J. (1983). The Soil Ecosystem. Pp 52-77 *In* Recher, H., Lunney, D. and Dunn, I. (eds). *A Natural Legacy.* Pergamon Press, Sydney.

70 Hurditch, W.J.,Charley, J.L. and Richards, B.N. (1980). Sulphur cycling in forests of Fraser Island and coastal New South Wales. Pp 237-250 *In* J.R. Freney and A.D. Nicholson (eds). *Sulphur in Australia.* Australian Academy of Science, Canberra.

71 Jones, R. (1969). Fire-stick Farming. *Australian Natural History* September 1969: 224-228.

72 Kailola, P.J., Williams, M.J., Stewart, P.C., Reichert, R.E., McKnee, A. and Grieve, C. (1993). *Australian Fisheries Resources.* Bureau of Resource Sciences, Department of Primary Industry and Energy, and Fisheries Research and Development Corporation, Canberra.

73 Kartzoff, M. (1969). *Nature and a City. The Native Vegetation of the Sydney Area.* Edwards and Shaw, Sydney.

74 Kelly, R., Robson, M., Birrell, M. (1993). *Establishment of a national reserve system in Australia.* ANZECC Meeting, 1993 (draft paper).

75 Kershaw, A.P. (1986). Climatic change and Aboriginal burning in north-east Australia during the last two glacials. *Nature* 322:47-49.

76 King, M. (1983). *Maori. A photographic and social history.* Heinemann, Wellington.

77 Kingdon, J. (1993). *Self-made Man and his undoing.* Simon and Schuster, London.

78 Krefft, G. (1862). The vertebrose animals of the lower Murray and Darling. *Transactions of the Philosophical Society of N.S.W.* 1862-1865:1-33.

79 Laitman, J. (1984). The Anatomy of Human Speech. *Natural History* 93:20-24

80 Lampert, R.J. (1979). Aborigines, Pp 81-89 *In* M. J. Tyler, C.H. Twidale and J.K. Kind (eds). *Natural History of Kangaroo Island.* Royal Zoological Society of South Australia.

81 Lavery, H.J. (ed.) (1978). *Exploration North. Australia's Wildlife from desert to reef.* Richmond Hill Press, Melbourne.

82 Leahy, M. and Crain, M. (1937). *The Land That Time Forgot.* Hurst and Blackett, London.

83 Lumholtz, C. (1889). *Among Cannibals.* J. Murray, London.

84 MacPherson, P. (1885). Some causes of decay of Australian forests. *Journal of the Royal Society of NSW* 29:85-96.

85 Majnep, I.S. and Bulmer, R. (1990). Kalam hunting traditions. Part 2. The truly arbo real kapuls. The Copper Ringtail and the giant rodents. *Working Papers in Anthropology, Archaeology, Linguistics and Maori Studies,* No 86. Department of Anthropology, University of Auckland.

86 Maning, F.E. (1887). *Old New Zealand. A tale of the good old times by a Pakeha Maori.* Golden Press, Auckland.

87 Marshall, L.G., and Corruccini, R.S. (1978). Variability, evolutionary rates and allome try in dwarfing lineages. *Palaeobiology* 4:101-119.

88 Martin H. (1993). *Published notes.* Lecture, Linnean Society of New South Wales.

89 Martin, P.S. (1984). Prehistoric overkill: the global model. Pp 354-403 *In* P.S. Martin and R.G. Klein (eds). *Quaternary Extinctions, a prehistoric revolution.* University of Arizona Press, Tuscon.

90 Martin, R. (1992). Of Koalas, tree-kangaroos and man. *Australian Natural History* 24:22-31.

91 May, R.M. (1977). Thresholds and breakpoints with a multiplicity of stable states. *Nature* 269:471-477.

92 McDowell, R.M. (ed) (1980). *Freshwater Fishes of Southeastern Australia.* A.H. and A.W. Reed, Sydney.

93 McKelvey, B. (1994). Cold war over warm ice. *Australian Natural History* Autumn: 49-53

94 Meehan, B. and Jones, R. (1986). Pp 15-32. *In* S.A. Wild (ed). Rom, *an Aboriginal*

Ritual of Diplomacy. Institute of Aboriginal Studies, Canberra.

95 Meslee, La, E.M. (1883). *The New Australia.* Reprinted Heinemann, Melbourne 1973.

96 Micco, M.H. (1971). King Island and the Sealing Trade. A translation of Chapters 22 and 23 of the narrative by François Péron published in the official account of the 'Voyage of Discovery to the Southern Lands'. *Roebuck Society Publication* No. 3.

97 Millener, P. (1991). The Quaternary avifauna of New Zealand. Pp 1317-1344 *In* Rich, P.V., Monaghan, J.M., Baird, R.F. and Rich, T.H. (eds). *Vertebrate Palaeontology of Australasia,* Pioneer Design Studio, Melbourne.

98 Miller, G.S. and Magee, J. (1992). Drought in the Australian outback: anthropogenic impact on regional climate. *Abstracts, A.G.U. Fall meeting,* 104.

99 Miller, R. (1983). *Continents in Collision.* Time-Life Books, Amsterdam.

100 Mitchell, T. (1838). *Three Expeditions into the Interior of eastern Australia, with descriptions of recently explored regions of Australia Felix, and the present colony of New South Wales.* T. and W. Boone, London.

101 Mitchell, T. (1848). *Journal of an expedition into the interior of tropical Australia.* London.

102 Morton, S.R. (1979). Diversity of desert-dwelling mammals: a comparison of Australia and North America. *Journal of Mammalogy* 60:253-264.

103 Nix, H.A. (1990). The environmental base. Pp 22-27 *In International Agriculture.* Special Issue of Agricultural Science.

104 O'Brian, P. (1987). *Joseph Banks. A Life.* Collins Harville, London.

105 Oneil, G. (1993). Beginnings. *Time* August 9, 1993.

106 Parris, H. (1948). Koalas on the lower Goulburn. *Victorian Naturalist* 64: 192-193.

107 Pascual, R., Archer, M., Ortiz Jaureguizar, E., Prado, J.L., Godthelp, H. and Hand, S.J. (1992). First discovery of monotremes in South America. *Nature* 356:704-706.

108 Phillips, H. (1990). A tale of three species: the stilt, the shrimp and the scientist. *Australian Natural History* 23:322-329.

109 Pierson, E.D., Sarich, V.M., Lowenstein, J.M., Daniel, M.J. & Rainey, W.E. (1986). A molecular link between the bats of New Zealand and South America. *Nature* 323:60-63.

110 Plomley, N.J.B. (1983). *The Baudin Expedition and the Tasmanian Aborigines 1802.* Blubber Head Press, Tasmania.

111 Powell, J.M. (1993). *Griffith Taylor and "Australia Unlimited".* University of Queensland Press.

112 Pyne, S.J. (1991). *Burning Bush. A fire history of Australia.* Allen and Unwin.

113 Ratcliffe, F. (1938). *Flying Fox and Drifting Sand.* Sirius Books, Sydney.

114 Recher, H. (1974). Colonisation and extinction. The birds of Lord Howe Island. *Australian Natural History* 18:64-69.

115 Reynolds, H. (1981). *The Other Side of the Frontier.* James Cook University, Queensland.

116 Robson, R.W. (1979). *Queen Emma. The Samoan-American Girl Who Founded an Empire in 19th Century New Guinea.* Pacific Publications, Sydney.

117 Rolls, E. (1969). *They All Ran Wild.* Angus and Robertson, Sydney.

118 Rolls, E. (1993). *From Forest to Sea.* University of Queensland Press.

119 Roth, H.L. (1897). *Ethnographical Studies Among the North-West-Central Queensland Aborigines.* Brisbane.

120 Schmid, M. (1981). *Fleurs et plantes de Nouvelle Calédonie.* Les Editions du Pacific,

Singapore.

121 Seargentson, S.W. and Hill, A.V.S. (1989). The colonisation of the Pacific: the genetic evidence. Chapter 8 *In* A.V.S. Hill and S.W. Searjentson (eds). *The Colonisation of the Pacific. A genetic trail.* Oxford University Press.

122 Singh, G. and Geissler, E.A. (1985). Late Cenozoic history of vegetation, fire, lake levels and climate at Lake George, New South Wales, Australia. *Philosophical Transactions of the Royal Society of London* 311:379-447.

123 Shephard, M. (1992). *The Simpson Desert, Natural History and Human Endeavour.* Royal Geographical Society of Australasia.

124 Souter, G. (1963). *New Guinea: the Last Unknown.* Angus and Robertson. Sydney.

125 Spencer, M. and Connell, A.W.J. (1988). *In* New Caledonia. *Essays in Nationalism and Dependency.* University of Queensland Press.

126 Spencer, M., Ward, A. & Connell, J. (eds). (1988). *New Caledonia. Essays in Nationalism and Dependency.* University of Queensland Press.

127 Steven, R. (1989). Land and white settler colonialism: the case of Aotearoa. Pp 21-34 *In* D. Novitz and B. Willmott (eds). *Culture and Identity in New Zealand.* GP Books, Wellington.

128 Szalay, F.S. (1982). A new appraisal of marsupial phylogeny and classification. Pp 621-40 *In* M. Archer (ed.) *Carnivorous marsupials.* Royal Zoological Society of New South Wales.

129 Taçon, P. (1993). Art of the Land. Pp 158-159. *In* Burenhult, G. *The Illustrated History of Humankind.* Vol. 1. Harper, San Francisco.

130 Tasman, A.J. (1642). *Journal.* Reprinted 1964. Australian Heritage Press, Adelaide.

131 Tench, W. (1793). *Sydney's First Four Years.* Being a reprint of 'A Narrative of an Expedition to Botany Bay', and 'A Complete Account of the Settlement at Port Jackson'. Reprinted 1974, Angus and Robertson, Sydney.

132 Tilman, D., (1982). Resource Competition and Community Structure. *Monographs in Population Biology* 17. Princeton University Press, Princeton, USA.

133 Tunbridge, D. (1991). *The Story of the Flinders Ranges Mammals.* Kangaroo Press, NSW.

134 Vickers Rich, P. and Rich, T.H. (1993). *Wildlife of Gondwana.* Reed, Australia.

135 Wallace, A.R. (1876). *The geographical distribution of animals, with a study of the relationships of living and extinct faunas as elucidating past changes of the earth's surface.* Volume 1. Harper and Brothers, New York.

136 Wallace, C. (1984). *The Lost Australia of François Péron.* Nottingham Court Press.

137 Webb, L. (1966). The rape of the forests. Pp 156-205 *In* J. Marshall (ed.). *The Great Extermination.* Heinemann, Adelaide.

138 Webb, S. (1984). *Prehistoric Stress in Australian Aborigines.* Ph.D. thesis, Prehistory, Research School of Pacific Studies. Australian National University.

139 White, J. (1790). *Journal of a Voyage to New South Wales.* Republished 1962, Angus and Roberston, Sydney.

140 White, J.P. (1993). Australia: the different continent. Pp 207-227. *In* G. Burenhalt (ed.). *The Illustrated Encyclopaedia of Humankind.* Vol 2. San Francisco.

141 Whitley, G.P. (1974). From First Fleet to El Torito. *Australian Natural History* 18:38-44

142 Williams, L. (1993). Australia: the secret history. Pp 11-17. *Good Weekend* October 2.

143 Wright, R. (1986). New light on the extinction of the Australian megafauna. *Proceedings of the Linnean Society of New South Wales* 109:1-9.

INDEX

bold = illustration

Printed in the USA
CPSIA information can be obtained
at www.ICGtesting.com
JSHW022018250324
59876JS00001B/5